# PHARMACEUTICAL BIOTECHNOLOGY

## Second Edition

# PHARMACEUTICAL BIOTECHNOLOGY

## Second Edition

*Edited by*

# Michael J. Groves

Taylor & Francis
Taylor & Francis Group
Boca Raton   London   New York

A CRC title, part of the Taylor & Francis imprint, a member of the
Taylor & Francis Group, the academic division of T&F Informa plc.

Published in 2006 by
CRC Press
Taylor & Francis Group
6000 Broken Sound Parkway NW, Suite 300
Boca Raton, FL 33487-2742

International Standard Book Number-10: 0-8493-1873-4 (Hardcover)
International Standard Book Number-13: 978-0-8493-1873-3 (Hardcover)
Library of Congress Card Number 2005043707

### Library of Congress Cataloging-in-Publication Data

Pharmaceutical biotechnology / edited by Michael J. Groves.—2nd ed.
    p. cm.  Includes bibliographical references and index.
  ISBN 0-8493-1873-4 (alk. paper)
  1.  Pharmaceutical biotechnology. I. Groves, M. J. (Michael John)

RS380.P475 2005
615'.19--dc22                                        2005043707

# Preface

In 1992 Mel Klegerman and I edited a textbook with the title *Pharmaceutical Biotechnology: Fundamentals and Essentials*. The present volume is effectively a second edition of this textbook and is designed to provide an update of a subject that has developed into a major component of current pharmaceutical research.

The decision to proceed with this second edition was prompted to some extent by the realization that the original book was out of print in the United States but was being photocopied and circulated in a number of countries, especially in the East and Far East. It is a pure coincidence that a number of my friends who contributed chapters to this volume have emanated from Turkey but one can only hope that copyrights will be respected this time!

Although it is true that the subject has developed over the past decade, some aspects have not changed significantly. Some sections from the original book have been developed and expanded. However, subjects such as fermentation, production, and purification have not changed very much over the years and are not discussed here. Interested readers are referred to the original volume for what remain excellent reviews of these subjects. In addition, the previous emphasis on biological response modifiers seems out of place here and the function and activity of proteins, for example, is better discussed elsewhere.

The chapter on the formulation of proteins has been expanded but the contents will reflect my own personal scientific interests, and the current literature will, inevitably, contain additional information on different approaches to the subject. Two additional chapters are included since some proteins, such as albumin or gelatin, can serve as drug delivery systems in their own right. In addition, proteins and phospholipids undergo characteristic interactions and deserve closer and separate attention.

The major expansion in this present volume concerns the subjects of proteomics and gene therapy, both of which offer so much promise for the future. Pulmonary administration is another likely route of delivery for the future and this is reviewed separately. Conventional wisdom suggests that proteins cannot be delivered orally but there is strong evidence suggesting that this is not always true and this is another exciting area that is reviewed here. The earlier review of vaccines has been expanded considerably since this is another area of current interest with potential for wider future application.

The question of whom this book is written for needs to be addressed. The previous volume was intended for industrial application by the publishers although the editors had the academic market in mind and even provided test questions and answers. This dichotomy has been avoided here since the intention is that students and graduate students would be most likely to gain the most benefit from a review of this type. Of course, scientists without a pharmaceutical background coming into the biotechnology industry for the first time will also find much to interest them, and it is hoped that a much wider audience both here and abroad will benefit.

The other question that needs to be answered in these days of the Internet is why a book is even necessary when so much information can be gleaned by simple use of a keyboard. In part, this is exactly why a textbook is needed. There is a real danger of students getting "information overload," with so much information being readily available that they are unable to digest and assimilate it. For example, a recent Google search for "chitosan stability" achieved over 22,000 "hits," most of which were useless since, for some reason, they were connected with cosmetic applications. Try searching under the term "emulsions" and be prepared for an astronomical response!

Sending a student to the library to carry out an old-fashioned book search in order to learn what has been achieved in a particular subject area is generally unpopular, but how else can the subject be appreciated from what might be called a classical perspective? The vast majority of new students do not seem to appreciate that there was any science prior to 1990. This is about as far back as most Internet databases can reach so, in effect, the student is denied access to the pioneering work that produced the advances in the first place.

Another related issue is the fact that groups of scientists are now putting the results of their research on the Internet which makes for a more rapid publication and, in some cases, this is of interest. However, what most neophytes do not appreciate is that this work has often not been peer-reviewed prior to publication and this can lead to uncertainty since the research may be flawed or even downright incorrect.

Uncritical acceptance of information is and remains the real issue with the so-called information age. The value of textbooks such as the present one is that access to the classical work is provided and much of the information has been digested to an acceptable format. With this in mind, individual chapters have suggested reading lists in an attempt to provide a bridge between the present omnivorous information bases and the subject basics.

I would like to thank the many friends and colleagues who have helped and encouraged me in this project. However, errors are mine and I take full responsibility.

**M. J. Groves**
*Chicago*

# Acknowledgments

I would like to acknowledge with gratitude the help I have received from friends and colleagues in putting this collection of essays together. Specifically I would like to recognize Ms. Helena Redshaw, CRC Press, for the help received in the assembly of the draft and Ms Alina Cernasev for encouragement and assistance with preparation of some of the figures. In addition I acknowledge with pleasure the discussions I have had with Dr. Simon Pickard (University of Illinois at Chicago) on the subject of measuring clinical outcomes. The results described in Chapter 15, however, are mine, and responsibility for errors that may have occurred lies with me.

# The Editor

**Michael J. Groves**, a pharmacist with a doctorate in chemical engineering, has spent much of his career working in industry and academe. Now retired, his scientific interests include dispersed drug delivery systems and quality control issues for parenteral drug products. Editor or joint editor of a number of books, he has published 400 research papers, patents, reviews, and book reviews. He is a Fellow of the Royal Pharmaceutical Society of Great Britain, the Institute of Biology, and the American Association of Pharmaceutical Scientists.

# Contributors

**H. O. Alpar**
Head of Centre for Drug Delivery
  Research
The School of Pharmacy
University of London
Brunswick Square, London, UK

**Kadriye Ciftci**
Temple University
School of Pharmacy
Philadelphia, Pennsylvania, USA

**Richard A. Gemeinhart**
Assistant Professor of Pharmaceutics
  and Bioengineering
Department of Biopharmaceutical
  Sciences
College of Pharmacy
University of Illinois at Chicago
Chicago, Illinois, USA

**David J. Groves**
Independent Consultant
Gimblett's Mill
Laneast, Launceston, Cornwall, UK

**Michael J. Groves**
(Retired) Professor of Pharmaceuticals
Department of Biopharmaceutical
  Sciences
College of Pharmacy
University of Illinois at Chicago
Chicago, Illinois, USA

**Anshul Gupte**
Temple University
School of Pharmacy
Philadelphia, Pennsylvania, USA

**Nefise Ozlen Sahin**
Mersin University
Faculty of Pharmacy
Department of Pharmaceutics
Yenisehir Campus
Mersin, Turkey

**Wei Tang**
Temple University
School of Pharmacy
Philadelphia, Pennsylvania, USA

**Charles P. Woodbury, Jr.**
Department of Medicinal Chemistry
  and Pharmacognosy
College of Pharmacy
University of Illinois at Chicago
Chicago, Illinois, USA

**Camellia Zamiri**
Department of Biopharmaceutical
  Sciences
College of Pharmacy
University of Illinois at Chicago
Chicago, Illinois, USA

# Table of Contents

# 1 Introduction

## Michael J. Groves, Ph.D., D.Sc.

In 1992 Mel Klegerman and I edited a text book called *Pharmaceutical Biotechnology: Fundamentals and Essentials* (Interpharm Press, Buffalo Grove, IL). We intended this book to encapsulate as much as possible the area identified as pharmaceutical biotechnology; and in the introductory chapter, we tried to provide a suitable definition of the term *at that time*.

Biotechnology simply means the technology of life or the industrial use of life forms. In principle it is a venerable term since we are all familiar with biotechnological processes, such as wine, beer, honey, bread and cheese making, and the development and processing of cereals. These connections can be traced back to the dawn of human civilization. In addition, exploitation of naturally derived products includes the use of wood, paper, rubber, leather, and turpentine, as well as objects woven from silk, wool, or cotton. Obviously, the definition of *pharmaceutical biotechnology* must then include all materials made or derived from natural materials that can be used as drugs for human or animal medicine. Accordingly, in our definitions, we attempted to provide a consensus, taking all these factors into account, such as

1. Use of biological systems to produce drug substances
2. Exploitation of modern technologies to produce drug substances from biological systems

More than a decade later, the word *biotechnology* has changed meaning in a very subtle way and now encompasses the use of material manipulation using DNA in living cells. If we add *pharmaceutical* to the key word we can suggest that the subject is about the development of drug substances, in particular those large protein and polypeptide molecules involved in some way or another in the disease states humans (and animals) suffer from. The inference now is that DNA in cells is manipulated so that the cells will produce a required molecule, usually proteinaceous, that has desirable properties in terms of the treatment of disease. If this term is accepted one can note that other areas of scientific endeavor not directly involved with drugs (such as DNA fingerprinting, used to uniquely identify individuals in forensic cases) are included but these need not be of concern here.

Some drugs situations are frankly confusing and need to be explored in greater depth. For example, tumor necrosis factor (TNF) is produced in the body as a natural defense against the many spontaneous tumors that arise in the body and has been suggested as a treatment for cancer, which may very well be effective under some conditions. However, TNF is also believed to be involved in the pain associated with rheumatoid arthritis, and monoclonal antibodies with specific activity against TNF are now on the market. The use of TNF is therefore a complex issue and needs to be evaluated with considerable care.

It should be pointed out that some of these applications of modern biotechnology are by no means free of controversy; the best-known example being the use of genetically modified (GM) cereals. In the United States GM products seem to be entering the marketplace except in specific areas; but they are banned in Europe and some other countries. These crops were simply an extension of the traditional processes used since the beginning of civilization to improve plants by selecting desirable features; and the modern process involving DNA manipulation is under better control, is more accurate, and is much faster.

The current marketing situation of GM cereals seems to be a result of hysteria in the popular press and, in the case of Europeans, a political move to boycott American imports. Taking a long-term and pragmatic view point, these issues may well disappear when time demonstrates they are incorrect in the first place and unfounded in the second.

A decade ago sales of pharmaceutical biotechnological products were estimated at about $2.4 billion annually worldwide and the future looked good (Klegerman and Groves, 1992). Stock markets looked favorably on biotech stocks and all seemed set for a bright future. Suddenly, companies with a promising product and future started going out of business and for several years there was a marked downturn in the value of biotech stocks as investors turned elsewhere. What had happened? Of course this downturn occurred during a depression in the value of stocks in general, but there were more subtle reasons underlying the situation in biotechnology. The scientific facts were oversold to investors who did not understand the science and were usually not interested in anything other than a return on their money. Companies faced stark reality when the FDA refused to recognize the often superficial work carried out on what were intended to be human drugs, and when they attempted to comply they found, all of a sudden, that compliance came with a high price tag. Some of the bigger companies could afford to hire the skilled employees and the expensive equipment required to get a drug onto the marketplace; but the smaller ones could not and went out of business.

This period of upheaval lasted most of the decade and to a diminished degree is still going on. Investors have become wiser and more knowledgeable and pharmaceutical biotechnology has survived in a marginally different form. Ten years ago there were 16 approved drugs, with another 20 or so awaiting approval, and perhaps as many as 135 in various phases of development. The total market was estimated to be $2.5 billion in 1991; in 2003 it had risen to $500 billion with just 500 drugs on the market and perhaps as many as 10,000 drug targets. This has to be considered in the context of healthcare costs overall, with an estimated market of nearly $2,100 billion.

The industry has stabilized and the future appears to be assured. Perhaps confidence for the future has prompted a revision of this present book. Many issues remain to be resolved, but the excitement and challenge keeps our interest for a long time ahead.

Returning to the genetically engineered drugs, the word *pharmaceutical* also implies that these drugs need to be prepared, formulated, and delivered to the patient in accordance with the more traditional technologies associated with small molecule drugs. With this in mind the present book takes the earlier edition a step further and provides pharmaceutical methodology on the properties and formulation of these newer types of drugs, including information on the analytical procedures involved.

Some aspects of the subject are not dealt with here, for example, large-scale manufacturing methodology, since this has remained much as it was 10 years ago, albeit with some improvements. The student searching for this information will find it covered in two chapters of the first edition, and I take responsibility for not including the same material in this edition. However, it might be noted here that one of the main issues at present is the limited amount of fermentation capacity throughout the world. Many of the drug candidates are prepared by manipulation of growing cultures of yeast or *Escherichia coli* cells that have been suitably modified, excreting the required protein or polypeptide into their liquid growth medium. On a large scale, big and very expensive fermentation tanks are required for this process, and, at least until recently, there was a shortage of this type of equipment, which tended to decrease availability of some desirable proteinaceous drug substances.

## THE CHAPTER SUBJECTS IN THIS BOOK

Ten years ago monoclonal antibodies (MAb) were of considerable interest but faded after clinical and commercial failures. Further clinical and laboratory testing revealed that only humanized MAb were likely to be of any benefit in a human patient. From a practical standpoint this research developed, and humanized MAb were launched onto the market, with success in the treatment of several diseases including rheumatoid arthritis. These developments justify a chapter on the subject in this book.

The amount of investigational work on the subject of formulation also needs further review and amplification. Basic formulation studies have suggested that proteins could be used to form drug delivery systems (e.g., microparticles); the interactions of proteins with phospholipids is of sufficient interest to justify a short chapter.

Genetic engineering and genomics have also seen developments over the past decade and these subjects need their own chapters, especially since the much lauded completion of the Human Genome Project. Although the expenditure on this project was vast and the science complicated beyond imagination, we can be tempted to look at the results and ask, "So what?". The public was educated to expect good things (often undefined) from the project but, looking with the advantage of hindsight, it may have been oversold. Almost every day now we are told that scientists have isolated the gene responsible for this or that disease—but where are the cures for these diseases? For example, the genetic variant responsible for the terrible Huntington disease (chorea) in which nerve cells in the brain are irreversibly killed, resulting in mental deterioration and spasmodic muscle movements, was identified in the early 1980s. Twenty years later we are probably no closer to finding a cure, although diagnosing the disease is now easier. Some of this frustration is described in Chapters 5 and 13 in this present edition.

The reason for this dissatisfaction in most cases is that the gene may have been identified but not necessarily the associated protein(s). Protein identification has been achieved in a few limited cases, but it is more difficult to produce a protein in sufficient quantities for testing or, more critically, for delivery to the patient. That these proteins will be found, characterized, and produced in sufficient quantities for human trial and evaluation cannot be doubted; just how long this will take in individual situations is an entirely different matter.

Indeed, it is often the case that a disease is experienced by only a few unfortunate individuals, who collectively cannot afford the very real costs associated with the development and production of a drug. This topic will be discussed in the final chapter; but the sad truth of the matter is that expectations may have been raised with no real expectation of reality. All the hype in the press has led the public to expect miracle cures for cancer, as an example. Moreover, many companies were formed to exploit the available technology or to develop new technology. They have found to their cost that the market was often simply not big enough to justify the expense of running a business, and many of these companies have quietly disappeared.

# 2 Proteins

*Charles P. Woodbury, Jr., Ph.D.*

## CONTENTS

## PROTEIN STRUCTURE

### CLASSIFICATION OF PROTEINS

Proteins are often put into one of two categories, globular or fibrous, on the basis of their overall structure. Globular proteins have compactly folded structures and tend to resemble globes or spheroids in overall shape. Some are soluble in water and can function in the cytosol or in extracellular fluids; other globular proteins are closely associated with lipid bilayers, being buried in part or in whole in the biological membranes where they function. Globular proteins include enzymes, antibodies, membrane-bound receptors, and so on. Fibrous proteins, in comparison, are generally insoluble in water or in lipid bilayers, and they have extended conformations, often forming rodlike structures or filaments. In connection with their solubility, fibrous proteins are generally found as aggregates. Their major biological role is a structural or mechanical one, to support or connect cells or tissues. These proteins include keratin (a major component of skin and hair), collagen (in connective tissue), fibroin (the protein found in silk), and many others.

Another way of classifying proteins is by their biological function. Viewed this way, proteins can be put into one of several categories. Below are some of these categories and one or more examples for each category.

- Structural support and connection: Collagen, elastin
- Catalysis: Enzymes such as alcohol dehydrogenase, acetyl cholinesterase, or lysozyme
- Communication: Peptide hormones, hormone receptors
- Defense: Antibodies, toxins from snake venom (these toxins may themselves be enzymes)
- Transport: Hemoglobin, albumin, ion "pumps"
- Energy capture: Rhodopsin
- Mechanical work: Actin, myosin, tubulin, dynein

## AMINO ACIDS

The building blocks of proteins are the 20 different naturally occurring amino acids. These amino acids each have a central carbon atom, the *alpha* ($\alpha$)-carbon, to which are attached a carboxyl group, an amino group (or in the case of proline, an imino group), a hydrogen atom, and a side chain. The side chain is the feature that distinguishes one amino acid from another; the other three groups about the $\alpha$-carbon are common to all 20 amino acids. The $\alpha$-carbon is a chiral center, which means that each amino acid can exist in either of two enantiomeric forms (denoted by D and L); the L form predominates in nature.

Based on the properties of the side chains, the 20 amino acids can be put into six general classes. The first class contains amino acids whose side chains are aliphatic, and is usually considered to include glycine, alanine, valine, leucine, and isoleucine. The second class is composed of the amino acids with polar, nonionic side chains, and includes serine, threonine, cysteine, and methionine. The cyclic amino acid proline (actually, an imino acid) constitutes a third class by itself. The fourth class contains amino acids with aromatic side chains: tyrosine, phenylalanine, and tryptophan. The fifth class has basic groups on the side chains and is made up of the three amino acids lysine, arginine, and histidine. The sixth class is composed of the acidic amino acids and their amides: aspartate and asparagine, and glutamate and glutamine.

The exact ionic state of the side chains in the last two classes will depend on the pH of the solution. At pH 7.0 the side chains of glutamate and aspartate have ionized carboxylates, and the side chains of lysine and arginine have positively charged, titrated amino groups. Since the pKa of the imidazolium side chain of histidine is about 6.0, we expect to find a mixture of uncharged and charged side chains here, with the uncharged species predominating at pH 7.0.

Amino acids are commonly represented by three-letter abbreviations, for example Pro for proline. There is also an even more succinct one-letter abbreviation or code to represent each amino acid. These abbreviations, the structures of the amino acids, and their isoelectric points (the pH at which the amino acid has no net electrical charge, i.e., a balance of positive and negative charge has been struck) are summarized in Table 2.1.

## PRIMARY PROTEIN STRUCTURE

In a protein the amino acids are joined together in a linear order by amide linkages, also known as peptide bonds. The sequence of the covalently linked amino acids is referred to as the primary level of structure for the protein. In writing the sequence of amino acids in a chain, it is conventional to orient the chain so that the amino acid on the left is the one with a free amino group on its α-carbon, while the last amino acid on the right is the one whose α-carbon carboxylate is free. In other words, the amino- or N-terminus of the peptide chain is written on the left, and the carboxyl- or C-terminus is written on the right. One more convention: the term "backbone" for a protein refers to the series of covalent bonds joining one α-carbon in a chain to the next α-carbon.

In a peptide bond, the carboxylate on the α-carbon of one amino acid is joined to the amino group on the α-carbon of the next amino acid. The peptide bond that joins the amino acids in a protein has partial double bond character. In a standard representation of an amide group, the carbonyl oxygen and carbon share a double bond, while the amide nitrogen carries a lone pair of electrons. Actually, there are more ways than this to share the electrons among the atoms involved. The result is a shift of electron density into the region between the carbon and nitrogen atoms, with development of a partial positive charge on the nitrogen and partial double bond between the carbon and nitrogen. Because of the electronegativity of the oxygen in the carbonyl group, another result is a shift in electron density toward that atom and development of a partial negative charge (Figure 2.1).

The partial double bond character of the carbon–nitrogen bond requires co-planarity of the carbonyl carbon and oxygen, the amide nitrogen and its hydrogen, and the α-carbons of both amino acids involved—a total of six atoms. The double bond character of the linkage would be disrupted by rotation about the bond joining the amide carbon and nitrogen, and so the peptide linkage resists torsional rotation and tends to stay planar. Because of the planarity of the peptide link, there are essentially only two different configurations allowed here, *cis* and *trans*. In the *cis* configuration the α-carbons of the first and second amino acids are closer together than they are in the *trans* configuration. This leads to steric repulsion between the side chains attached to these two atoms, and so the *cis* configuration is energetically much less favorable than the *trans*. The great majority of amino acids in proteins are thus found to be joined by peptide linkages in the *trans* configuration. Proline is the only major exception to this rule. Because of steric constraints on the cyclic

## TABLE 2.1

AROMATIC AMINO ACIDS

Tyrosine
( Tyr, Y)
pI = 5.7
pK ( sidechain) = 10.1

Phenylalanine
( Phe, F)
pI = 5.5

Tryptophan
( Trp, W)
pI = 5.9

BASIC AMINO ACIDS

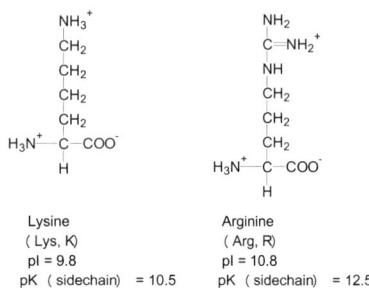

Lysine
( Lys, K)
pI = 9.8
pK ( sidechain) = 10.5

Arginine
( Arg, R)
pI = 10.8
pK ( sidechain) = 12.5

ACIDIC AMINO ACIDS AND THEIR AMIDES

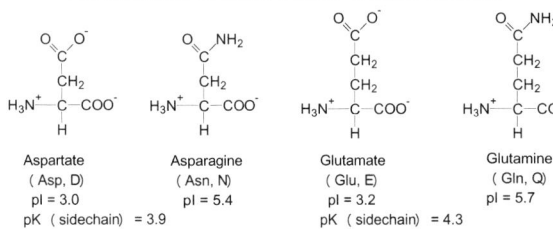

Aspartate
( Asp, D)
pI = 3.0
pK ( sidechain) = 3.9

Asparagine
( Asn, N)
pI = 5.4

Glutamate
( Glu, E)
pI = 3.2
pK ( sidechain) = 4.3

Glutamine
( Gln, Q)
pI = 5.7

Histidine
( His, H)
pI = 7.6
pK ( sidechain) = 6.0

POLAR AMINO ACIDS

Serine
( Ser, S)
pI = 5.7

Cysteine
( Cys, C)
pI = 5.0

Threonine
( Thr, T)
pI = 5.6

Methionine
( Met, M)
pI = 5.7

ALIPHATIC AMINO ACIDS

Glycine
( Gly, G)
pI = 6.0

Valine
( Val, V)
pI = 6.0

Alanine
( Ala, A)
pI = 6.0

Leucine
( Leu, L)
pI = 6.0

Isoleucine
( Ile, I)
pI = 6.0

CYCLIC AMINO ACID

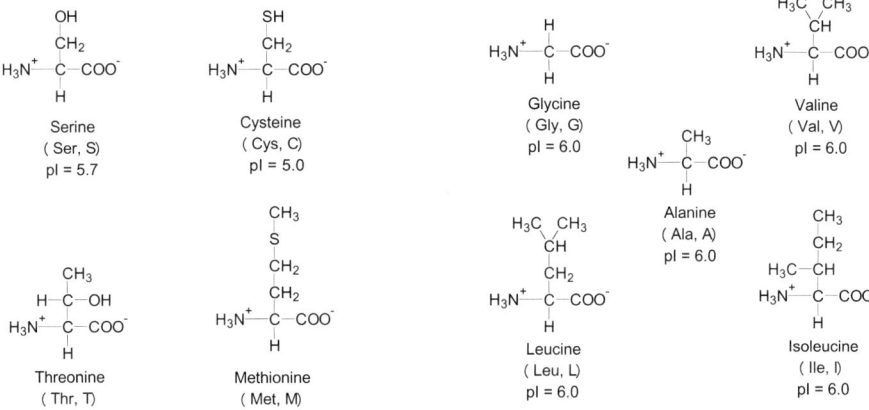

Proline
( Pro, P)
pI = 6.3

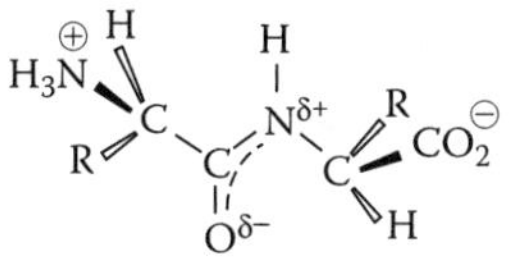

**FIGURE 2.1** The peptide bond, showing the partial double bond character of the amide linkage and the development of partial charges on the nitrogen and oxygen atoms.

side chain in proline, the *cis* and *trans* isomers are more nearly equal in terms of energy, and the *cis* isomer occurs relatively frequently.

The constraints on rotation are much weaker for the carbon–carbon bond joining an amino acid's $\alpha$-carbon and its carbonyl carbon; this bond enjoys a great deal of rotational freedom, especially by comparison to the highly restricted amide bond. Rotational freedom is also enjoyed by the nitrogen–carbon bond that joins the amide nitrogen and the adjacent $\alpha$-carbon. (The exception here is again proline because of its ring structure.) Thus, of the three covalent bonds contributed by each amino acid to the protein backbone, one is strongly constrained in terms of rotation while the other two bonds are relatively free to rotate. This rotational freedom in each amino acid allows protein chains to wind about in a large number of conformations. Some of these conformations will be energetically preferred over others, however, leading to the formation of secondary and higher levels of structure for the polypeptide chain.

The side chains of cysteine residues contain a terminal thiol (–SH) group. These thiols are sensitive to oxidation/reduction reactions, and can form covalent disulfide (-S-S-) bridges among themselves. The two joined cysteine residues are then said to be combined into a single *cystine* unit. There may be one or several of these disulfide bridges present in a polypeptide. These disulfide bridges can be formed between cysteine residues that may be separated by tens or hundreds of residues along the polypeptide chain, and they can thus bring two distant regions of the polypeptide chain into close spatial proximity, a factor that may be quite important in determining the overall shape of the protein. Disulfide bridges can also be formed between cysteine residues on separate polypeptide chains, and can serve to hold two chains together covalently.

For naturally occurring proteins the polypeptide chains will have between 50 and 2000 amino acids joined together. On the average an amino acid residue in these chains will have a molecular weight of about 110, so the typical polypeptide chain will have a molecular weight between 5500 and 220,000.

## PROTEIN SECONDARY STRUCTURE

Secondary structure refers to regularities or repeating features in the conformation of the protein chain's backbone. Four major types of secondary structure in proteins are: (1) the *alpha* ($\alpha$) helix, formed from a single strand of amino acids; (2) the *beta* ($\beta$) sheet, formed from two or more amino acid strands (from either the same chain or from different chains); (3) the *beta* ($\beta$) bend or reverse turn, in a single strand; and (4) the collagen helix, composed of three strands of amino acids.

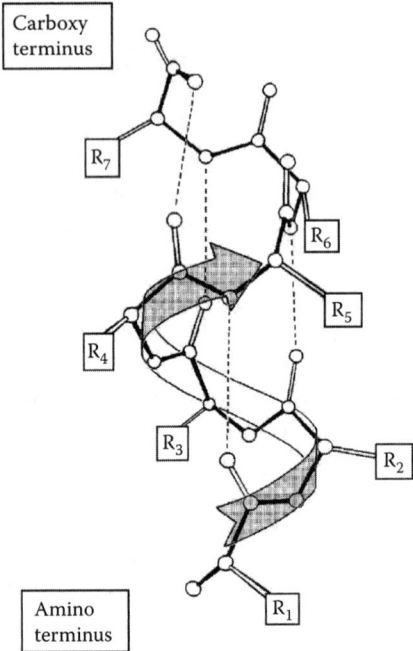

**FIGURE 2.2** The alpha ($\alpha$) helix, showing the pattern of intrachain hydrogen bonds that stabilize the structure, and the radial extension of amino acid side chains from the helix axis. (Adapted from Richardson, J.S. [1981]. The anatomy and taxonomy of protein structure. *Advances in Protein Chemistry,* 34, 167–339, copyright 1981, with permission from Elsevier Science.)

The $\alpha$ helix most commonly found in proteins is represented in Figure 2.2. This is a right-handed helix built from L-amino acids. The $\alpha$ helix is stabilized principally by a network of hydrogen bonds. These hydrogen bonds link amino acids that are otherwise separated along the protein chain. Specifically, an amino acid residue at position $i$ in the chain will be hydrogen bonded to the amino acid at position $i + 3$; the hydrogen bond is between the carbonyl oxygen of residue $i$ and the amide hydrogen of residue $i + 3$, and the bond is nearly parallel to the long axis of the helix. This interaction among neighboring residues compacts the polypeptide chain by comparison to a fully extended conformation. Functional groups on the amino acid side chains do not participate in the network of stabilizing hydrogen bonds. Instead, the side chains extend radially outward as the backbone forms the helical coil.

Some amino acids are found more frequently in $\alpha$ helices than are others. Some amino acids often found in $\alpha$ helices are alanine, phenylalanine, and leucine, since it is relatively easy for these to rotate into the proper conformation without steric clash or other repulsions between the side chains. Some other amino acids are rarely found in $\alpha$ helices, primarily because of such unfavorable interactions; these include arginine and glutamate. Also, proline is not found in the middle of helices because the nitrogen–carbon bond in the pyrrolidine ring of proline cannot rotate into the proper conformation to keep up the hydrogen bonding network (proline can sometimes be found at the end of an α helix, however).

β sheets come in two types, and both involve almost complete extension of the protein backbone, in contrast to the α helix. In the parallel β sheet, two or more peptide chains align their backbones in the same general direction such that the hydrogen bonding moieties of the amide linkages on adjacent chains line up in a complementary fashion with donors opposing acceptors. Again, there is a hydrogen bonding network involving carbonyl oxygens and amide hydrogens, but the bonding is between chains, not along the same chain as was the case with the α helix. The antiparallel β sheet is quite similar to the parallel type, with adjacent chains having parallel but opposite orientations. Frequently, a long peptide chain will fold back on itself in such a way that two or more regions of the chain will line up next to each other to form either a parallel or an antiparallel β sheet structure. For either kind of β sheet structure the side chains will project away from (up or down from) the plane of the sheet. The two types of β sheet are represented in Figure 2.3.

As with the α helix, some amino acids are found more frequently in β sheets than are others. Because of packing constraints, small nonpolar amino acids are the most frequently found here, including glycine, alanine, and serine along with some others. Proline is sometimes found in β sheets, but tends to disrupt them. This is because the pyrrolidine ring of proline constrains the protein backbone from adopting the almost completely extended conformation required to form the β sheet. Amino acids with bulky or charged side chains also disrupt the packing and alignment

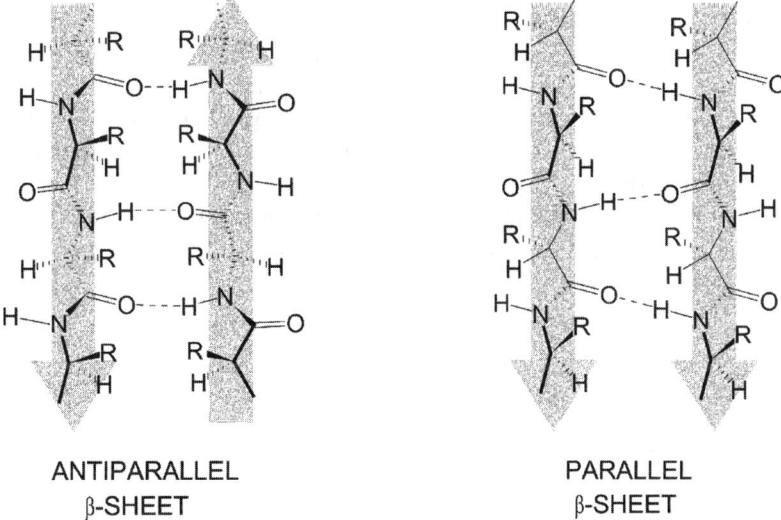

ANTIPARALLEL
β-SHEET

PARALLEL
β-SHEET

**FIGURE 2.3** The antiparallel and the parallel beta (β) sheet or ribbon, showing the pattern of interchain hydrogen bonds and the protrusion out of the ribbon's plane of the amino acid side chains (denoted by R). The *arrows* indicate the relative directions of the peptide chains.

needed to form a β sheet. Thus glutamate, aspartate, arginine, tryptophan, and some others are not common in either type of β sheet.

The third major type of secondary structure is the β bend or reverse turn, where the polypeptide chain turns back on itself (for example, to form an antiparallel β sheet). There are actually several different subtypes of reverse turn that differ in details of the bond angles in the participating amino acids. Such turns typically extend over four adjacent amino acids in the chain, with hydrogen bonding of the carbonyl oxygen of the first residue in the turn with the amide hydrogen of the fourth residue. Proline is often found as the second residue in these turns; its rigid imino ring helps to bend the backbone of the chain. Glycine is often found as the third residue in β bends, because its hydrogen side chain offers little steric repulsion to the tight packing required in this region of the turn.

Fibrillar collagen, the fibrous protein found in connective tissue, has its own characteristic primary and higher levels of structure. A strand of fibrillar collagen is composed of three peptide chains. Each chain typically contains glycine in every third position, and contains a high proportion of proline, lysine, hydroxyproline, and hydroxylysine residues. (Hydroxyproline is a modified proline with a hydroxy group on the ring at the 4 position, while hydroxylysine carries a hydroxyl group on the number 5 carbon.) The primary structure can be written as Gly-X-Y, where the X position often contains proline. The hydroxylated amino acids are formed after biosynthesis of the peptide chain by enzymatic hydroxylation of the corresponding unmodified amino acids. Enzyme specificity restricts these unusual amino acid derivatives to the Y position in the Gly-X-Y sequence.

In fibrillar collagen, each participating amino acid chain forms a left-handed helix (unlike the α helix, which is right-handed), then three of these chains wind about each other to form a right-handed super helix. This triple-stranded super helix is stabilized by hydrogen bonds between adjacent peptide strands (again different from the α helix, where the hydrogen bonds are all among residues along the same strand), and by covalent chemical cross-links that are formed after the individual chains have wound around each other. These cross-links are Schiff base linkages between the side chains of unmodified and modified lysine residues of adjacent peptide strands. Overall, the combination of crosslinking and superhelical structure makes for a protein that is mechanically very strong, one that is rigid and resists bending and stretching— very desirable characteristics in a protein used in connective tissue.

## TERTIARY AND QUATERNARY STRUCTURE IN GLOBULAR PROTEINS

Tertiary structure is defined by the packing in space of the various elements of secondary structure. A globular protein may have several α helices and two or more regions of β sheet structure that are all tightly packed together in a way characteristic of the protein. This regularity in packing of α helices and β sheets, together with the β bends, constitutes the third level of structure of proteins. Parts A, B, and C of Figure 2.4 compare the primary, secondary, and tertiary levels of structure for a polypeptide chain.

In large proteins, tertiary structures can often be divided into domains. A domain is a region of a single peptide chain with a relatively compact structure; it has folded

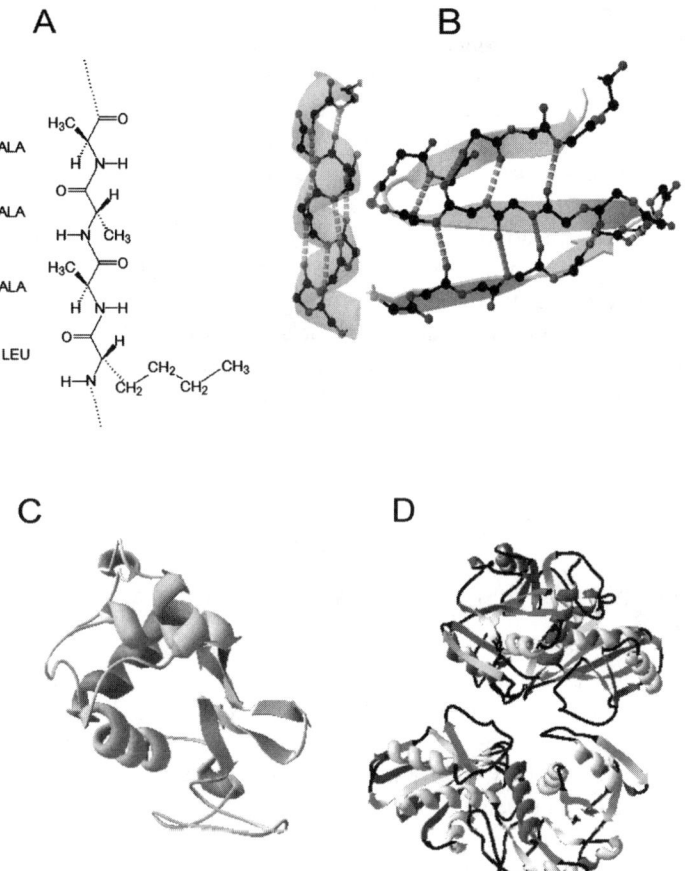

**FIGURE 2.4** The levels of protein structure. A. The primary level, showing the peptide backbone and side chains. B. Elements of secondary structure, an α helix and three strands of an antiparallel β sheet. C. The tertiary structure of hen egg white lysozyme, showing the packing of α helices and β sheet structures. D. An example of quaternary structure: the dimer of glycerol phosphate dehydrogenase from *E. coli*. (Images for B, C, and D created using the Swiss protein data bank viewer spdbv version 3.7 and protein data bank files 1HEW and 1dc6.)

up independently of other regions of the peptide chain. Domains are of course still connected one to another by the peptide backbone. Once formed, two or more domains may pack together in a characteristic fashion that defines the overall tertiary structure of that region of the protein.

Quaternary structure refers to the specific aggregation or association of separate protein chains to form a well-defined structure. Part D of Figure 2.4 compares the quaternary structure of a dimeric protein (two polypeptide chains) to the lower levels of protein structure. The separate protein chains are often referred to as subunits or monomers; these subunits may be identical or may be of quite different sequence and structure. The forces holding the subunits together are weak, noncovalent interactions. These include hydrophobic interactions, hydrogen bonds, and van der Waals interactions. Because of the large number of such possible interactions between two protein surfaces, the aggregate can be quite stable, despite the weakness of any single noncovalent interaction. Furthermore, the association can be quite specific in matching protein surfaces to one another: because of the short range of the stabilizing interactions, surfaces that do not fit closely against one another will lack a large fraction of the stabilization enjoyed by those protein surfaces that do fit snugly together. Thus, mismatched subunits will not form aggregates that are as stable as subunits that are properly matched.

It is common practice to include under the umbrella of quaternary structure other kinds of complexes between biopolymers. For example, the complex of DNA with histones to form nucleosomes may be said to have quaternary structure, the DNA also being regarded as a component subunit.

## PROTEIN BIOSYNTHESIS

### MESSENGER RNA AND RNA POLYMERASE

The primary repository of genetic information in the cell is DNA. For this information to be expressed in the form of an enzyme, an antibody, or other protein, an intermediate between the protein and the DNA is used. This intermediate is composed of RNA, and (after some intermediate processing) is called messenger RNA, or mRNA. The general flow of information, from DNA to RNA and finally to protein, was summarized by Francis Crick as the "Central Dogma" for molecular biology. Figure 2.5 presents schematically the main features of the Central Dogma as applied to a eukaryotic cell with a nucleus.

For a given gene, the corresponding mRNA is complementary to (matches the base pairing of) the strand of the DNA where the information in the gene is stored. The conventions on nucleic acid nomenclature, nucleic acid structure, and the Watson-Crick rules for complementary base-pairing are given in Chapter 3.

The copying of the template DNA into a strand of RNA is called transcription, and the RNA that results is referred to as a transcript. It is conventional to write double-stranded DNA sequences from left to right, with the top strand having its 5′ end on the left and its 3′ end on the right (the numbering convention here refers to particular carbons of the nucleotide sugar moieties). Also by convention, the bottom DNA strand is the template for the RNA transcript; this is the DNA strand that is complementary in sequence to the RNA transcript. The RNA transcript is identical in sequence to the top (or "coding") DNA strand, except that the RNA contains the base uracil where the coding DNA contains thymine, and the sugar deoxyribose of DNA is replaced by ribose in RNA.

Enzymatic synthesis of RNA from a DNA template runs along the template strand toward its 5′ end. In a gene, this corresponds to movement in the 5′ to 3′ direction along the complementary coding strand. The 5′ to 3′ direction on the coding

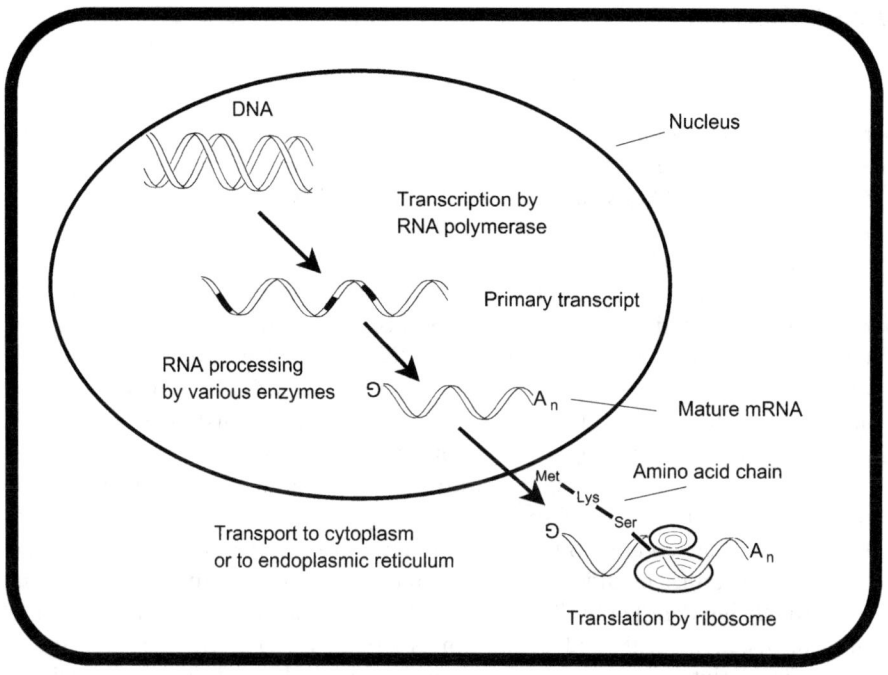

**FIGURE 2.5** The "Central Dogma": Double-stranded DNA is transcribed to messenger RNA (in eukaryotes, with processing of the transcript), which in turn is translated by the ribosome into the chain of amino acids making up a protein.

strand is referred to as "downstream" in the gene, while the 3′ to 5′ direction along the coding strand is described as "upstream."

The process of transcription runs under quite complicated control mechanisms. Among other things, these mechanisms direct the RNA synthesis to start and stop in precise places on the DNA, and they control the rate of mRNA synthesis. In this way gene expression can be integrated with the metabolic state of the cell, and with the cell cycle for cellular replication and differentiation.

The synthesis of mRNA is catalyzed by the enzyme RNA polymerase. Bacteria have only one type of RNA polymerase, which is responsible for virtually all bacterial RNA polynucleotide synthesis. Higher organisms have three types of RNA polymerase. Type II RNA polymerase transcribes genes into mRNA molecules and synthesizes certain small RNAs found in the nucleus. This enzyme is the one closest in general function to the bacterial RNA polymerase. Type I RNA polymerase in eukaryotes transcribes genes for certain RNA species that are a part of the ribosome (ribosomal RNA, or rRNA; see below). Type III RNA polymerase transcribes genes for certain types of small RNA molecules used in RNA processing and in protein transport. It also transcribes certain genes involved in protein synthesis, including the genes for a small ribosomal RNA species and for all the transfer RNA species.

## mRNA Processing

In prokaryotes (single-cell organisms lacking organelles), the mRNA can immediately be used to direct protein synthesis. This is not so for eukaryotes (higher organisms consisting of nucleated cells), where the mRNA receives extensive processing before it is ready to be used in protein synthesis. This processing includes removal of certain RNA sequences, the addition of RNA bases at the ends of the molecule (these additional bases are not directly specified by the transcribed gene), and the chemical modification of certain bases in the RNA.

In eukaryotes, transcription takes place in the cell's nucleus (Figure 2.5). The primary RNA transcript often includes copies of DNA regions that do not code for protein sequences; these DNA regions are the so-called intervening regions, or introns, found in many eukaryotic genes. The DNA regions that actually code for amino acid sequences are known as exons (expressible regions). The parts of the RNA transcript corresponding to the introns must be removed before the RNA can faithfully direct synthesis of the protein from the copy of the exons. The process of removing the intron RNA is referred to as *splicing*, and it occurs before the transcript leaves the nucleus. Figure 2.6 presents this series of operation in RNA processing.

There are further modifications of the transcript, however. To protect the transcript from being digested by RNA nucleases outside the nucleus, the RNA gains a run of polyadenylic acid residues (known as a poly A tail) at the 3' end. At the 5' end, the transcript gains a "cap" of a guanine residue attached via an unusual 5-5 triphosphate link (Figure 2.7). Furthermore, this guanine is usually methylated at the N7 position, and in some cases the sugar residues of the first one or two bases in the original transcript are methylated at the 2' hydroxyl group. Besides protecting the RNA transcript against degradation, the cap serves as a binding point for the ribosome. Finally, after all these processing steps, the transcript RNA is ready to be

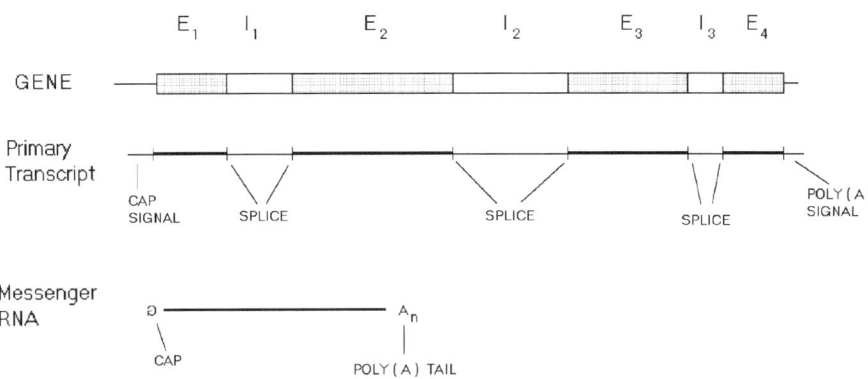

**FIGURE 2.6** Processing of mRNA in the eukaryotic cell, emphasizing the splicing of the messenger with excision of intervening sequences (introns) away from the RNA that codes for amino acid sequences (exons). The reversed G indicates the methylated guanine cap found at the 5' end of processed messenger RNA (mRNA), and the symbol An indicates the run of adenine residues that form a tail at the 3' end of the mRNA.

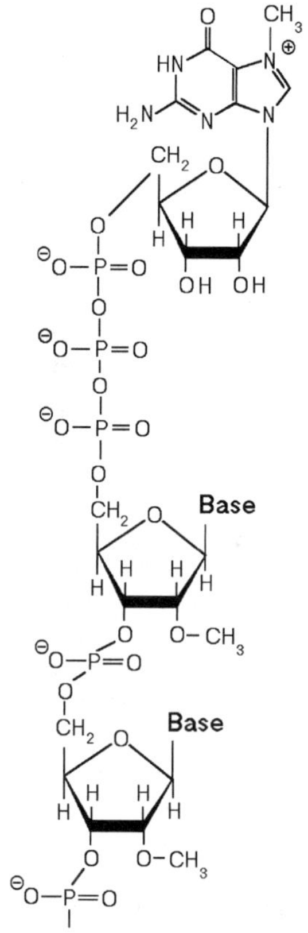

**FIGURE 2.7** The "cap" structure found in eukaryotic cells, showing the modified terminal nucleotide containing N7-guanine, the unusual linkage of this nucleotide to the rest of the chain, and the methylation of two neighboring sugar groups.

used in protein biosynthesis, and it passes out of the nucleus to the endoplasmic reticulum or into the cytosol as a fully processed messenger RNA.

## THE GENETIC CODE

The way in which the coding strand of DNA specifies the sequence of amino acids in a protein is known as the genetic code. The code is made up of triplets of nucleic acid bases, known as codons. A given series of three bases in the coding strand DNA of a gene will unambiguously specify a particular amino acid and no other.

There are 64 possible triplet codons ($64 = 4^3$; there are four choices of base at each of the three positions in a codon). However, only 61 codons are used to designate amino acids; three codons are used to signal the end of the amino acid sequence for a protein, and these three codons are called termination or stop codons. With 20 amino acids used in most proteins and 61 codons to specify amino acids, some amino acids are coded for by more than one codon, i.e., there are synonymous

**TABLE 2.2**
**The Genetic Code**

| First Base (5'-end) | Second Base U | Second Base C | Second Base A | Second Base G | Third Base (3'-end) |
|---|---|---|---|---|---|
| U | Phe | Ser | Tyr | Cys | U |
|   | Phe | Ser | Tyr | Cys | C |
|   | Leu | Ser | STOP | STOP | A |
|   | Leu | Ser | STOP | Trp | G |
| C | Leu | Pro | His | Arg | U |
|   | Leu | Pro | His | Arg | C |
|   | Leu | Pro | Gln | Arg | A |
|   | Leu | Pro | Gln | Arg | G |
| A | Ile | Thr | Asn | Ser | U |
|   | Ile | Thr | Asn | Ser | C |
|   | Ile | Thr | Lys | Arg | A |
|   | Met | Thr | Lys | Arg | G |
| G | Val | Ala | Asp | Gly | U |
|   | Val | Ala | Asp | Gly | C |
|   | Val | Ala | Glu | Gly | A |
|   | Val | Ala | Glu | Gly | G |

codons for some amino acids. This is sometimes referred to as "degeneracy" in the genetic code.

Since a messenger RNA is an exact copy of a gene's coding strand (with uracil replacing thymine), the mRNA carries a copy of the codons that determine the amino acid sequence in the protein as specified by the gene. The codons here are in RNA form, not in DNA form. This genetic information can be read from the mRNA by a ribosome, codon by codon, as the ribosome covalently joins together the corresponding amino acids, in exactly the order of the codons on the original coding strand of DNA.

A standard version of the genetic code for the triplet codons on a messenger RNA is given in Table 2.2.

## ACTIVATED AMINO ACIDS AND TRANSFER RNA

Before amino acids can be joined together by the ribosome, they must be activated by attachment to a special species of RNA known as transfer RNA (tRNA). The activation serves two purposes: first, activation chemically prepares the amino acid for forming an amide linkage to another amino acid; and second, its attachment to a tRNA helps to direct the proper sequential joining of the amino acids in a protein.

Transfer RNA comes in many different types that can be distinguished by details in their nucleotide sequence. Regardless of sequence differences, all tRNA molecules fold up into the same general structure with several short double-helical regions

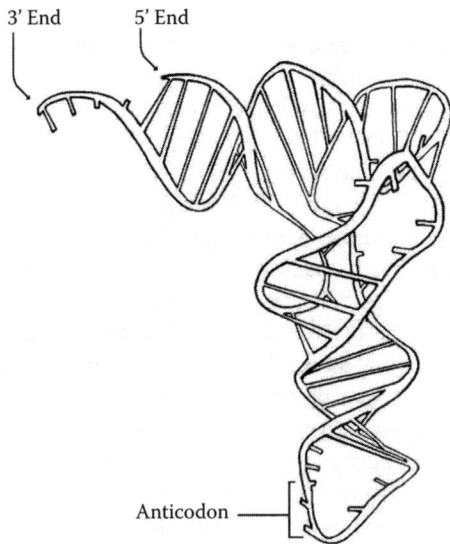

3' End          5' End

Anticodon

**FIGURE 2.8** The general structure of transfer RNA (tRNA) showing the phosphodiester backbone as a ribbon, with bases or base pairs projecting from the backbone. (Adapted from Rich, A. [1977]. Three-dimensional structure and biological function of transfer RNA *Accounts of Chem. Res.,* 10, 388–396, copyright 1977, American Chemical Society, with permission from *Accounts of Chemical Research.*)

(called stems) and short single-stranded loops at the ends of most of the stems. The overall spatial shape resembles a lumpy L (Figure 2.8).

At one end of the tRNA strand there is a short stem with a protruding single-stranded region that has a free 3' hydroxyl group. This is the site where an amino acid is attached covalently via its carboxyl group. The joining of the proper amino acid to its corresponding tRNA is catalyzed by a special enzyme, one of the aminoacyl-tRNA synthetases. There is a specific synthetase for joining each type of amino acid and its corresponding tRNA. This avoids the problem of having, say, an alanine residue attached to the tRNA for glutamine, which might lead to insertion of the wrong amino acid residue in a protein and the consequent loss of activity or specificity in that protein.

On each of the tRNA molecules, one of the single-stranded loops contains a trinucleotide sequence that is complementary to the triplet codon sequence used in the genetic code to specify a particular amino acid. This loop on the tRNA is known as the anticodon loop, and it is used to match the tRNA with a complementary codon on the mRNA. In this way the amino acids carried by the tRNA molecules can be aligned in the proper sequence for polymerization into a functional protein.

## THE RIBOSOME AND ASSOCIATED FACTORS

The ribosome is a huge complex of protein and nucleic acid that catalyzes protein synthesis. There are differences between prokaryotic and eukaryotic ribosomal structure;

for simplicity this discussion will be centered on the synthesis of proteins in the bacterium *Escherichia coli*. The general organization and functions are the same, however, in both prokaryotic and eukaryotic systems.

All ribosomes have two subunits, and each subunit contains several protein chains and one or more chains of RNA (ribosomal RNA, or rRNA). In the ribosome from *E. coli*, the smaller of the two subunits is known as the 30S subunit and the larger is referred to as the 50S subunit. (The unit S stands for Svedberg, a measure of how rapidly a particle sediments in a centrifuge.) The two subunits combine to form the active 70S ribosomal assembly. The special RNA molecules that are a part of the ribosome are quite distinct from messenger or transfer RNA molecules, and they play important roles in forming the overall ribosomal quaternary structure and in aligning mRNA and tRNA molecules during protein biosynthesis.

At one point or another during protein synthesis, several other proteins will be associated with the ribosome. These include factors that help in initiating the synthetic process, others that help in elongating the peptide chain, and yet others that play a role in terminating the synthesis of a peptide chain. Beyond this, there is also the mRNA to consider, as well as the aminoacylated tRNA molecules. Finally, since protein biosynthesis consumes energy, there is the hydrolysis of ATP and GTP to AMP and GDP, respectively, by the ribosome.

## MESSAGE TRANSLATION AND PROTEIN SYNTHESIS

Protein synthesis has three main stages: initiation, elongation, and termination. In the initiation stage, the 30S subunit binds a mRNA molecule, then binds a 50S subunit. The mRNA is aligned for proper translation by complementary hydrogen bonding to a portion of a rRNA molecule found in the 30S subunit. The assembly now binds a particular tRNA that has been aminoacylated with methionine (more precisely, with formyl-methionine). This tRNA pairs up with the first codon on the mRNA, and so guarantees that the first amino acid in the peptide chain will be methionine. In *E. coli*, three different proteins, called initiation factors (IF-1, IF-2, and IF-3), also aid in forming this initiation complex.

After initiation comes peptide chain elongation. In this stage, the mRNA molecule is read by the ribosome from its 5′ end toward its 3′ end. Activated amino acids are added step by step to the growing peptide chain, the sequence of addition being governed by the order of codons on the mRNA and the binding and alignment of the appropriate aminoacylated tRNA species. This phase of peptide synthesis involves the formation of peptide bonds between the amino acids. The peptide chain is elongated from amino terminus to carboxy terminus, i.e., the first codon on the mRNA specifies the amino acid at the N-terminus of the protein to be synthesized. During chain elongation two additional proteins help in the binding of tRNA and in peptide bond formation (elongation factors EF-Tu and EF-Ts in *E. coli*). The ribosome reads the mRNA at the rate of about 15 codons per second, so that the synthesis of a protein with 300 amino acids takes about 20 seconds.

The last stage of peptide chain synthesis is termination. The genetic code specifies three stop codons, indicating the termination of a coding sequence. When the ribosome encounters one of these stop codons on the mRNA, certain release factors

(RF-1, RF-2, and RF-3 in *E. coli*) help to dissociate the newly synthesized protein chain, the mRNA and the ribosomal subunits from one another. The mRNA is free to be used in another cycle of protein synthesis, as are the ribosomal subunits.

Protein synthesis can be carried out by ribosomes free in the cytosol. In eukaryotes, ribosomes also carry out protein synthesis while bound to the surface of the endoplasmic reticulum. In addition, a given mRNA molecule usually has more than one active ribosome translating it into protein; an assembly of several ribosomes on a single mRNA is called a polyribosome, or polysome for short.

## PROTEIN MODIFICATION

### Types of Modification

During protein synthesis and afterwards, proteins can undergo substantial covalent modification. The types(s) and extent of modification will depend on the protein, and often play an important role in the biological function of the protein. In general, protein modifications can be divided into two major classes: reactions on the side chains of the amino acids, and cleavages of the peptide backbone.

The side chains of the amino acids may be modified by

- Hydroxylation (e.g., proline to hydroxyproline)
- Methylation (e.g., serine or threonine, at the hydroxyl group)
- Acylation (e.g., attachment of a fatty acid, such as stearic, myristic or palmitic acid, to one of several different amino acid side chains)
- Acetylation (at the side chain amino group of lysine residues in histones)
- Phosphorylation (of serine, threonine, or tyrosine)
- Carbohydrate attachment (to serine, threonine, or asparagine; this glycosylation gives rise to "glycoproteins")

One very important type of modification involves the joining of sulfhydryl groups from two cysteine residues into a disulfide bridge, as mentioned earlier in the discussion of the primary level of protein structure. The pairing of the cysteines for bridge formation is done quite specifically, with help from specialized enzyme systems. Disulfide bridge formation frequently occurs during the folding of the protein. Some are even formed on an incompletely synthesized chain during translation. Bridge formation is an important factor in guiding the folding process toward the correct product. The bridges help to stabilize the protein against denaturation and loss of activity. Proteins with mispaired cysteines and incorrectly formed disulfide bridges are often folded in the wrong fashion and are usually biologically inactive.

Backbone cleavages may involve removal of the N-terminal formyl moiety from the methionine, the removal of one or more amino acids at either the N- or C-terminus of the chain, and cuts in the backbone at one or more sites, possibly with the removal of internal peptide sequences.

Besides these modifications, certain proteins can acquire prosthetic groups (e.g., hemes, flavins, iron-sulfur centers, and others). The prosthetic groups may be attached covalently (usually to amino acid side chains) in some cases, noncovalently

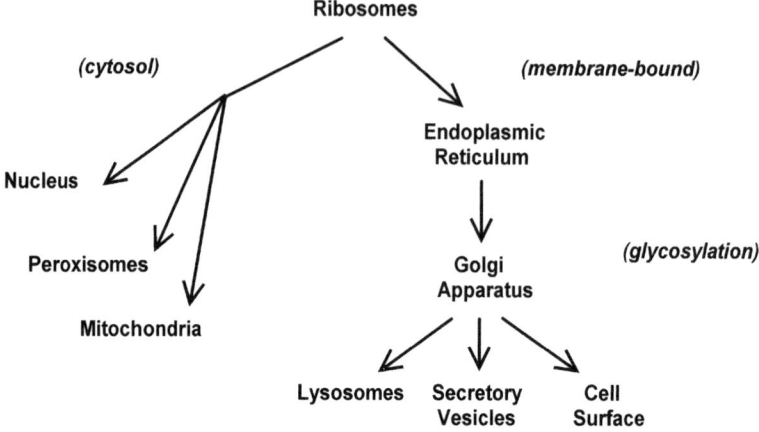

**FIGURE 2.9** The protein processing pathway in eukaryotes.

in others, but in either case they cause a substantial change in the properties of the protein.

## PROTEIN TRANSPORT AND MODIFICATION

As the ribosome synthesizes a new peptide chain, the chain usually begins to fold, creating regions of secondary and even incomplete tertiary structure. Enzymes then act on these folded residues to modify them. As noted above, an important modification that occurs at this stage is the formation of disulfide bridges. Another is the cleavage of the peptide backbone at specific sites, which may be important for the transport of the protein across membranes in the cell. These modifications occur during the process of translation, and so they are described as cotranslational modifications.

In eukaryotes, the mature form of a protein may be separated by several membrane barriers from its site of synthesis. For example, proteins secreted from the cell must pass through the membrane of the endoplasmic reticulum (ER) into its lumen, then through channels of the ER (where glycosylation usually occurs) to the Golgi complex. Inside the Golgi complex, there may be further glycosylation; there may also be removal of certain carbohydrate residues or proteolytic cleavage of the protein. Finally the protein passes to secretory vesicles which release the protein outside the plasma membrane of the cell. Proteins destined for other locations inside the cell will receive different types and levels of glycosylation and proteolytic cleavage. Figure 2.9 summarizes the overall pathway for protein processing.

## PROTEIN STABILITY

### THE IMPORTANCE OF NONCOVALENT INTERACTIONS

Although a protein's primary structure is determined by covalent bonds along the peptide backbone, the secondary and higher levels of structure depend in large part on relatively weak, noncovalent interactions for stability. In terms of standard-free energy change, the strength of these interactions is typically in the range of 0 to 7 kcal/mol (0 to 30 kJ/mol), much less than that involved in covalent bond formation. Furthermore, such weak interactions can be individually disrupted fairly easily by thermal agitation under physiological conditions, unlike covalent bonds which seldom break under these conditions.

The persistence of secondary and higher levels of structure is due to the large number of these weak interactions in a folded protein. While it is possible to disrupt one or a few of these weak interactions at a given moment, in a folded protein there are many more undisrupted noncovalent interactions that maintain the folded structure. Only on raising the temperature, perturbing the solution pH, or otherwise changing conditions away from the physiological, will one see a significant shift in the equilibrium from the native to the denatured conformation.

On the other hand, the weakness of these interactions does give a protein in solution an appreciable amount of conformational flexibility. The solution conformation of a protein is not absolutely rigid. Instead, because of momentary disruptions of weak interactions, most proteins in solution are constantly flexing, stretching, and bending, while still maintaining their overall shape. The result is that the protein molecules are distributed in a dynamic equilibrium over a host of closely related conformations, all very nearly the same in stability. This flexibility may be quite important in the biological functioning of proteins. Enzymes, for example, often undergo conformational changes on binding substrate, and a number of membrane-bound receptor proteins change conformation on binding their respective chemical messengers.

## TYPES OF NONCOVALENT INTERACTIONS

The fundamental noncovalent interactions of interest here include: (1) so-called exchange interactions, the very short range repulsion between atoms due to the Pauli exclusion effect; (2) polarization interactions, due to changes in the electron distribution about the interacting atoms; (3) electrostatic interactions, including interactions among charged groups and dipoles; and (4) the hydrogen bond, where two electronegative atoms partially share a hydrogen atom between them. The term van der Waals interaction is often used to summarize interactions in classes 1 and 2 above. Figure 2.10 presents these different interactions as they might occur between two polypeptide chains.

Exchange interactions are very short ranged, on the order of 1.2 to 1.5 Å (0.12–0.15 nm) for carbon, nitrogen, and hydrogen; these interactions essentially define what we usually regard as atomic and molecular shapes. Exchange interactions are strong enough to block any substantial interpenetration of atoms, and they are primarily responsible for what is called steric repulsion.

The hydrogen bond is a fairly short-range attractive interaction. The hydrogen atom involved lies between two electronegative atoms, usually oxygen, nitrogen, or a halogen species, although sulfur can also participate. The hydrogen atom usually has a strong covalent attachment to one of the electronegative atoms, and thus it lies closer to this atom than to the other. This group is referred to as the donor of the hydrogen bond, and the other electronegative center is the acceptor group. Hydrogen

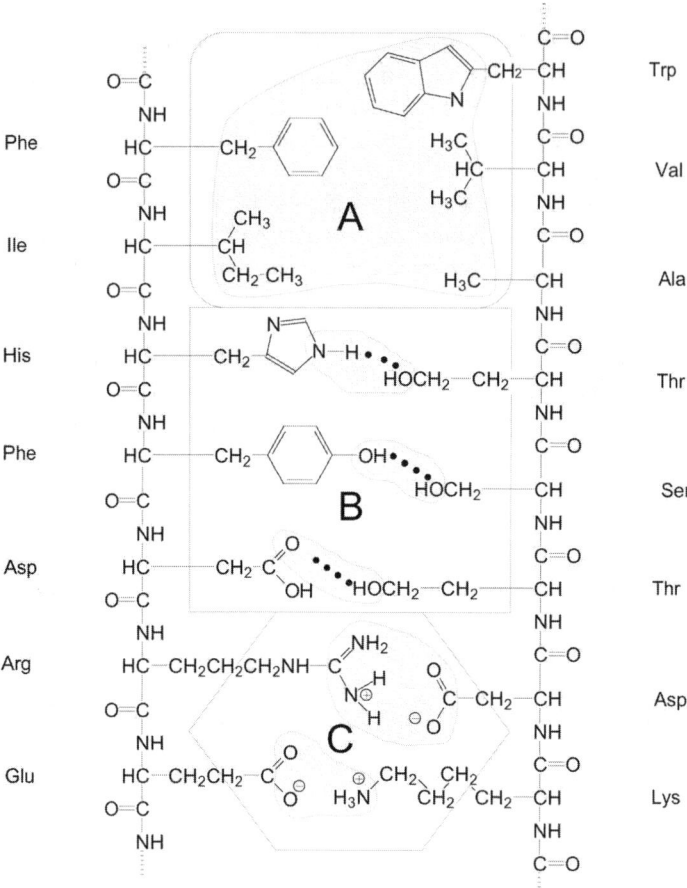

**FIGURE 2.10** Weak interactions that stabilize protein structures. A, Hydrophobic interactions; B, hydrogen bonding; C, ionic interactions.

bonds have a range around 2 Å (0.2 nm). The three atoms participating in the interaction typically are collinear with a bond angle of 180°, though distortions of the bond angle of 20°–30° are common. A hydrogen bond would typically cost 2 to 9 kcal/mol (8 to 38 kJ/mol) to disrupt.

Electrostatic interactions can be either attractive or repulsive, depending on the charges and/or the orientation of dipoles involved; charges of opposite sign attract and like charges repel one another. The strength of the interaction depends on the local dielectric constant, and this in turn is strongly dependent on the nature and organization of the medium surrounding the interacting species. For example, in the interior of a protein the dielectric constant may vary from 2 to 40, so that electrostatic interactions are very strong there, while in an aqueous environment the interactions (with the same species, distances, and orientations) can be up to 40 times weaker because the dielectric constant is around 80. The interaction energy for a pair of singly charged ions in water at 298 K separated by 1 nm is about one-half of a kcal/mol (about 2 kJ/mol), while in the interior of a membrane, with a dielectric constant around 2, it would be about 17 kcal/mol (70 kJ/mol).

Polarization interactions for atoms and small molecules or functional groups are much weaker than the other interactions listed above. For example, in vacuum the attractive energy between two methyl groups is only about 0.15 kcal/mol (0.6 kJ/mol) at a separation of 0.4 nm. However, polarization interactions are additive, so that for large bodies with many individual polarization interactions (e.g., a protein binding a large substrate molecule) the overall contribution may be 10 to 20 kcal/mol (40–80 kJ/mol). Furthermore, these interactions will be present for both nonpolar and polar (even ionic) groups.

## THE HYDROPHOBIC EFFECT

Nonpolar molecules tend to have low solubilities in water, and large nonpolar solutes tend to form aggregates in aqueous solution. In the past these tendencies were sometimes explained by invoking a special "hydrophobic bond" between nonpolar groups. However, "bond" is a misnomer here, and it is better to refer to an "effect," because there is no exchange of bonding electrons involved in either of the tendencies noted above. Instead, the hydrophobic effect is a combination of several of the fundamental noncovalent interactions, and it involves details of the organization of water molecules around nonpolar solute molecules.

In the bulk phase, water molecules are hydrogen bonded to one another and attract one another by polarization interactions. A single water molecule can participate in up to four hydrogen bonds at the same time, with the bonds oriented in tetrahedral fashion about the central oxygen atom. In ice, this leads to a rigid three-dimensional lattice of molecules. In liquid water, the hydrogen bonding interactions are broken up enough so that there is no longer any long-range order or rigidity; but on average, water molecules will still have about four hydrogen-bonded partners around them and the local (as opposed to long-range) structure will still resemble that of the ice lattice.

When a small nonpolar solute molecule is dissolved in water, the local network of hydrogen bonds will rearrange to make room for the new solute. Polarization interactions among the water molecules will also be disrupted. There will be, however, new polarization interactions between the solute and the surrounding water molecules, and for the most part these new polarization interactions will almost exactly balance the lost ones. The rearrangement of the hydrogen bonding network is more important. Because of the strength of the hydrogen bond, there will be a strong tendency for the water molecules immediately surrounding the nonpolar solute to reorient themselves so as to each keep four hydrogen bonding partners. This results in a cage-like water structure about the solute which can fluctuate in shape, and which mostly maintains the favorable hydrogen bonding energies. The local water structure is, however, somewhat more organized than the loose structure in pure liquid water.

Elementary thermodynamics provides a framework to interpret these changes in energy and structural organization. The key points can be summarized briefly as follows. The energy changes in a system at constant temperature and pressure are conveniently represented by $\Delta H$, the change in enthalpy, while the changes in order or organization can be represented by $\Delta S$, the change in entropy. As is well known, $\Delta H$ and $\Delta S$ can be combined into a single thermodynamic function, the free energy change $\Delta G$:

$$\Delta G = \Delta H \; T\Delta S$$

where T is the absolute temperature in Kelvins. The quantity $\Delta G$ determines whether a process will occur spontaneously; it can also be used to determine the point of equilibrium in chemical and biochemical systems. Any process where $\Delta G$ is negative will occur spontaneously, and equilibrium is reached when $\Delta G$ for the process goes to zero.

The thermodynamic relations above can be applied to the dissolution of nonpolar solutes in water. The enthalpic contribution to the free energy change is rather small, because of the compensations in hydrogen bonding and polarization interactions. The organization of water molecules around the nonpolar solute (an increase in the system's order) lowers the entropy of the system, that is, $\Delta S$ is negative for this process. This makes $T\Delta S$ positive, which results in an unfavorable contribution to the free energy change for dissolving the nonpolar solute. The magnitude of the $T\Delta S$ term for such processes is usually much larger than that of the $\Delta H$ term, and entropy changes dominate the thermodynamics. Since the entropy change is unfavorable, the overall free energy change is unfavorable, and nonpolar solutes tend to have low solubilities.

The tendency for large nonpolar solutes to aggregate can be explained on the same basis. There will be a sheath of water about each isolated solute molecule, and the water in the sheath is again more organized than the water in the bulk solution (though its organization at an extensive nonpolar surface is different than the cagelike structure formed around small nonpolar solutes). When two such large hydrated molecules form an aggregate, some of the organized water that was between the solutes will be released to the bulk solution as the solutes make contact with one another. Again, there will be relatively small changes in the energy (enthalpy) because the hydrogen bonding and polarization interactions are largely compensated, but there will be a major change in the entropy. The water molecules released from the hydration sheaths will pass to a more disorganized state, which is entropically favorable. This entropic contribution dominates the enthalpic term in the free energy of aggregation, and makes the process favorable overall.

## PROTEIN FOLDING AND NONCOVALENT INTERACTIONS

The folding of proteins into their characteristic three-dimensional shape is governed primarily by noncovalent interactions. Hydrogen bonding governs the formation of $\alpha$ helices and $\beta$ sheets and bends, while hydrophobic effects tend to drive the association of nonpolar side chains. Hydrophobicity also helps to stabilize the overall compact native structure of a protein over its extended conformation in the denatured state, because of the release of water from the chain's hydration sheath as the protein folds up. (Hydration of the folded protein is limited mainly to the protein's surface and its crevices and corrugations.) Exchange interactions put an upper limit on how tightly the amino acids can be packed in the interior of a folded protein, and of course play a role in the constraints on secondary and tertiary structure. Solvation effects favor the placement of highly polar and ionic groups where they are exposed to solvent and not buried in the (generally nonpolar) interior of a protein.

## THERMODYNAMICS AND KINETICS OF PROTEIN FOLDING

Calorimetric studies of protein denaturation have revealed the following general points. First, proteins generally are folded into the most stable conformation available; the native structure has the lowest free energy among all conformations. Second, for small globular proteins, the transition from native to denatured form can be adequately represented as a two state process. That is, the $\Delta H$, $\Delta S$, and $\Delta G$ values are consistent with a model where there are only two macroscopic states, the native and the denatured. This in turn implies that the transition is highly cooperative ("all-or-none"); the individual amino acids do not switch states independently of one another, but instead tend to switch states in a concerted way. Third, for large proteins with several structural blocks or domains, the native structure is lost in stages, each stage corresponding to the all-or-none denaturation of individual structural blocks. Fourth, these cooperative transitions are accompanied by characteristic increases in the heat capacity of the system. Furthermore, the change in heat capacity is strongly correlated with an increase in the surface area of the protein chain that is exposed to solvent. This points to the importance of the contact of nonpolar groups with water in determining the thermodynamics of the unfolding or folding of proteins. The increase in heat capacity upon denaturation is likely due to ordering of water about the nonpolar groups that are exposed as the protein unfolds, and the order of these water molecules decreases more rapidly than that of bulk water as the temperature rises.

Table 2.3 presents thermodynamic data on the denaturation of selected proteins that have a single compact globular shape and that fit the pattern of an all-or-none transition from native to denatured form. In general, both $\Delta H(298\ K)$ and $\Delta S(298\ K)$ for the transition are positive, and $\Delta H$ and $T\Delta S$ nearly cancel one another. This means that $\Delta G$ for denaturation is in general small, though positive, at 298 K. A typical free energy of denaturation at 298 K would be around 20 to 50 kJ/mol of protein. Experiments and theory are consistent with a temperature of maximum stability between 273 and 298 K (0–25°C) for most proteins. As might be expected, raising the temperature destabilizes the native form of the protein with respect to the denatured form. Interestingly, cooling the protein below the temperature of maximum stability also destabilizes the native form, and can lead to the curious phenomenon of cold denaturation.

An important topic of current research is how the sequence of amino acids in a newly synthesized protein can direct the folding of the chain into a precise, biologically active shape. Can the amino acid sequence be used to predict the final three-dimensional shape of the protein? The short answer to this question is, "Not completely, not yet." Present computer-aided predictions are about 70% accurate with

**TABLE 2.3**
**Thermodynamic Data on the Denaturation of Selected Proteins[a]**

| Protein | Molecular Weight | $\Delta H$ (298 K) | $\Delta S$ (298 K) | $\Delta C_P$ (323 K) |
|---|---|---|---|---|
| Ribonuclease | 13,600 | 2.37 | 6.70 | 43.5 |
| Carbonic anhydrase | 29,000 | 0.80 | 1.76 | 63.3 |
| Lysozyme (hen) | 14,300 | 2.02 | 5.52 | 51.7 |
| Papain | 23,400 | 0.93 | 1.60 | 60.1 |
| Myoglobin | 17,900 | 0.04 | − 0.80 | 74.5 |
| Cytochrome $c$ | 12,400 | 0.65 | 0.90 | 67.3 |

[a] $\Delta H$ in kJ/mol of amino acid residue; $\Delta S$ in J/K per mole of amino acid residue; $\Delta C_P$ in J/K-mol of amino acid residue.

Data adapted from Privalov, P.L., and Gill, S.J. (1988).

respect to secondary structure, and the prediction of tertiary structure is often a good deal less accurate. But a great deal of progress has been made and a general mechanism for protein folding has emerged.

Biophysical studies have shown that many denatured proteins can spontaneously refold *in vitro,* upon removal of a denaturing agent (urea, detergent, acid, and so on). Certain enzymes and other proteins can accelerate the folding process *in vitro*, and it has been concluded that their role *in vivo* is to prevent misfolding or aggregation. However, these protein factors serve more to facilitate the folding process than to specify a particular spatial shape for the product.

The biophysical studies have also shown that folding is a relatively rapid process, with folding being completed in a matter of seconds to perhaps a few minutes in most cases. This rules out a "trial and error" approach to folding, where the protein randomly folds into all possible conformations in a search for the most stable conformation. For a typical protein of 100 residues, with about 10 different conformations for each amino acid residue, the total number of conformations would be around $10^{100}$, far too many to be searched in a reasonable amount of time.

Instead of proceeding at random, protein folding follows a stepwise process, with three main steps. These are: (1) formation of unstable, fluctuating regions of secondary structure; (2) aggregation or collapse of these embryonic structures into a more compact intermediate structure or structures; and (3) rearrangement or adjustment of the intermediate(s) into the final conformation (or family of conformations) that we identify as the native structure. From *in vitro* studies it is known that the first step is very fast, typically occurring within 0.01 seconds. The second step is fast, taking about 1 second for most proteins; while the third step can be rather slow *in vitro*, running from a few seconds up to 2500 seconds. The slowness of this last step is often connected with isomerization of proline from the *cis* to the *trans* isomer, and what happens *in vivo* may be considerably faster, thanks to facilitation by proline isomerase enzymes.

The solution of the protein folding problem will have wide ramifications. For example, the completion of the Human Genome Project, with the full DNA sequence of the human genome, has generated the sequences of many genes whose protein products are as yet uncharacterized. It would be a great help in deducing the function of each of these gene products if it were possible to predict the overall three-dimensional folding of the protein, and thus relate it to possible enzymatic, structural, or signalling functions in the cell. Eventually, improved diagnostic tools or therapies would result from this. Also, molecular biology laboratories are developing novel artificial proteins, not derived from any preexisting gene, in attempts to obtain more efficient enzymes, enzyme inhibitors, and signalling factors, for use in medicine and in industrial chemical processes. An effective way of predicting the final folded structure from a peptide's sequence would greatly increase the efficiency of these research and development efforts.

## FURTHER READING

Berman, H.M., Goodsell, D.S., and Bourne, P.E. (2003). Protein structures: From famine to feast. *American Scientist,* 90, 350–359.

Dill, K.A. and Chan, H.S. (1997). From Levinthal to pathways to funnels. *Nature Structural Biol.,* 4, 10–19.

King, J., Haase-Pettingell, C., and Gossard, D. (2002). Protein folding and misfolding. *American Scientist,* 90, 445–453.

Privalov, P.L. and Gill, S.J. (1988). Stability of protein structure and hydrophobic interaction. *Adv. Protein Chem.,* 39, 191–234.

Stryer, L. (1995). *Biochemistry,* 4th ed. W. H. Freeman, New York.

# 3 Recombinant DNA Basics

*Charles P. Woodbury, Jr., Ph.D.*

## CONTENTS

## THE USES OF RECOMBINANT DNA

In the 1970s a new technology emerged allowing molecular biologists to isolate and characterize genes and their protein products with unprecedented power and precision. One important result was the ability to recombine segments of DNA from diverse sources into new composite molecules, or recombinants. In developing and applying this technology, the molecular biologists drew on several different areas of basic research, including the enzymology of nucleic acid and protein synthesis, the details of antibiotic resistance in bacteria, and the mechanisms by which viruses infect bacteria and how those bacteria protect themselves against such infections. The collection of techniques for manipulating DNA and making recombinant molecules is known today as recombinant DNA technology or genetic engineering.

Recombinant DNA technology is now in routine use in basic and clinical research. It has led to new therapeutic agents, such as pure human insulin for diabetics that avoids immunological side reactions to insulin derived from animal sources, and pure human growth hormone in sufficient quantities for routine therapy of dwarfism. New diagnostic agents have been derived that allow early and precise characterization of infections and genetic diseases, with improved prospects for successful therapy as a result. Therapy at the level of repair of the genetic defect in a few genetic diseases is not far away, and genetic engineering is playing a central role in the development of new antiviral and anticancer agents. Finally, genetically engineered microorganisms are in increasing use in industry, in the production of organic chemicals (including pharmaceuticals), and in the detoxification of chemical waste and spills.

To understand how these modern methods work, it is necessary first to review some general laboratory techniques ubiquitous in genetic engineering. Among the most important are gel electrophoresis of nucleic acids, nucleic acid hybridization assays, and the polymerase chain reaction.

## GEL ELECTROPHORESIS

When an electric field is applied through electrodes to a solution of charged molecules, the positively charged molecules will move toward the negative electrode, while the negatively charged molecules will move toward the positive electrode. The motion of these molecules in response to the electric field is referred to as electrophoresis. DNA molecules of course carry a high negative charge, and so they will electrophorese toward a positive electrode. For large DNA molecules in free solution both the electrical force per unit length and the hydrodynamic drag per unit length are constant (the DNA behaves as a free-draining polymer). Thus, for a given DNA molecule, the magnitudes of both forces are each proportional to the length of that molecule. However, the electrophoretic mobility is proportional to the *ratio* of these forces, and so the electrophoretic mobility of these large DNA molecules is *independent* of size. The result is that free solution electrophoresis is not a useful way to separate large DNA molecules by length. For small DNA molecules, which behave hydrodynamically as rods, there is a (weak) logarithmic dependence of the electrophoretic mobility on the length of the DNA, but this is not enough to make free solution electrophoresis convenient for separating small DNA molecules.

Instead of open solutions, nowadays one uses gels of agarose (a polysaccharide extracted from certain algae) or polyacrylamide (a synthetic polymer) that are cross-linked to form a network of pores in the gel. The gel matrix provides a sieving effect, retarding the migration of large molecules more than of the small molecules. The result is that small DNA molecules electrophorese through the gel more rapidly than

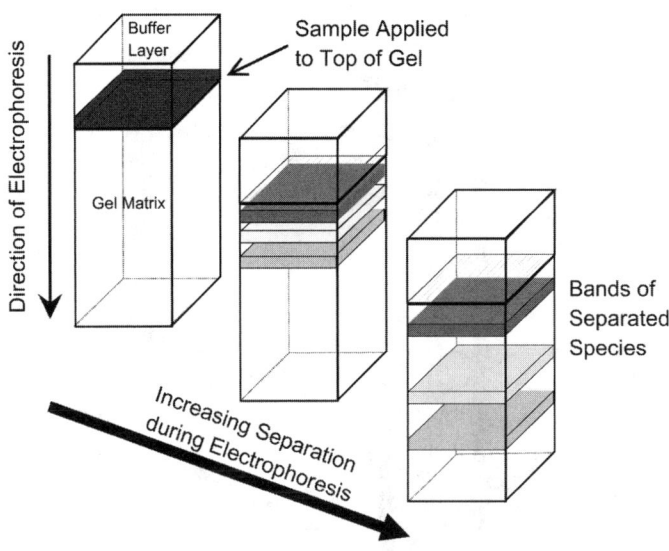

**FIGURE 3.1** The principle of separation of DNA by gel electrophoresis. A DNA sample is applied to the top of the gel, and the gel acts as a sieve to separate the DNA by size.

large ones. The use of the gel matrix also reduces any convective mixing, an undesirable effect that would degrade resolution. By proper choice of the gel concentration and of the degree of cross-linking, the pore size can be adjusted to optimize the size resolution. It is possible to resolve DNA species that differ in length by only one base pair, in molecules that may be hundreds of base pairs in overall length. Electrophoresis can also be used with single-stranded DNA or RNA, with the same sort of resolution.

DNA molecules of the same length will move through the gel at the same rate. Assuming that the DNA sample is applied to the gel in a small volume, so that the applied sample forms a very thin layer of liquid at one end of the gel, then the DNA molecules will migrate through the gel as sharp bands, each band having DNA molecules of the same length. A DNA sample with many different sizes of molecules will thus yield many DNA bands upon electrophoresis (Figure 3.1), while a purified DNA sample whose molecules are all of the same length will show only one band.

In a typical experimental setup, the agarose or polyacrylamide is cast as a slab (a slab gel); the slab may be cast vertically, between two glass plates, or (for agarose but not polyacrylamide) it may be cast horizontally onto a single plate. Before the gel is actually poured for casting, a plastic template with multiple teeth, which resembles a coarse comb, is inserted at the top of the slab. This is done so that the solidified gel will keep the impression of those teeth as a series of indentations or wells that can hold individual DNA samples. The different DNA samples can then be run side-by-side on the same gel, which facilitates comparison of the DNA sizes from sample to sample. A typical slab gel might have 10 to 40 such sample wells, so that a large number of DNA samples may be compared simultaneously (see Figure 3.2).

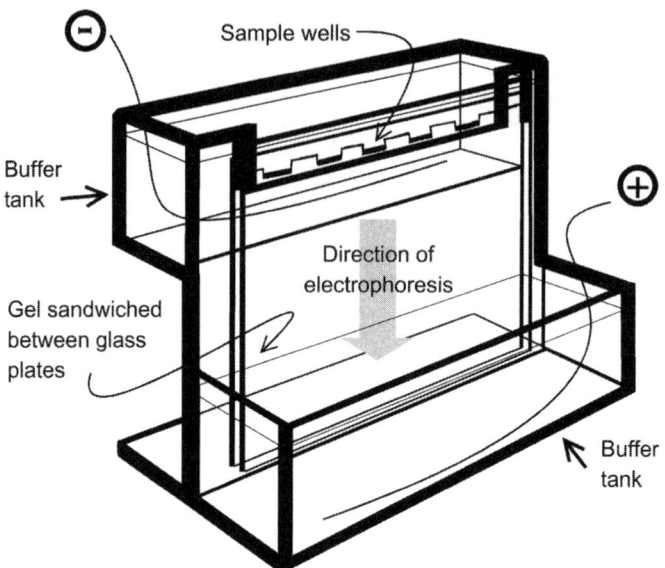

**FIGURE 3.2** A slab gel electrophoresis apparatus. A voltage source generates an electric potential difference between the upper and lower buffer chambers, causing the applied DNA sample to migrate through the gel toward the positive electrode.

After a suitable period of electrophoresis, the DNA samples on the slab gel can be visualized by staining. A common stain is the dye ethidium bromide. This dye binds to DNA and becomes highly fluorescent when ultraviolet light is shined on it. The regions on the gel where there is DNA will then appear as bright orange-red bands under a blacklight, and it is easy to mark their positions by photography. Alternatively, if the DNA is radioactively labeled, then the gel can be dried and placed in the dark next to a sheet of unexposed x-ray film. The x-ray film will be exposed by the radioactivity in the various DNA bands; the developed film will have dark bands exactly where the DNA bands were located in the gel. This process of using x-ray film to locate radioactivity is referred to as *autoradiography*.

## NUCLEIC ACID HYBRIDIZATION ASSAYS

### DNA STRUCTURE AND COMPLEMENTARY BASE PAIRING

DNA in cells exists mainly as double-stranded helices. The two strands in each helix wind about each other with the strands oriented in opposite directions (antiparallel strands). The bases of the nucleotides are directed toward the interior of the helix, with the negatively charged phosphodiester backbone of each strand on the outside of the helix. This is the famous B-DNA double helix discovered by Watson and Crick (Figure 3.3).

In this structure, the bases on opposite strands pair up with each other in a very specific way: adenine (A) pairs with thymine (T), and guanine (G) pairs with

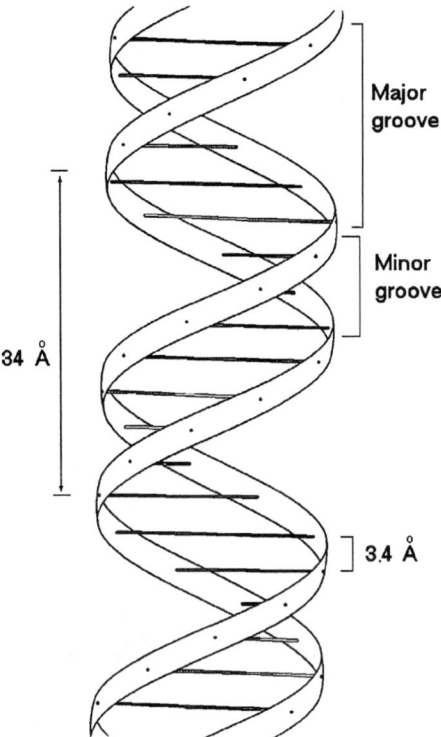

Major
groove

Minor
groove

34 Å

3.4 Å

**FIGURE 3.3** The B form of the DNA double helix, showing major and minor grooves.

cytosine (C). This is referred to as Watson-Crick pairing of complementary bases (Figure 3.4). The term "complementary" is further used to describe the match of one strand of the helix with the other, when all the bases obey Watson-Crick rules of pairing (Figure 3.5).

The specificity of pairing is governed by hydrogen bonding and steric fit of the bases with each other. To fit into the interior of the helix, a purine base must match up with a pyrimidine base; purine-purine or pyrimidine-pyrimidine pairs simply will not fit within the steric constraints of the double helix. If forced into a double-stranded molecule, such pairings will distort and destabilize the helix. It is possible to mispair a purine with a pyrimidine (e.g., adenine and cytosine, or guanine and thymine), but these pairings lack the proper match of hydrogen bond donors and acceptors, and they destabilize the native structure appreciably.

## DNA Renaturation, Annealing, and Hybridization

Double-stranded DNA molecules with one or more mispairs are more easily disrupted or denatured than are properly and fully paired double helices; the more mispairing, the less the stability of the molecule. If a DNA molecule, with or without mispairs, is too unstable, then the two strands will dissociate from one another to

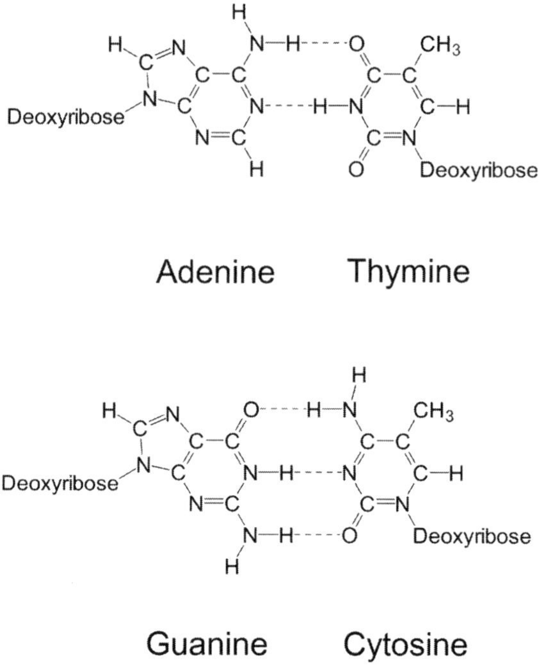

**FIGURE 3.4** Watson-Crick base pairing in DNA. Adenine is complementary to thymine, and guanine is complementary to cytosine.

form two single-stranded DNA molecules. In general, low salt concentration or high temperature will cause this so-called helix-coil transition.

It is possible to renature the separated strands of a DNA molecule by the process of annealing (Figure 3.6). Here, the complementary DNA strands in a hot solution will pair up with one another as the solution slowly cools, and eventually the complete double helix can be recovered with all the bases properly aligned and paired. (Rapid cooling of the solution would result in quenching, in which the pairing of strands would be incomplete, and unstable intermediates are kinetically trapped as the main product.) It is also possible to "hybridize" two strands of DNA that do not have entirely complementary sequences, provided there is substantial agreement in pairing. Such molecules are, of course, less stable than those with complete Watson-Crick pairing, and they will tend to dissociate at lower temperatures or higher salt concentrations than the corresponding correctly paired molecules. By manipulating the temperature and salt concentration the hybridization can be controlled to permit only correctly paired molecules to form; alternatively, less stringent conditions might be used to allow a certain level of mispairing in the duplexes formed.

## HYBRIDIZATION ASSAYS

A variety of techniques have been developed that take advantage of the ability of single-stranded DNA (or RNA) molecules to pair up with and hybridize to other

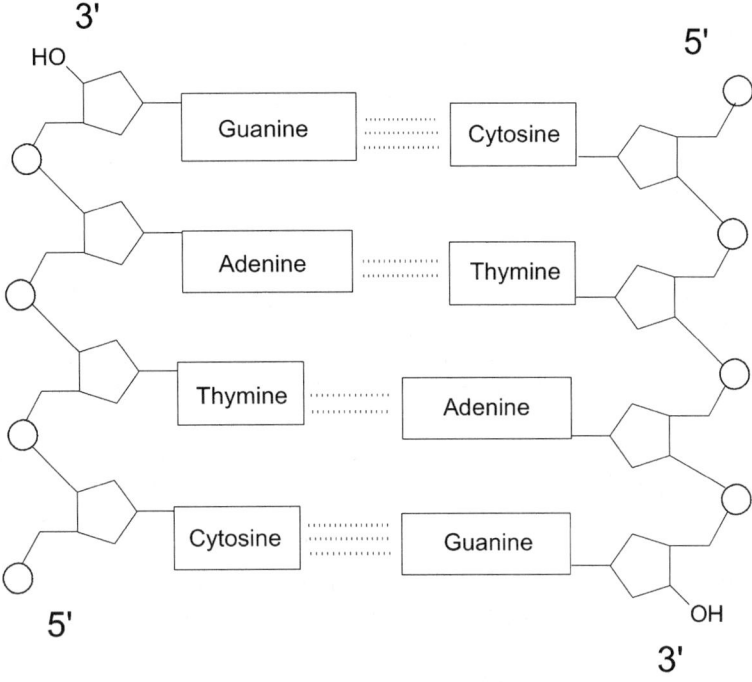

**FIGURE 3.5** Double-stranded DNA showing complementary pairing of bases along two antiparallel strands. Sugar moieties are represented by *pentagons*, phosphate moieties by the *circles*, and hydrogen bonds by the *dotted lines* connecting the complementary bases.

single-stranded nucleic acid chains. These hybridization assays can be used to determine whether a particular DNA or RNA sequence is present in a sample that contains a wide variety of nucleic acid fragments. This can be useful in diagnosing a viral infection, in detecting bacterial pathogens in biological samples, in screening for different genetic alleles, and in analyzing forensic samples.

The general procedure is as follows. A sample of nucleic acid, which may contain a particular sequence of interest (the target sequence), is denatured and affixed to a

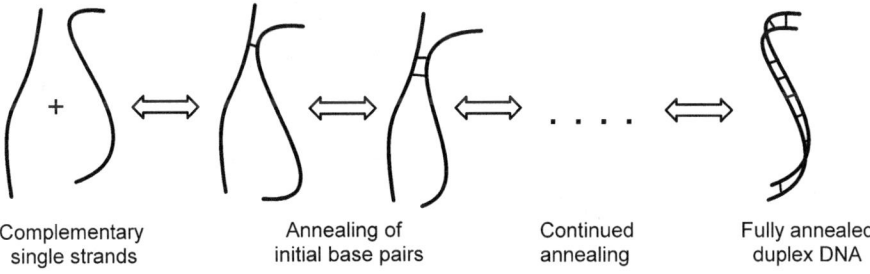

Complementary single strands     Annealing of initial base pairs     Continued annealing     Fully annealed duplex DNA

**FIGURE 3.6** Annealing of two complementary strands of DNA to form a fully paired double-stranded molecule.

suitable membrane. Then the membrane is washed with a labeled oligonucleotide (the probe) whose sequence matches a part of that of the target. A positive signal indicates that the probe DNA found a complementary sequence on the membrane to hybridize with, and that consequently the target sequence was present in the sample.

Since hybridization assays are so ubiquitous in recombinant DNA work, it is worthwhile to look at some of the details of the method and some commonly used variants.

The dot-blot assay is perhaps the simplest variant. DNA is prepared from a biological sample thought to contain a selected target sequence. For example, this target might be a key part of a viral gene that would identify that virus, and the assay used to determine the presence or absence of that virus in the sample. Alternatively, the target might be the region of the human hemoglobin gene that determines the sickle cell characteristic, and the assay used to check for the presence or absence of this allele. In any case, the sample DNA is extracted, denatured by alkali, then simply applied in a small spot or dot to a membrane of nitrocellulose or nylon; the term dot-blot derives from this method of applying DNA to a membrane. The DNA is fixed in place on the membrane by heating (for nitrocellulose membranes) or by cross-linking with ultraviolet light (for nylon membranes). The membrane is then bathed with a solution containing the probe sequence, under selected conditions that promote specific hybridization of the probe to the target. The membrane is washed to remove excess probe, then dried.

The probe species is often radioactively labeled, or it may carry a fluorescent tag, or some other chemical or enzymatic moiety to generate a positional signal. For radioactive labeling, a common choice of radioisotope is phosphorus-32 (or $^{32}$P), because it can be incorporated as phosphate into DNA or RNA relatively easily, and it emits energetic beta particles that are easy to detect. The radioactivity on the membrane can be used to expose an adjacent x-ray film in a pattern corresponding to the radioactive spots on the membrane. After a suitable exposure time, one develops the film and studies the location and intensity of the images of the radioactive spots to deduce the position and degree of probe hybridization on the membrane.

The Southern blot method is often used in research work, particularly in constructing genetic maps. It is used to detect DNA fragments that carry a specific target sequence and that have been separated into bands of identical molecules by electrophoresis on a slab gel (Figure 3.7). The DNA in the slab gel is denatured by alkali, which exposes the single strands to potential hybridization partners. Then a sheet of nitrocellulose membrane is laid over the slab gel, and this is then covered with many layers of absorbent paper. The stack of absorbent paper acts as a wick to draw the moisture from the slab gel. As the water passes from the gel to through the membrane it carries with it the denatured DNA molecules. The DNA is bound by the nitrocellulose to form a replica of the original pattern of DNA bands on the slab gel, or, in other words, a "blot" of the slab gel (Figure 3.8). After sufficient time for the blotting to reach completion, the membrane is dried by baking, to fix the DNA molecules in place. Next, the membrane is incubated with a solution containing a labeled probe oligonucleotide, whose sequence is complementary to the target sequence. Then excess probe is washed away and the membrane inspected for signs

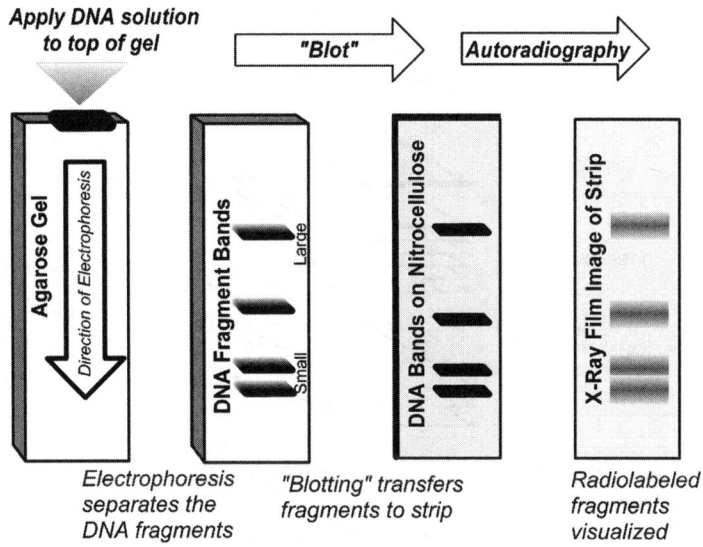

**FIGURE 3.7** Visualization by autoradiography of electrophoresed, radiolabeled DNA. After electrophoresis the DNA bands are blotted onto a nitrocellulose membrane, and x-ray film is placed next to the membrane. The positions of labeled DNA bands appear when the film is developed.

of hybridization. $^{32}$P-labeled probes are commonly used, along with autoradiography to detect hybridization.

Another variant of the hybridization assay is the Northern blot. Here it is RNA, not DNA, that is separated on a slab gel and transferred to a membrane. In the original version of the method, a special chemically treated cellulose membrane was used to hold the RNA, since nitrocellulose does not normally bind RNA. However, conditions have now been found where nitrocellulose will indeed retain RNA molecules. Nylon membranes can also be used. Radiolabeled DNA probes and autoradiography are then employed as above in the Southern blot method. The method is often useful in studying how levels of RNA species in a cell vary with stages of development and differentiation.

In all of the hybridization assays, the conditions of salt and temperature must be carefully selected to permit proper hybridization. Typically, hybridization conditions are used that are stringent enough to allow only perfectly complementary sequences to renature. If the conditions are not sufficiently stringent, then one has a problem with the probe binding to too many different sequences on the membrane. In this situation there will be a large amount of background noise from the nonspecifically bound probe, which may hide a weak positive signal from the probe molecules that have bound specifically. Thus, great care must be taken in performing the hybridization if its sensitivity is to be preserved.

The amount of probe hybridization onto a membrane-bound sample is, of course, limited by the amount of complementary target nucleic acid present in that sample;

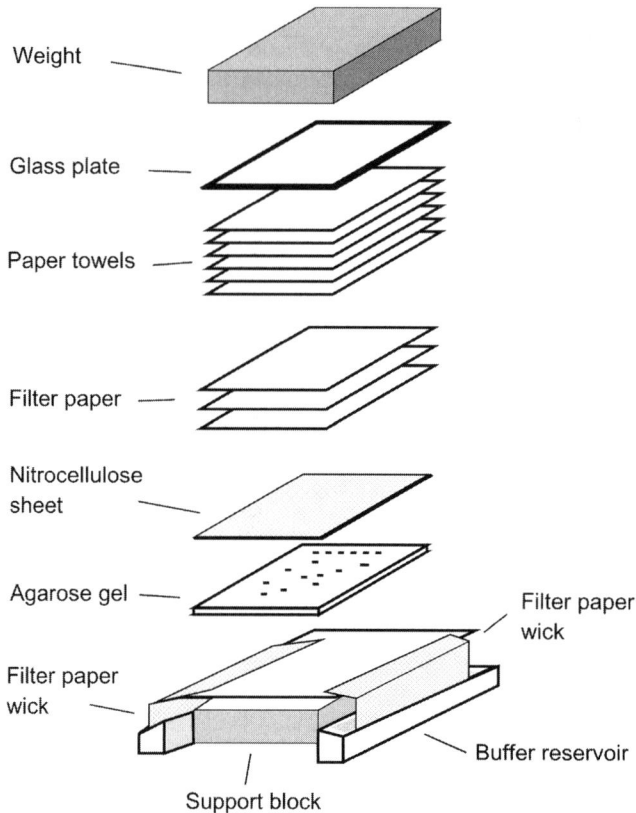

**FIGURE 3.8** Setup for a Southern blot. The agarose gel containing the separated DNA is placed in contact with a nitrocellulose sheet, then pressed with filter paper and paper towels. The buffer in the wet gel moves by capillary action into the filter paper and towels; the lower reservoir of buffer helps promote the migration of the DNA from the gel onto the nitrocellulose sheet.

the greater the amount of complementary DNA (or RNA) in the sample, the stronger the signal from the hybridization assay. A single target molecule in the sample is usually not enough for reliable detection unless the probe is extremely radioactive, and for a reliable assay it is desirable to have many copies of the target sequence present in a sample. One way to do this is to "amplify" the target sequence, using the polymerase chain reaction.

## THE POLYMERASE CHAIN REACTION

One area of basic biochemical research that has paid unexpected dividends is DNA replication. Enzymological work here has characterized the various DNA polymerases in bacterial and eukaryotic cells. With progress in the biochemical characterization of these enzymes, new applications have been found for them in research and medicine. One of the most important and exciting is their use in the so-called polymerase chain reaction (PCR). PCR is a technique that produces millions of exact copies of a selected DNA sequence out of a mixed population of DNA molecules. The DNA amplification possible with PCR is now being used in forensic tests to identify criminal suspects or to determine paternity, in characterizing genetic disorders, and in recombinant DNA research.

## TEMPLATES, PRIMERS, AND DNA POLYMERASE

PCR is based on the ability of single-stranded DNA to pair up with a complementary single-stranded molecule to form a double-stranded molecule. This annealing or hybridization of single-stranded DNA into duplexes occurs spontaneously under certain well-understood conditions of temperature, pH, and salt (see above the section on hybridization assays). With the proper choice of sequence, one can readily anneal a short oligonucleotide to a much longer molecule and expect to have exact base pairing so that the oligomer attaches to the long DNA at only the exact complementary sequence. The resulting DNA molecule will have at least one duplex region flanked by single-stranded regions (Figure 3.9).

If one then adds DNA polymerase and sufficient amounts of all four deoxyribonucleotide triphosphates to a solution containing the hybridized DNA, the DNA polymerase will bind to the duplex region and start to synthesize new DNA that is complementary in sequence to the single-stranded region. The DNA polymerase synthesizes new DNA by stepwise attachment of the 5′ phosphate of an incoming

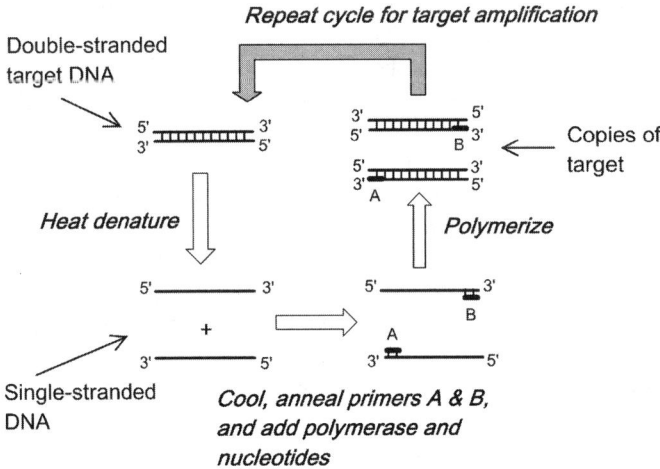

**FIGURE 3.9** The polymerase chain reaction (PCR). A target sequence is cut at either end, denatured, and annealed with two oligonucleotide primers; the primers are complementary to regions flanking the target sequence, on opposite strands as shown. DNA polymerase then copies the exposed single-stranded region, which includes the target sequence. The duplex molecules are denatured, annealed with more primers, and polymerized again. Repeated cycles of denaturation, primer annealing, and polymerization lead to large numbers of identical DNA molecules.

nucleotide to an existing 3′ hydroxyl group on the end of the duplex region; the incoming nucleotide triphosphate is hydrolyzed to the monophosphate, and it is the nucleotide monophosphate that is actually incorporated into the new DNA. Synthesis proceeds in only one direction, thus only the 3′ end of the oligomer receives the new DNA. In this process, the oligomer that is elongated is referred to as a primer and the longer polynucleotide, whose sequence is complementary to the new DNA, is referred to as the template. Under appropriate conditions, the DNA polymerase is capable of duplicating DNA for several hundred to a few thousand bases, so quite long single-stranded regions can be duplicated.

## THE PCR AMPLIFICATION PROCESS

Suppose one were interested in detecting a particular DNA sequence, say a viral gene in an infected cell culture, and that by other means one already knew the sequence of the gene. The problem is to distinguish the particular viral DNA sequence in a mixture containing DNA from all the other viral genes as well as from the host cell. With PCR, one can amplify the desired DNA sequence to the point where the rest of the DNA can be ignored, and so the viral gene's presence can be easily detected, for example, by using a hybridization assay.

To do this, one must first chemically synthesize two sets of DNA oligomers to use as primers for DNA polymerase. The sequence of each of the primers is carefully chosen to match the viral DNA sequence just within—or just outside, if possible— the gene of interest, one primer for each end of the gene. The primers' sequences are chosen so that they will hybridize to opposite strands, with their 3′ ends pointing toward one another (refer to Figure 3.9 again).

Next, the mixture of viral and host DNA is denatured by heating, and a great excess of the two sets of primers is added. On cooling, the primers hybridize to their respective complementary sequences on the viral DNA, and a heat-stable DNA polymerase is added along with sufficient deoxynucleotide triphosphates for DNA synthesis to proceed. (The biological source of the commonly used DNA polymerase is the thermophilic bacterium *Thermus aquaticus*, and the polymerase is known as Taq polymerase.) The timing and concentration of reagents are controlled so that DNA synthesis will extend from one primer sequence past the sequence complementary to the other primer. The reaction is halted by heating, which dissociates the newly synthesized DNA from the template DNA.

As the mixture cools, the original single-stranded viral DNA will bind primer oligomers again; more important, the newly synthesized DNA can also bind primers, because it was extended from one primer past the sequence complementary to the other primer. The DNA polymerase again synthesizes DNA, this time using as templates not only the original viral sequence but also the newly synthesized DNA. After sufficient time, the reaction is again quenched by heating, and the DNA duplexes dissociate. Repeating this cycle—reaction, heating and dissociation, then cooling and hybridization of fresh primer—leads to more and more copies of the DNA sequence that lies between the two primer sequences. Since the newly synthesized DNA is itself a template for more DNA synthesis, copies will be made of copies in an ever-widening chain of reactions, hence the term *polymerase chain*

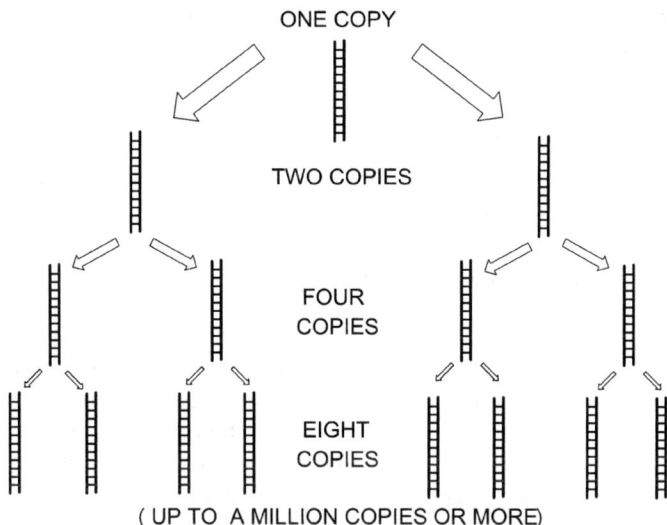

ONE COPY

TWO COPIES

FOUR COPIES

EIGHT COPIES

( UP TO  A MILLION COPIES OR MORE)

**FIGURE 3.10** Polymerase chain reaction can amplify a single molecule of DNA into millions of identical copies.

*reaction* (Figure 3.10). The number of copies of the selected region doubles with each cycle (ignoring slight inefficiencies in the process). After 20 cycles, a mixture containing a single copy of the selected sequence will have nearly $2^{20}$, or about one million copies of that sequence. Virtually all the DNA now present is the desired sequence, with the other viral and host DNA sequences constituting only a tiny fraction of the total DNA. In this way, a DNA sample originally containing only a single copy of the sequence of interest can be treated to enrich the sample with an overwhelming number of copies of that sequence.

## APPLICATIONS OF PCR

The polymerase chain reaction has numerous applications in basic research, and its use as a diagnostic tool is expanding. The main limitation to the method is that, to synthesize appropriate oligonucleotide primers, the sequence of at least part of the molecule to be amplified must be known beforehand.

Basic research applications include virtually any operation in which one would like to produce many copies of a particular nucleic acid sequence, even though the original may be only a single molecule in a vast mixture of all manner of sequences. It is relatively easy to amplify with PCR a single copy of a gene to the point where DNA sequencing can be done on it, or to make sufficient quantities of the gene's sequence for cloning into bacteria for studies on expression of the gene and characterization of the protein product, if any.

Diagnostic applications of PCR include relatively rapid and accurate identification of pathogens of various types, including bacteria, viruses, and parasites. DNA can be amplified from samples of tissue or from biological fluids such as blood.

To make a diagnosis, one chooses the proper primer sequences to amplify the pathogen's DNA; this can be done with quite dilute samples that contain a relatively high level of "background" DNA from the patient's own cells. PCR is also a valuable tool in identifying various genetic diseases. Again, the DNA sample can come from tissue or biological fluid. This time, the primer sequences are chosen to flank the DNA sequence of the genetic site in question; the resulting amplified DNA is then tested for a match to known alleles of the gene.

## CONSTRUCTING RECOMBINANT DNA MOLECULES

With recombinant DNA technology, one can use bacteria, yeast, or cultures of special types of cells from higher organisms to obtain large amounts of a protein that the cell would not otherwise normally produce, for example, human growth hormone production in *E. coli*. Of course, the production of these foreign proteins means that the cultured organism must have acquired the DNA for the gene for the particular protein. Molecular biologists now have a very successful general strategy for doing this. The details will differ from case to case, but the general plan is usually the same.

The steps in this strategy are

1. Make or isolate the DNA for the gene of interest
2. Cut and join this DNA to another type of DNA molecule to prepare it for transfer into the production organism
3. Introduce the joined DNA into the cell
4. Check the cells for the acquisition of the DNA and expression of the foreign gene
5. Grow the cells in large amounts for peptide production

### DNA CLONES

The term "clone" originally referred to a population of genetically identical cells or organisms, all having the same single ancestor and all being derived by asexual reproduction. "Cloning" in recombinant DNA work now usually refers to the insertion and multiplication of identical molecules of DNA in a host organism.

### Sources of DNA for Cloning

By using the genetic code, one can take the amino acid sequence of a given protein and come up with a series of trinucleotide codons that would direct that protein's synthesis. Then, thanks to advances in nucleic acid chemistry, the DNA with those codons in the proper sequence can readily be made in a laboratory. In short, one can synthesize an artificial gene.

For short peptides, one could simply make a single-stranded oligonucleotide for one strand of the gene DNA, then make the complementary strand and hybridize the two strands together. (Actually, because of the redundancy of the genetic code, usually several different DNA sequences would code for the same amino acid sequence.) For very long amino acid sequences, the direct chemical synthesis of the corresponding gene might have to be done in stages. A series of single-stranded

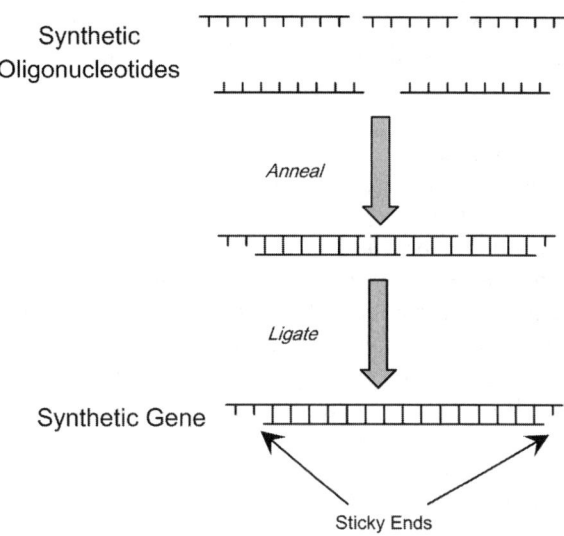

**FIGURE 3.11** Construction of a synthetic gene from chemically synthesized oligonucleotides.

oligonucleotides are made that overlap in sequence, so that they can partially hybridize to their neighbors (Figure 3.11). The oligonucleotides' sequences are chosen to represent alternating strands. Once hybridized, the complete double-stranded DNA gene is made by using DNA polymerase and the necessary deoxynucleotide triphosphates to fill in the single-stranded regions. Finally, DNA ligase is used to reseal any nicks (broken phosphodiester linkages) in the DNA backbone.

For proteins of high molecular weight, the direct chemical synthesis of the corresponding gene is fairly tedious. Moreover, the direct synthesis strategy requires knowing the entire amino acid sequence of the protein to construct the proper nucleotide sequence in the gene. For large or rare proteins, this information may be unknown. A general alternative strategy is then to construct a gene library or gene bank, and then screen or select the DNA for the gene of interest from this library. Methods for screening or selecting genes are discussed below; here the focus is on the construction of the library.

One method is to extract the entire chromosomal DNA, then to cut the DNA randomly into many pieces. The cutting can be done by mechanical shearing of the DNA solution; the conditions are typically chosen to produce random fragments that are 20,000 to 40,000 base pairs in length. This fragment size is long enough to contain most or all of a typical gene, but short enough to be cloned by present-day techniques.

Another method is to extract the DNA, then treat it with an enzyme that cleaves double-stranded DNA at specific sequences (a restriction enzyme; see below). By partially digesting the DNA, large fragments can be generated; moreover, these fragments will contain identical overlapping sequences at their ends that will help later in the cloning process.

How many clones must be made to be reasonably sure that a library will contain a particular gene? This depends on the size of the genome studied and the size of the fragment cloned; the larger the fragment or the smaller the genome, the fewer the number of clones required for the library. Depending on the degree of certainty demanded, a library adequate for *E. coli* might contain only 700 clones, but a human genome library might require over 4 million clones.

A third alternative starts with an extract of RNA, not DNA. Mature eukaryotic mRNA contains a long run or tail of adenine residues at its 3 end. The poly(rA) tail can be hybridized with an oligomer of thymine residues, and the oligo(dT) can then be used as a primer for a particular kind of DNA polymerase known as reverse transcriptase. This enzyme, a polymerase associated with retroviruses, will use RNA as a template to make a complementary DNA copy of the RNA, creating a DNA-RNA double-stranded hybrid. In another round of synthesis, the enzyme can replace the RNA strand entirely with DNA, so that the RNA-DNA hybrid is completely converted to double-stranded DNA containing an exact copy of the original RNA sequence. This DNA molecule is known as cDNA because it has a strand that is complementary to (or a copy of) the original RNA.

Unless the mRNA for a particular gene has been purified away from all the other mRNA species, the result of using reverse transcriptase is to make cDNA copies of all the different mRNA species originally present, including that for the gene of interest. This mixture of cDNA species can be cloned in much the same way as genomic DNA to give a gene library, but with the important difference that only DNA from expressed genes (i.e., those for which mRNA is made) will be present in the library. Gene control sequences and unexpressed genes may be missing from the library, which may limit its usefulness. For those genes that are expressed, however, the cDNA cloning route is often an attractive one, since it reduces the number of clones that must go into the library.

## CUTTING AND JOINING DNA

Early basic research on viral infections of *E. coli* led to the discovery of enzyme systems that protect the bacterium against viral infection. These restriction systems, so-called because they restrict the growth of the virus, were found to be of two general types, differing in their enzymology. Type II systems are now used in recombinant DNA work.

Type II systems have two enzymes, a DNA methylase and a DNA endonuclease. Both enzymes recognize and act on the same particular sequence of DNA, generally 4 to 6 base pairs in length, which is known as the recognition sequence or restriction site. Methylation of one or two bases in the site by the methylase will usually block action by the endonuclease at that site; if the sequence is unmethylated, the endonuclease hydrolyzes both strands of the DNA phosphodiester backbone to cleave the DNA. Bacterial DNA is methylated soon after it is synthesized, and thus it is protected against cleavage. However, the DNA of an invading virus usually is not methylated properly, and so the restriction endonuclease cleaves it into fragments to block the infection.

The recognition sequences of type II restriction endonucleases are usually symmetric; they have a twofold axis of symmetry so that they read the same in the 5'-to-3' sense along opposite—but, of course, complementary—strands. Such sequences are

## TABLE 3.1
### Properties of Selected Restriction Endonucleases

| Enzyme | Source | Recognition Sequence |
|---|---|---|
| *Eco*RI | *Escherichia coli* | G↓AATTC |
| *Bam*HI | *Bacillus amyloliquifaciens* H | G↓GATCC |
| BglII | *Bacillus globigii* | A↓GATCT |
| SalI | *Streptomyces albus* | G↓TCGAC |
| PstI | *Providencia stuartii* 164 | CTGCA↓G |
| HpaII | *Hemophilus parainfluenzae* | C↓CGG |
| HindII | *Hemophilus influenzae* $R_d$ | GTPy↓PuAC |
| HindIII | *Hemophilus influenzae* $R_d$ | A↓AGCTT |
| HaeIII | *Hemophilus aegyptius* | GG↓CC |

called *palindromes*. An example of an English palindrome is A MAN, A PLAN, A CANAL, PANAMA. An example of such a palindromic DNA sequence is the recognition sequence for the restriction endonuclease *Eco*RI: the sequence is

(5′) GAATTC (3′)
(3′) CTTAAG (5′)

The *Eco*RI endonuclease cleaves this hexanucleotide sequence on each strand between the G and the first A residue (reading 5′-to-3′); this point of cleavage is marked by arrows in the sequences in Table 3.1. Notice that the points of cleavage are offset, or staggered, for *Eco*RI. Not all restriction endonucleases make staggered cuts in DNA; some cleave the DNA without any offset, to produce blunt-ended cuts.

Cleavage by the *Eco*RI enzyme produces a single-stranded tail of four nucleotides on the resulting DNA fragments. These tails are, of course, complementary in sequence, and they can be paired up again under the proper conditions (hence they are often called "sticky ends"). The breaks, or nicks, in the DNA backbone can be re-sealed into covalent bonds by the action of the enzyme DNA ligase and an intact DNA molecule is thus recovered. A key point here: the joining of the ends depends only on the sequence of the single-stranded tails on the DNA. As long as the tails are complementary, any two DNA fragments can be ligated. This makes it possible to join together into a single molecule DNA fragments from quite different sources (e.g., animal and bacterial cells) and to rearrange DNA segments to produce novel genetic combinations that create new enzymes, receptors, or other proteins of interest (Figure 3.12).

As mentioned before, DNA for cloning is sometimes prepared by shearing high molecular weight chromosomal DNA. This produces random double-strand breaks in the DNA, so that a wide range of sizes of randomly broken DNA results. These molecules will lack the standardized sticky ends generated by restriction enzymes that make the annealing of different DNA molecules so easy. Also, cDNA will lack sticky ends, because there is no reason for mRNA to have sequences at its 5′ and 3′ ends that will lead to convenient restriction sites. How are such DNA molecules prepared for cloning?

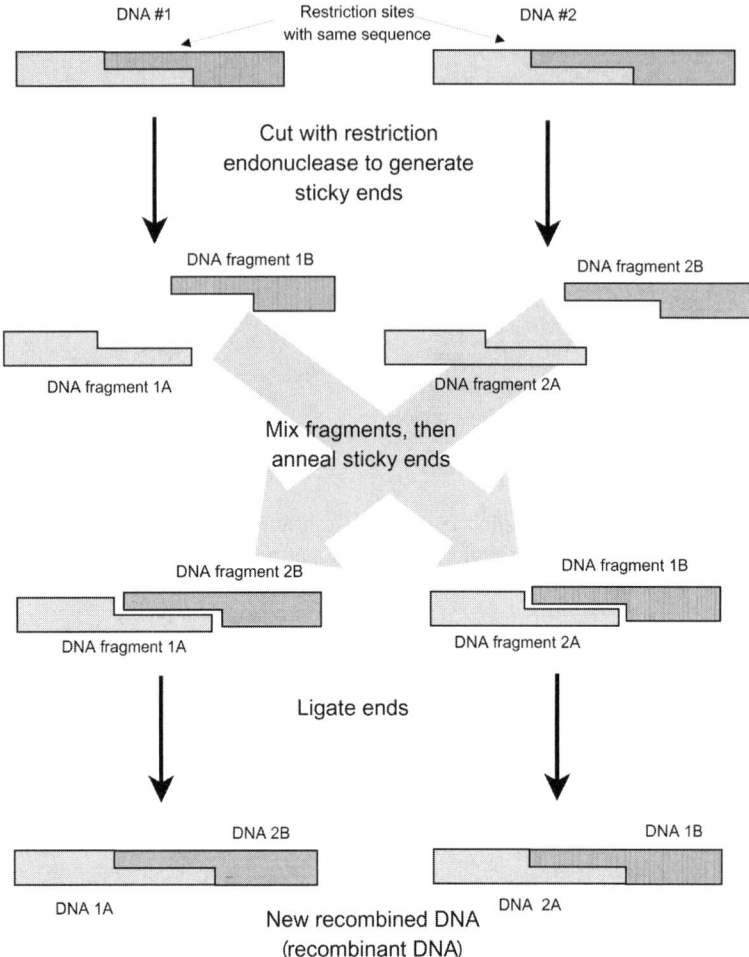

**FIGURE 3.12** Using "sticky ends" generated by the action of a restriction endonuclease to recombine DNA sequences.

One method is to attach single-stranded homopolymer tails (e.g., a run of G residues) to the 3′ ends of all the double-stranded DNA to be cloned, and complementary homopolymer tails (e.g., a run of C residues) to the 5′ ends of the cloning vehicle DNA (see the following the section on DNA vectors). These tails are synthesized and attached enzymatically, using the enzyme terminal deoxynucleotidyl transferase. Since the tails are complementary, the fragments will anneal by their tails to form a nicked or gapped region of double-stranded DNA. DNA polymerase can then be used to fill in the gaps in the annealed molecules, and ligase can be used to seal up the nicks, yielding completely joined molecules, suitable now for cloning.

Another method is to enzymatically add suitable restriction sites to the blunt ends of the DNA. For this, a specially designed blunt-ended synthetic oligonucleotide species, known as a "linker," is added and enzymatically ligated to the larger DNA.

This oligonucleotide contains sequences for one or more restriction sites, and after digestion with the proper restriction enzyme, the whole molecule will now have the proper sticky ends.

Finally, DNA ligase can be used to join two blunt-ended fragments of DNA. This is not very efficient, however, since the complex of enzyme with two blunt-ended DNA molecules is not stabilized by pairing of sticky ends, and it is therefore unlikely to form. When joining DNA, it is usually more efficient to take advantage of the stability of the complex offered by complementary sticky ends.

## DNA VECTORS

DNA vectors are special types of DNA that help in transferring foreign DNA into a cell, replicating that DNA once it is transferred, and expressing the gene(s) on that DNA.

Vectors for cloning DNA come in several different types. Some are special strains of virus that have been genetically altered to introduce desirable properties and eliminate undesirable traits. Other vectors are plasmids, small circular double-stranded DNA molecules that can replicate inside the cell independent of the host cell's own replication, and which frequently carry on them one or more genes for antibiotic resistance. In general, the vector must have at least one cloning site, a sequence that can be cleaved by a restriction endonuclease to permit its joining with a similarly cleaved foreign DNA fragment. Table 3.2 lists some commonly used cloning vectors, the type of host cell, and some of the properties of the vector.

**TABLE 3.2**
**Properties of Selected Cloning Vectors**

| Name | Type | Host Cell | Remarks |
|------|------|-----------|---------|
| PBR322 | Plasmid | *E. coli* | Resistance to ampicillin, tetracycline. General purpose vector |
| pUC8 | Plasmid | *E. coli* | Resistance to ampicillin. *lac* screening. General purpose vector |
| pEMBL8 | Plasmid | *E. coli* | Resistance to ampicillin. *lac* screening. Gene expression, sequencing, mutagenesis |
| pBluescript | Plasmid | *E. coli* | Resistance to ampicillin. *lac* screening, RNA transcripts. cDNA cloning |
| M13mp18 | Filamentous | *E. coli* | DNA sequencing, mutagenesis. *lac* screening bacteriophage |
| λgt10 | Bacteriophage | *E. coli* | cDNA cloning |
| λgt11 | Bacteriophage | *E. coli* | cDNA cloning, gene expression |
| λZAP | Bacteriophage | *E. coli* | Resistance to ampicillin. *lac* screening. cDNA cloning, gene expression, automatic excision into plasmid pBluescript SK |
| λEMBL3A | Bacteriophage | *E. coli* | Genomic library construction. Reduced chance of escape from laboratory; biological containment |
| YEp24 | Plasmid | *E. coli,* yeast | Ampicillin resistance. Yeast shuttle vector |

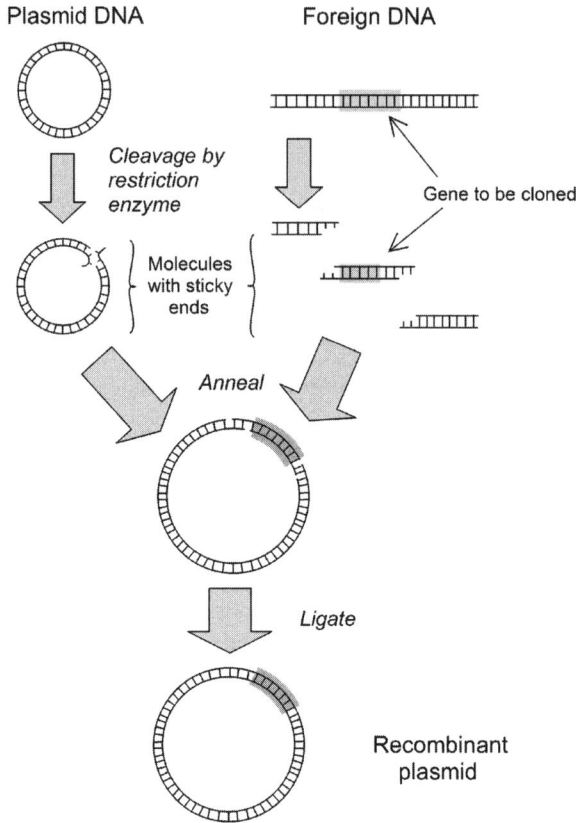

**FIGURE 3.13** Insertion of a foreign DNA sequence into a plasmid cloning vehicle, using restriction endonucleases and ligase.

Assuming the vector DNA and the DNA fragment to be cloned have been suitably prepared by one or another of the methods described earlier in this section, the two species are joined to one another by annealing their ends and sealing the nicks with DNA ligase (Figure 3.13).

The process of actually getting the recombinant DNA into a cell will depend on the type of vector used. If the vector is a plasmid, the DNA is now ready for transfection into the host cells. If the vector is a virus, it may be necessary to package the DNA into infectious virus particles before it can be introduced into the host cells. Infection is usually a fairly efficient means of introducing the recombinant DNA into the host, while transfection tends to be quite inefficient. Transfection involves special chemical and enzymatic treatment of the host cell to make its cell membrane permeable to the DNA, and such treatment tends to reduce the viability of the host.

There is then the problem of separating transformed cells, or transformants, that have acquired the correct recombinant DNA from cells that have not. This leads to the processes of screening and selecting transformants that have the desired characteristics, processes dependent both on the nature of the host cell and on the type of DNA newly introduced into the cell.

## Host Cells for Recombinant DNA Work

Bacterial, yeast, and mammalian cells are all commonly used in recombinant DNA work. The choice of the type of cell, and the particular strain of that cell type, will depend on the type of product sought, the quantity of product desired, and the economics of production. For example, the cell type that produces the highest level of a particular protein may not be the best objective choice for production, if that cell type does not properly fold and modify the protein for full activity. Usually, each new product must be evaluated on its own when choosing a host cell line.

Some factors to consider when choosing a host cell line include (1) the growth patterns of the cells (slow or fast); (2) the cost of growing the cells (special media or supplements, type of growth vessel, special handling, etc.); (3) the level of expression of the product that the cell type can achieve; (4) whether the cell can secrete the product to the surrounding medium, for ease of purification; (5) whether the cells can properly fold and modify a protein product for full activity; and (6) the availability of suitable vectors for transforming the host cells and for "shuttling" pieces of DNA between different cell types. With this variety of factors to consider, most choices end up as compromises over several factors.

The advantages of using bacteria include their rapid growth on low-cost media. Scaleup from laboratory to industrial production volumes is usually straightforward, too. With the proper choice of vector, one can obtain excellent levels of production of protein. The workhorse bacterium *E. coli* is genetically well characterized and is the most popular choice here, although strains of *Salmonella* and *Bacillus* are in use as well. Disadvantages of using bacteria include the lack of proper protein glycosylation and other types of modification, their tendency to deposit as insoluble intracellular aggregates any protein at high levels of expression, the lack of a good means of secreting "foreign" proteins to the medium, and the improper processing of transcripts from eukaryotic genes that contain intervening, noncoding sequences, or introns.

Yeast offers certain advantages over bacteria, notably the capacities to process RNA transcripts to remove introns, and to glycosylate proteins (this latter capacity may occasionally be a disadvantage, since yeast often hyperglycosylate foreign proteins). Other advantages of yeast are their relatively rapid growth and the low cost of the media. Yeast also can secrete protein products to the medium more readily than can *E. coli*. Recently, certain plasmid types have been developed that can replicate in either yeast or *E. coli*, and that have restriction sites suitable for cloning work. These so-called shuttle vectors—they can shuttle DNA between the two host species—can be used to combine the genetic engineering and expression advantages in both yeast and *E. coli*.

One disadvantage of yeast as host cells is the possible difference in glycosylation pattern for proteins made in yeast versus the original eukaryotic source; furthermore, the glycosylation may vary among batches of yeast-produced protein, depending on the growth conditions used. Another problem is that refolding of overexpressed and aggregated proteins is often needed. Finally, the level of protein production in yeast is often appreciably lower than that obtainable with bacteria.

Eukaryotic cell types that are frequently used include Chinese hamster ovary (CHO) cells, baby hamster kidney (BHK) cells, and various tumor-derived cell lines. Cell lines derived from insects are also used. These eukaryotic cells grow much more slowly than yeast or bacteria (a doubling time of 18 to 24 hours versus 20 to 90 minutes for bacteria), and they are more sensitive to shearing forces and to changes in pH, temperature, oxygen level, and metabolites. Culturing these cells is thus a considerable problem, particularly for large-scale production. Levels of protein expression can be variable, as well. However, two major advantages of using these cell types are their ability to process RNA transcripts properly for gene expression, and their ability to properly modify the expressed protein by cleavage, glycosylation, refolding, and so on. One further advantage is that, with the proper control sequences, the protein product can be secreted to the medium after processing, which protects it from various proteases inside the cell and also helps in the purification process.

## METHODS FOR SCREENING AND SELECTING TRANSFORMANTS

A screen is basically a test of an individual cell line for the presence or absence of a specific trait, for example, the production of a certain nonessential enzyme. A selection operates on an entire population of cell lines simultaneously, to permit the growth and propagation of only those cells with the desired trait. A selection might involve, for example, antibiotic resistance or the ability to grow without certain nutrient supplements. Generally speaking, selection of transformants is a more powerful and efficient method than screening, but sometimes an appropriate selection method is not available for the specific trait of interest in a transformant, and one must resort to screening.

For genetic engineering with *E. coli*, a popular plasmid cloning vector, pBR322, carries the genes for resistance to ampicillin and tetracycline. Thus, a transformant that has acquired a copy of pBR322 DNA would also have acquired the traits of resistance to both antibiotics. Such a transformant could then easily be selected by growing the bacterial population in a medium that contains one or both of the antibiotics; bacteria that were not transformed simply do not grow.

Now suppose that this vector carries a piece of foreign DNA that is inserted into the plasmid DNA in such a way that it does not affect either of the genes for antibiotic resistance. Selection for antibiotic resistance would not be enough to guarantee that the selectees also carry the foreign DNA, since the selection did not operate on traits derived from that DNA. One must now use a screen of one type or another to find the desired clones.

If the foreign DNA is a gene that codes for, say, an enzyme not present in the host cell, then it might be possible to screen clones by looking for that enzyme's activity. Alternatively, certain immunological techniques could be used to detect the presence of the unique protein being produced by the desired clones. However, unless the enzyme assay is easy and fast, or suitable antibodies are available, it may not be feasible to use such screens. Moreover, there is no guarantee that a bacterial host would properly express a eukaryotic gene and make a detectable protein, since bacteria do differ considerably in expression mechanisms from other organisms. A better type of screen is one that would detect the foreign DNA directly without relying on any protein production. Several such screening methods have been developed with this in mind.

One method involves purifying the plasmid DNA from each antibiotic-resistant bacterial isolate, and digesting that DNA with a battery of restriction endonucleases. The resulting fragments can be separated from one another by gel electrophoresis and the fragmentation pattern studied for each isolate. By comparing this pattern with the pattern for plasmid DNA without the foreign DNA, one could then deduce which isolates had indeed received the recombinant plasmid as opposed to those that had picked up only the ordinary plasmid. A variation on this method is to use PCR, choosing primer sequences to match the DNA sequence on either end of the foreign DNA (this assumes that one already knows those sequences, of course). One then attempts to amplify the DNA between the primer sequences in the usual way. Clones containing the foreign DNA will yield easily detected amounts of amplified DNA, while clones without the foreign insert will not respond to amplification and so can be readily identified.

Another DNA screening method uses a type of hybridization assay. In this method, bacterial colonies are grown at low density in petri dishes on nutrient agar. A membrane of nitrocellulose is then lightly pressed onto the agar surface to pick up a trace of bacteria from each colony, so that the membrane carries a replica of the pattern of colonies on the agar. The bacteria on the membrane are killed and broken open, and their DNA denatured. The membrane now has spots or dots of single-stranded DNA, each spot representing one of the original bacterial colonies. The DNA in the spots is fixed in place by heat treatment. Next, the spotted nitrocellulose membrane is bathed in a solution containing a labeled probe oligomer whose sequence is complementary to a portion of gene of interest. After rinsing off excess probe and drying the membrane, one can use either autoradiography or a special scanning apparatus to detect spots on the membrane where the probe hybridized, and so to identify colonies carrying the proper recombinant DNA.

This method is often used in screening gene libraries made from sheared genomic DNA or from cDNA. There is, however, the technical difficulty of deciding on the sequence to be used in the probe, if the cloned gene's sequence is not yet known. Fortunately, there are some ways around this problem.

First, there may already be available information on the amino acid sequence of the protein product of the gene. By using the genetic code in reverse, one can guess a suitable sequence of codons and use this as the basis for the probe sequence. Second, the gene of interest may have a homolog in a different organism, and the sequence of this homologous gene (assuming that it is known) can be used to direct synthesis of a probe. Third, the gene may be member of a family of related genes in the same organism, and the sequence of one or more of these related genes can be used in choosing the probe's sequence. Fourth, the gene may be expressed at a high level in the original organism. Presumably there would be abundant mRNA in the organism corresponding to transcripts of that gene. A cDNA library made from the extracted mRNA would likely have more clones of the highly expressed gene than of any other gene. One could then pick a clone at random, purify the recombinant DNA, label it, and use it as a probe. By doing this with many different clones, then picking the one that hybridizes to a large fraction of the clones in the library, one has a good chance of obtaining the cDNA clone for the gene of interest.

## EXPRESSION OF FOREIGN GENES

To use host cells to produce a foreign protein, it is not enough to transform the cells with the foreign gene's DNA inserted into a vector molecule. The gene must be properly expressed in the host as well. This means that the foreign gene DNA must have the proper sequences to direct (1) transcription of the gene, (2) translation of the resulting mRNA, and (3) the proper processing of the peptide product, if necessary. These sequences include a promoter sequence for the binding of RNA polymerase and the control of transcription, a ribosome-binding sequence and a start codon (the triplet ATG, specifying the amino acid methionine) along with a stop codon for proper translation, and sequences in the resulting peptide product to direct proteolytic cleavage, glycosylation, and other types of protein modification.

One common strategy to accomplish all of this is to use a vector that has the necessary promoter and ribosome-binding sequences next to the restriction site where the foreign DNA is inserted. Another strategy is to insert the foreign gene into an existing gene on the vector, to produce a fused mRNA that contains the codons for the vector's gene joined to the codons for the foreign gene. In either case, the reading frame for the inserted sequence must be properly aligned to get synthesis of the correct amino acid sequence. A popular choice for control sequences and for fused mRNA production in *E. coli* involves the gene for beta-galactosidase, an enzyme involved in sugar metabolism in the bacterium. The promoter for this gene generally directs high levels of production of beta-galactosidase mRNA, so one can expect good gene expression for the foreign gene.

When the fused-gene method is used, one recovers a "fusion polypeptide," in which the foreign protein has extra amino acids (from the vector's gene, e.g., beta-galactosidase) at one end. This run of extra amino acids, now covalently linked to the foreign polypeptide, is sometimes a useful feature for stabilizing, protecting, and marking the foreign gene product for later purification. After recovering the fusion polypeptide from the cell culture, it may be necessary to cleave it to purify the desired foreign gene product away from the vector's gene product. This can be done with cyanogen bromide (CNBr), which cleaves polypeptides specifically at methionine residues. Since the foreign polypeptide usually has a methionine as its first amino acid, CNBr can neatly cleave the fusion polypeptide exactly at the point where the two peptides join. The CNBr cleavage method is useful for foreign polypeptides that do not have any methionine residues anywhere except at the N-terminus. (If there are any internal methionine residues, they will be cleaved along with the one at the fusion point, which could inactivate the foreign gene product one is trying to recover.) The desired polypeptide can now be separated from the mixture of cleavage products by standard methods, for example, by chromatography.

# ENGINEERING PROTEIN SEQUENCES

One of the major applications of recombinant DNA technology has been to produce large amounts of commercially relevant proteins, including enzymes, receptors, and peptide messengers of various sorts. The sequences of these proteins, at least in the initial stages of investigation and production, have been those found in nature, so that the structure and function of the protein products of cloning would be the same as those of "natural" proteins extracted from tissue, serum, and so forth.

With the advances in chemical synthesis of DNA, in DNA sequencing, and in the cloning of DNA segments, scientists can now alter the DNA coding sequence in a cloned gene at precise locations. They can now correlate those changes with gene expression and with the structure and function of the protein gene product to understand better the biological functioning of the gene and its product *in vivo* and perhaps to obtain "improved" genes or protein products.

## SITE-DIRECTED MUTAGENESIS

There are several strategies for producing changes, that is, mutations, in a DNA sequence, some of them considerably more specific than others. Mutations at random locations (random mutagenesis) can be produced by any number of chemical or physical agents that cause DNA damage and subsequent changes in the sequence during its repair or replication. Random mutagenesis is fairly easy to do, but because it is random it is not an efficient way to study, for example, the connection of an enzyme's amino acid sequence and its catalytic efficiency or specificity.

A more efficient approach to this sort of study would be to pick out particular amino acids (perhaps those in the enzyme's catalytic site) and replace them with different amino acids in a systematic way. This would of course require mutations in the DNA sequence at particular sites. This process is known as site-directed or site-specific mutagenesis, and there are several different methods of doing it. Two of the more direct methods are described in the following.

If the amino acid sequence of a target protein is already known, then a synthetic gene can be constructed for the protein by using the genetic code in reverse and chemically synthesizing the indicated DNA sequence. To produce a new gene with substitutions in one or another of its amino acids, one can simply synthesize and clone the appropriately altered DNA sequence. In general, however, this method is limited to genes for small peptides where the corresponding nucleotide sequences are short and relatively easy to make. This process will be described below in more detail in connection with the cloning of the human somatostatin gene.

Another method of producing changes at a particular site in a cloned gene is to use heteroduplex mutagenesis (Figure 3.14). Assuming that the DNA of interest has been cloned in a plasmid, one can enzymatically nick the DNA and then digest one or another of the strands in the plasmid to produce a "gapped" duplex, the gap being the single-stranded region on the plasmid. Next, one chemically synthesizes an oligonucleotide that is (almost) complementary to the cloned gene over a short region where the mutation is wanted. This synthetic oligonucleotide will differ from complete complementarity just at the codon for the amino acid to be replaced; at that

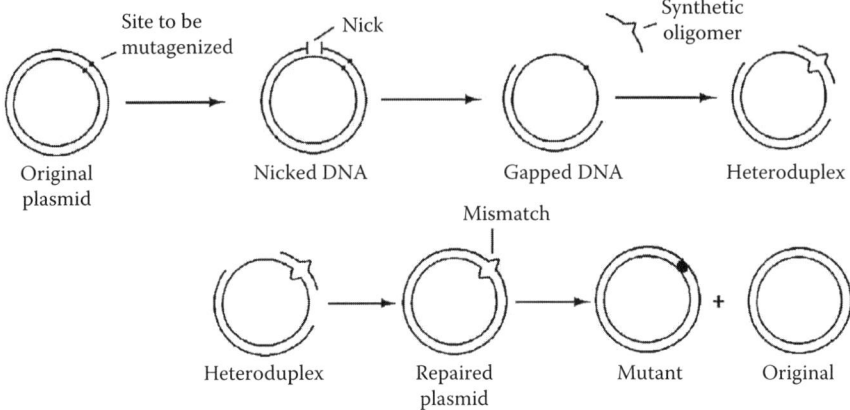

**FIGURE 3.14** Site-specific mutagenesis. A plasmid containing a gene to be mutagenized is nicked and the nick enlarged enzymatically. A synthetic oligomer containing the desired altered sequence is annealed to the exposed or gapped DNA to form a heteroduplex with a mismatch in base pairing at the site where the sequence is to be altered. Enzymatic repair of the gaps and subsequent transfection into the host and replication of the DNA will give bacteria carrying the mutated gene.

point a different codon is used, so that one, two, or three of the bases will not match those on the gene cloned in the plasmid. The synthetic oligonucleotide is annealed to the gapped duplex and enzymatically incorporated into a fully duplex molecule by using a polymerase, ligase, and added nucleotide triphosphates.

This duplex with mismatches can be transferred into *E. coli* just as any other plasmid, and there it will replicate. Half of the progeny plasmids will have the original DNA sequence, but the other half will be the result of copying the strand with the altered sequence, and so these progeny will code for the mutant protein. The mutant progeny can be identified by screening with a hybridization probe, and the bacteria harboring them can be used as a source of the mutant protein.

## EXAMPLES OF APPLIED RECOMBINANT DNA TECHNOLOGY

One of the principal applications of recombinant DNA technology is to the industrial production of proteins and polypeptides that have a high intrinsic value. These products would include, for example, the pharmaceutically valuable peptide hormones; there is also much interest in enzymes with industrial applications (e.g., lipases to improve the performance of laundry detergents, or glucose isomerase, used in the bulk conversion of starch to the widely used sweetener fructose). Another principal application is the engineering of microorganisms to perform industrial chemical processes better (e.g., to increase antibiotic yield from bacterial fermentations) or to reduce the use of toxic feedstocks or intermediates in the production of bulk quantities of high-value small organic compounds (e.g., catechols, benzoquinones, etc.). A third application, of great interest to those involved in drug discovery and development, is the functional expression of mammalian drug receptors in bacteria or yeast, for improved throughput when screening compound libraries. Here we give a few case histories of such applications.

## CLONING OF HUMAN SOMATOSTATIN IN *E. COLI*

Somatostatin is a polypeptide hormone involved in regulating growth; it is important in the treatment of various human growth disorders. The hormone is small, only 14 amino acids in length, making it a good candidate for cloning. In fact, it was the first human protein cloned in *E. coli*.

The amino acid sequence of the hormone is $(NH_2)$-Ala-Gly-Cys-Lys-Asn-Phe-Phe-Trp-Lys-Thr-Phe-Thr-Ser-Cys-(COOH). Using the genetic code, it is straightforward to deduce an appropriate DNA sequence that would code for this polypeptide. For production of the hormone in *E. coli*, the gene fusion method was used.

To do this, a synthetic oligonucleotide containing the coding sequence was made, with some special features. First, two stop codons were attached to the C-terminal Cys codon, to arrest any translation beyond the synthetic gene. Second, a methionine codon was placed before the N-terminal Ala codon, to have a CNBr cleavage site in the expected fusion polypeptide. Third, nucleotide sequences for restriction sites for the *Eco*RI and *Bam*HI restriction endonucleases were placed on the ends, to generate appropriate sticky ends after cleavage, for ligation of the oligomer into a plasmid vector.

The double-stranded oligomer was then joined with a plasmid that carried a portion of the beta-galactosidase gene and its associated control sequences, with *Eco*RI and *Bam*HI restriction sites in the beta-galactosidase coding region. After transfection of a special strain of *E. coli* and selection of the proper recombinants, the fusion polypeptide was purified away from all the other bacterial proteins and subjected to CNBr cleavage. This released the 14 amino acid hormone fragment, which was readily separated from the fragments of beta-galactosidase.

## ENGINEERING BACTERIA FOR INDUSTRIAL PRODUCTION OF VALUABLE SMALL ORGANICS

Quinic acid is a small organic compound very useful in chemical syntheses as a chiral starting point; it also serves as a precursor to benzoquinone and hydroquinone, which are bulk commodity industrial chemicals. Hydroquinone is used as a developing agent in photography and for production of antioxidants, while benzoquinone is used as a building block for many different industrial organic compounds. A traditional source of quinic acid is the bark of *Cinchona* but the production of multiton quantities from this source, as would be needed for industry, is problematic. The standard industrial synthetic route to quinoid compounds (e.g., benzoquinone and hydroquinone) starts with benzene, derived from petroleum feedstocks, and involves toxic intermediates; benzene is a well-known carcinogen. It is therefore desirable to produce compounds like quinic acid more cheaply, to avoid or reduce the use or production of toxic compounds, and in the long run, to reduce dependence on petroleum feedstocks by changing to renewable carbon sources such as corn and biomass. To achieve these aims, the bacterial (*E. coli*) production of quinic acid from glucose was undertaken. *E. coli* does not naturally produce quinic acid, so this process involved considerable genetic engineering, as well as an appreciation of details of microbial metabolism.

The bacterium *Klebsiella pneumonia* can use quinic acid as a carbon source for its growth; the first step here is oxidation of quinic acid to 3-deoxyhydroquinate (DHQ), which is catalyzed by the enzyme quinic acid dehydrogenase. Actually, thermodynamics predicts that the reverse reaction is favored. So if a second bacterium, which made DHQ, had inserted into it the gene for the dehydrogenase, and if this second bacterim did not normally metabolize quinic acid, then this would result in the second organism synthesizing quinic acid from DHQ.

Accordingly, a strain of *E. coli* was first engineered to produce elevated levels of DHQ by increasing the levels of certain key enzymes: transketolase, 3-deoxy-D-arabino-heptulosonic acid 7-phospate (DAHP) synthase, and DHQ synthase. Also, the strain has reduced levels of DHQ dehydratase, which if present would divert some of the metabolic flow into the biosynthesis of aromatic amino acids; its blockage results in higher production of quinic acid.

In the second stage of the engineering process, the enzyme from *Klebsiella* was inserted into this *E. coli* strain, and the bacterium grown on glucose as carbon source. A notable conversion of 80 mM glucose into 25 mM quinic acid was achieved. This is a promising route to the production of valuable aromatic compounds from simple, renewable feedstocks like glucose, while avoiding toxic starting materials or intermediates.

## CLONING RECEPTORS IN BACTERIA

Many important pathways of drug action involve membrane-bound G-protein–coupled receptors (GPCRs). These receptors bind agonist ligands (hormones, neurotransmitters, etc.) with their extracellular regions; they then transduce and amplify the external chemical signal by interaction through their intracellular regions with heterotrimeric G-proteins. The intracellular G-proteins then modulate the activity of a wide variety of enzymes and ion channels, and so affect or regulate many different cellular functions. GPCRs are thus an important class of targets for drug action.

Drug screening relies on the availability of large quantities of receptors. The traditional source of GPCRs is membrane preparations from mammalian tissue. However, this is an expensive source which may not be convenient for drug-screening operations. In some cases, mammalian cell lines can be engineered to overproduce the desired GPCR, but the cell lines are difficult and slow to grow, and are often genetically unstable, losing the ability to overproduce the desired receptor. For simple binding studies, such as might be used in a high-throughput screening operation, the receptor could more cheaply and simply be produced with genetically engineered bacteria, and the intact whole-cell bacterium used in the binding assay. However, bacteria lack endogenous signal transducers (e.g, the G-proteins) so that functional responses (as opposed to simple binding) generally cannot be measured. Still, bacteria are much cheaper to grow than mammalian cell lines, and they lack endogenous receptors, and so produce a low background in the binding assays.

An early example of cloning of a GPCR was the expression of human β1 and β2 adrenergic receptors in *E. coli*. The genes coding for these receptors were first isolated using a cDNA library of the human genome. The coding regions were fused to the 5′ end of a part of the bacterial *lam*B gene, as carried by the multicopy plasmid vector pAJC-264. The *lam*B gene product is a bacterial outer membrane protein responsible for the transport of maltose and maltodextrins; its fusion to the genes for the receptor could thus help to direct the fusion product polypeptide to the outer membrane of the bacterium, obviously a desirable location for a membrane-bound receptor. The plasmid vector pAJC-264 carries a gene coding for resistance to the antibiotic ampicillin, thus allowing for selection of transformed bacteria. It is also designed to express the cloned gene upon addition of isopropyl-thiogalactoside, or IPTG, to the growth medium. This is a common technique to postpone expression of a cloned gene until a sufficient bacterial growth level has been reached, whereupon the biosynthetic resources of the bacterium can be redirected toward synthesis of the recombinant polypeptide by inducing gene expression with IPTG. The multicopy feature of this plasmid allows for each bacterium to carry several copies of the same vector and thus to have several copies of the cloned gene; this can help raise the overall level of gene expression and increase yields of the desired protein product.

The plasmids (designated pSMLβ1 and pSMLβ2, respectively) were separately used to transform *E. coli*. After obtaining stable clones for each plasmid (there can sometimes be spontaneous loss of the entire plasmid or a part of the plasmid), synthesis of the *lam*B-β-adrenergic receptor fusion was induced with IPTG. Expression of the receptor was verified by a radioligand binding assay, using the β-adrenergic antagonist [125]I-iodocyanopindolol. Binding was specific and saturable, as would be expected for functional β-adrenergic receptors. Binding competition assays, using various β-adrenergic agonists and antagonists, showed the same order of binding affinity as for receptors prepared from mammalian tissue. Expression of the β2 receptor was higher than for the β1 receptor, with about 220 active β2 receptors per bacterium versus about 50 active β1 receptors per bacterium. The lower expression level of the β1 receptor was attributed to action of bacterial proteases at a sensitive site in this latter protein. However, the level of expression for the β1 receptor is still acceptable for pharmacological screening assays. Furthermore, background binding was quite low by comparison to assays with mammalian tissue or cell lines. Finally, one liter of bacteria, grown to an optimal cell density, could provide material for about 13,000 assays; this is quite significantly easier and cheaper than culturing mammalian cell lines in sufficient volume for the same number of assays.

## FURTHER READING

Nicholl, D.S.T. (2002). *An Introduction to Genetic Engineering*, 2nd ed., Cambridge University Press, Cambridge, UK.

Stryer, L. (1995). *Biochemistry*, 4th ed., W. H. Freeman, New York.

# 4 Monoclonal Antibodies

*David J. Groves, Ph.D.*

## CONTENTS

## WHAT ARE MONOCLONAL ANTIBODIES?

Antibodies, or immunoglobulins, are a group of glycoproteins (Figure 4.1) produced by B-lymphocytes in response to antigens. The antibodies bind to the antigens with varying affinities through noncovalent associations and also interact with other elements of the immune system so as to neutralize, or eliminate, the bound antigens

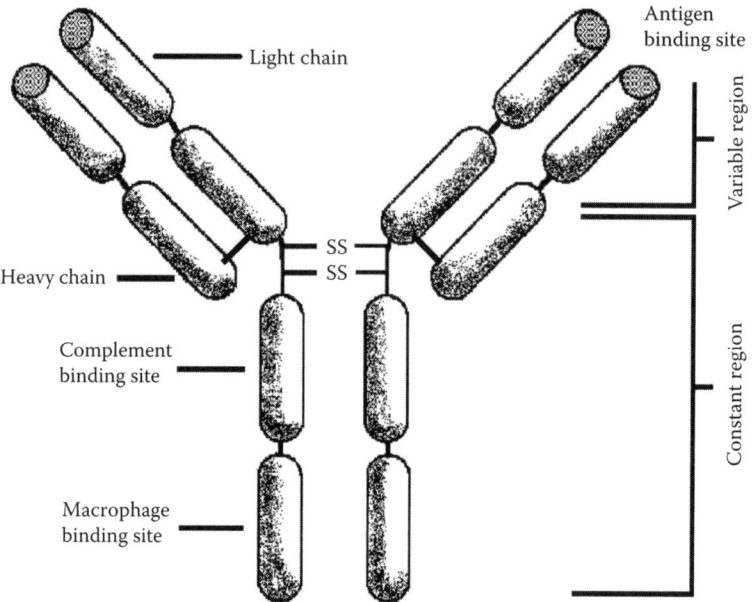

**FIGURE 4.1** Basic antibody structure.

from the body. Antigen receptors on the surface of each B lymphocyte recognize only one aspect (or epitope) of the antigen, although the total repertoire of antigens recognized by the immune system is enormous. Lymphocytes recognizing any particular antigen, therefore, form only a small subset of the total number. However, if a B lymphocyte recognizes and binds to an antigen, the cell is stimulated to proliferate, generating a large number of progeny, each producing antibody of identical specificity to the original surface receptor.

When antibodies are collected from the serum of an animal that has been exposed to antigenic challenge, they generally consist of immunglobulins produced by a range of B lymphocyte clones that have responded to different epitopes of the antigen along with antibodies produced to undefined, irrelevant, and past antigenic challenges. Antibodies produced in this way are referred to as polyclonal. They exhibit heterogeneity with respect to a number of properties. These may include binding avidity, specificity, and immunoglobulin class and isotype (genetically determined differences in the heavy chain constant domains). Although polyclonal antibodies can be purified extensively to limit the heterogeneity, this process can be time consuming, expensive, and unlikely to result in a preparation of antibodies with identical properties derived from a single B lymphocyte clone. Occasionally, subjects are identified with a B lymphocyte tumor, or myeloma, a single clone that produces large quantities of antibody (up to 10% of the total immunoglobulin). This can provide a highly enriched source of monoclonal immunoglobulin for the preparation of particular immunoglobulin classes or isotypes, but the specificity is rarely of interest.

B lymphocytes producing antibody *in vivo* are terminally differentiated and have a finite life span; therefore they cannot be readily cultured *in vitro*. In 1975 Köhler and Milstein successfully used inactivated Sendai virus to fuse antibody-producing B lymphocytes from a mouse spleen to a myeloma cell line from the same strain. The resulting progeny combined the immortality of the myeloma with the secretion of antibody. Antibody derived from such hybrid cell lines is referred to as monoclonal. The antibody forms a homogeneous preparation with respect to affinity, specificity, and chemical behavior. Additionally, under suitably controlled conditions, there is no batch-to-batch variation such as occurs with serum collected from different animals at different times.

The clonal derivation and uniform binding characteristics of monoclonal antibodies (MABs) can permit discrimination between chemically similar antigens with much greater specificity than that of polyclonal antisera raised to same immunogens. However, a MAB cannot discriminate between two antigens if the antibody is directed to an epitope common to the antigens. Polyclonal antisera may be able to discriminate, because they are responding to more than one of the epitopes and they are unlikely to bind in exactly the same manner to the different antigens.

The degree of specificity of the MAB may also present other disadvantages in assay development for some antigens. For instance, MABs to viral strains may be so specific that they do not cross-react with other minor strains of the same virus. The mixed response of polyclonal antibodies is likely to be directed to several antigenic determinants, one or more being present on each strain.

One other property of polyclonal antibodies that can be advantageous with respect to MABs is their potential for cooperatively binding to their respective antigens and so stabilizing their overall binding and mutually enhancing their affinities. Polyclonal antisera are also more likely to form precipitating complexes than MABs owing to the greater opportunity for network formation between antibody and antigen.

MABs exhibit two important advantages over polyclonal antisera; firstly, the potential to select for precisely defined characteristics. Specificity is normally the primary consideration; however, it is important to appreciate that MABs can also be selected on the grounds of their affinity for the antigen. High-affinity antibodies are required for the development of ultrasensitive assay systems or for use *in vivo* to neutralize circulating toxins, hormones, or drugs. In some cases, it may be advantageous to prepare antibodies of lower affinity. For purification purposes, a low-affinity antibody will enable the antigen to be eluted with less extreme conditions and thus avoid denaturation.

From a pharmaceutical perspective, the most important advantage of MABs over their polyclonal equivalents is the potential to generate large quantities of antibody under precisely controlled conditions.

## APPLICATIONS OF MONOCLONAL ANTIBODIES

In the quarter of a century since Köhler and Milstein's ground-breaking work, more than 100,000 MABs have been reported. Many researchers were quick to recognize the therapeutic potential of MABs and from the late 1970s academic and commercial laboratories rushed to produce MABs specific for a range of human diseases. This initial optimism was soon suppressed as clinical trials began to show that mouse immunoglobulins administered to patients were often neutralized by the patient's own immune system. Efficacy was transient, if seen at all, and the commercial boom that financed many of the early developments evaporated. Many of the companies involved in early developments folded or refocused. The idea of MABs was oversold to the general public, the medical profession, and commercial and financial sectors. The failure of the hype to be realized immediately resulted in a loss of confidence and developments were set back several years because of lack of support. This disappointment affected mainly pharmaceutical applications and work continued in other areas.

Most MABs have been produced using mice, an established and relatively inexpensive animal model. There are readily available myeloma cell lines and generally small amounts of immunogens are required. Not surprisingly, many early efforts were in the field of human antibodies and over 200 human antibodies have been developed since the first report of Olsson and Kaplan in 1980. As potential substitutes for human antibodies, and as model systems for the development of human fusions, primates have also received much attention and MABs have been developed in chimpanzees, macaque monkeys, and baboons. Rat and hamster MABs have been produced in the search for antibodies to endogenous murine antigens that produce poor responses in mice. Rabbits, which are routinely used for the production of polyclonal antisera, and guinea pigs, have proved difficult subjects for MAB production. Limited success has been reported by the use of interspecific fusions, although a rabbit plasmacytoma has recently been developed as a fusion partner. MABs have also been successfully derived from cattle, sheep, pigs, goats, horses, cats, dogs, and mink using interspecific fusions. This has generally been in the continuation of work on polyclonal antibodies or investigations of particular applications requiring nonmurine antibodies. However, pig and sheep MABs have both been suggested as candidates for human immunotherapy. Chicken MABs have been produced by a number of Japanese groups following the development of a chicken B lymphoblastoid line suitable for fusion.

Currently applications of MABs fall within four main areas: diagnosis, imaging, therapy, and purification. The primary focus is on human health care, but there have been many MABs generated for veterinary applications as well as nonmedical applications, such as environmental monitoring, and industrial processes, such as abzymes.

## DIAGNOSIS

Polyclonal antisera have been used in the diagnosis of disease since the end of the 19th century when bacterial antigens were used to test for pathogens such as typhoid and syphilis. The development of quantitative precipitation assays using purified antibodies in the 1930s, together with theories of antibody-antigen reactions, laid the groundwork for the first simple immunoassay to measure single antigens. The gel diffusion techniques of Oudin and Ouchterlony in the 1940s permitted the analysis of antigens and antisera. At the same time Coombs described the use of red cell agglutination as an indicator for a sensitive competitive immunoassay. Agglutination techniques are still used in many diagnostic tests, often with latex particles instead of blood cells as the solid phase. In the early 1960s the first radiolabeled immunoassays were reported, permitting analytical sensitivities in the pmol/liter range and resulted in increased demands on antisera quality. The chemical coupling of fluorogenic and enzyme labels to antibodies permitted the development of immunocytochemistry to localize antigens within tissue sections. In liquid phase systems, enzyme labels have been used in the generation of enzyme immunoassays, probably the most widely used immunodiagnostic system currently in use. The combination of cell sorting technology with labeled antibodies has produced a powerful analytical and diagnostic technique, Fluorescence activated cell sorting (FACS) is extensively used in analysis of blood cell composition, notably in the monitoring of immune responses in diseases such as AIDS and cancer.

Monoclonal antibodies are particularly useful in diagnostic areas where specificity is the most important criterion, for example, in the measurement of circulating steroid hormones or in discriminating between viral strains. They also permit analysis of disease phases, such as those associated with developmental stages of parasite life cycles characterized by alteration in the expression of cell surface antigens. The use of MABs in the analysis of cell surface antigens is also important in tissue typing for transfusion or transplantation purposes. Here MABs allow extensive analysis of the population and permit precise matching of the donor and the recipient. In both clinical and veterinary fields, MABs have been developed for the diagnosis of pregnancy using rapid, simple, and direct assays. In cattle breeding, MABs are also used in the sexing of embryos prior to implantation.

Perhaps the most important area clinically for MABs is in diagnosis, imaging, and treatment of tumors based on changes in cell surface antigens. Monoclonal antibodies to tumor-indicative markers are applied to biopsy specimens through immunocytochemical methods, to blood samples with fluorescence activated cell sorting, and to *in vivo* imaging using radiolabeling.

## IMAGING

Imaging is the use of specific, or semispecific, methods to identify and localize disease. Imaging can be considered to be intermediate to diagnosis and therapy. Antibodies, generally tagged with radioactive labels, are administered to the patient and therefore must fulfill safety criteria. The purpose is to search for and localize antigens, such as tumor markers, within the body, in order for treatment decisions to be made. Imaging is often used in the assessment of metastatic spread of tumors in cancer medicine. The first animal studies were performed in 1957 and demonstrated that external scintigraphy could be used to locate tumors in an animal following the administration of radiolabeled tumor-specific antisera. The first successful human studies were performed in 1967 using iodinated rabbit antisera to fibrinogen. Following the discovery of specific tumor-associated antigens, such as CEA and AFP, and MABs, research has accelerated in this field.

Imaging antibodies need to meet a stringent set of criteria—they must be stable during labeling and after administration. They must be of sufficiently high affinity to localize in small tumors and accumulate. There must be no inappropriate cross-reactivity *in vivo*. The target antigen must be carefully defined as no antigens have yet been described which are entirely tumor specific and they may occur on other tissues or at different times.

Radioactive labels are γ-emitters selected on the basis of half-lives, the energies emitted, decay products, ease of labeling, availability and expense. Iodine isotopes 121, 123, and 124, Indium 111, and Technetium 99 are the labels most widely used. The short half-lives of these labels (hours to days) means that radioimaging reagents are prepared immediately prior to treatment. Radioimaging of diseased tissue also provides useful information on the design of therapies that localize radioisotopes or toxins at tumor sites for therapy.

## THERAPY

The use of specific and nonspecific antisera in human medicine is well established and dates back to 1891 when Emil von Behring developed the first diphtheria antitoxin, but their use carries associated risks such as fluid overload and transmission of disease. The potential for MABs as therapeutic agents was quickly recognized and the first MAB was approved for therapeutic use in 1986 (Ortho Biotech's OKT3, a mouse MAB to CD3, for the reversal of transplant rejection). It became clear early on that the presence of, or appearance of, human antimouse antibodies (HAMA) in the patient, which neutralized subsequent treatments, often limited the efficacy of mouse MABs. More recently, chimeric, deimmunized, or fully human MABs have been developed.

At the time of writing 10 MABs have been approved by the FDA for therapeutic application (see Table 4.1). At least 80 more are currently in development.

MABs can be used directly to induce an inflammatory response to tumor cells, or to block receptors; they can be conjugated with a cytotoxic or radioactive molecule such as Iodine 131. By attaching liposomes containing cytotoxic drugs to single chain antibody constructs, anticancer drugs can be delivered effectively to the cell interior. Bi-specific MABs can be used to accumulate cytotoxic cells or molecules at the target site. The ready access of MABs to cells in the general circulation has resulted in many of the targets being hematological disease such as blood cancers. The neutralization of circulating immune cells or clotting components has also been achieved. Commercial and academic laboratories are developing antibodies to neutralize particular classes of T cells to control diseases such as psoriasis, and to limit tissue damage following cardiac arrest or stroke.

The treatment of solid tumors with such relatively large molecules as immunoglobulins presents certain challenges. Tumor regression in response to immune sera was first recorded in the 1950s and currently trastuzumab (Herceptin) and rituximab (Rituxan) are used successfully in the treatment of solid tumors. The antibody can act in several ways including the induction of apoptosis, the blocking of growth factor receptors, induction of cell- or complement-mediated cytotoxicity, or the blocking of angiogenesis. However, direct actions require the antibody to penetrate the tumor mass. This may be facilitated by the use of antibody fragments, but the combination of immunotherapy with surgery and radio- or chemotherapy has been demonstrated to be the most effective approach. Other therapeutic uses of MABs in both veterinary and clinical situations include the prevention of viral or

**TABLE 4.1**
**MAB Drugs Approved for Marketing, 2001**

| MABs | Antigen | Type | Target Disease | Company | FDA Approval |
|---|---|---|---|---|---|
| Orthoclone OKT3 | CD3 | Murine | Acute transplant rejection | Ortho Biotech/ Johnson & Johnson | 1986 |
| ReoPro | GP IIb/IIIa | Chimeric | Blood clotting during surgery | Centocor/Eli Lilly & Co. | 1994 |
| Rituxan | CD20 | Chimeric | Non-Hodgkin's lymphoma | Biogen IDEC/Hoffman La-Roche/Genentech | 1997 |
| Zenapax | Il-2 Receptor | Humanized | Acute transplant rejection | Protein Design Labs/Roche | 1997 |
| Herceptin | HER2 | Humanized | Breast cancer | Genentech/Roche | 1998 |
| Remicade | TNF | Chimeric | Rheumatoid arthritis and Crohn's disease | Centocor/ Schering-Plough | 1998 |
| Simulect | Il-2 Receptor | Chimeric | Acute transplant rejection | Novartis | 1998 |
| Synagis | RSV F Protein | Humanized | RSV infection | MedImmune | 1998 |
| Mylotarg | CD33 | Humanized | Acute myeloid leukemia | Celltech/Wyeth-Ayerst | 2000 |
| Campath | CD52 | Humanized | Chronic lymphocytic leukemia | Millennium Pharmaceuticals/Schering AG | 2001 |

bacterial infection by direct immunoneutralization (the blockading of viral entry by preventing binding to receptors) or the stimulation of endogenous antibody responses by using anti-idiotypic MABs to mimic the protective antigens of the infective agent. MABs can also be used to remove a specific substance from the circulation, including the acute elimination of venoms or toxins following, for example, snakebite. Accidental or deliberate overdoses of drugs or poisons, such as paraquat, can also be treated. In veterinary medicine, MABs have been used as contraceptives by removing endogenous hormones from the circulation, and as fertility enhancers by eliminating hormonal feedback during ovulation. MABs to steroids have also been tested for their effect on eliminating hormone-related taints in pig meat prior to slaughter.

### PURIFICATION AND OTHER APPLICATIONS

MABs provide highly selective agents for the affinity purification of antigens, allowing the recovery of antigen from crude mixtures or removing particular contaminants from a preparation. By using selected MABs for purification, it is possible to isolate the original antigen although only crude preparations are available for immunization. MABs have been proposed as possible reagents for industrial separations, for cleanup of waste streams, and as components of consumer goods. The technology is available for all these applications and the major obstacle to fulfillment is that of inexpensive production techniques at an industrial scale. Plant-based production techniques may offer the most realistic source of material, especially as glycosylation and purity are likely to be much less critical than in the production of clinical material.

Alongside these areas, MABs also find extensive, and occasionally esoteric, uses in basic research, especially in the investigation and analysis of the immune system itself. For instance, MAB technology allows the preparation of large amounts of homogenous antibodies of the different isotypes, previously only obtainable by the extensive fractionation of serum proteins or by identification of suitable myelomas. As it develops, MAB technology will allow the comparison and exploitation of the immune systems of many different species. The application and elaboration of this work has resulted in the MAB occupying a key position in the research, diagnosis, prophylaxis, and treatment of a wide range of diseases, in both human and veterinary medicine.

# THE GENERATION OF MONOCLONAL ANTIBODIES

The production of all MABs follows the same basic pathway (Figure 4.2). Depending on the antibody and the application question, it may not be necessary to proceed to all stages of the pathway.

### IMMUNIZATION

The preparation of any MAB commences with the generation of lymphocytes sensitized to the antigen of choice. Conventionally this is done by the immunization of a mouse against the antigen followed by repeated booster injections until a suitable serum response has been obtained. Optimal results are obtained if the immune

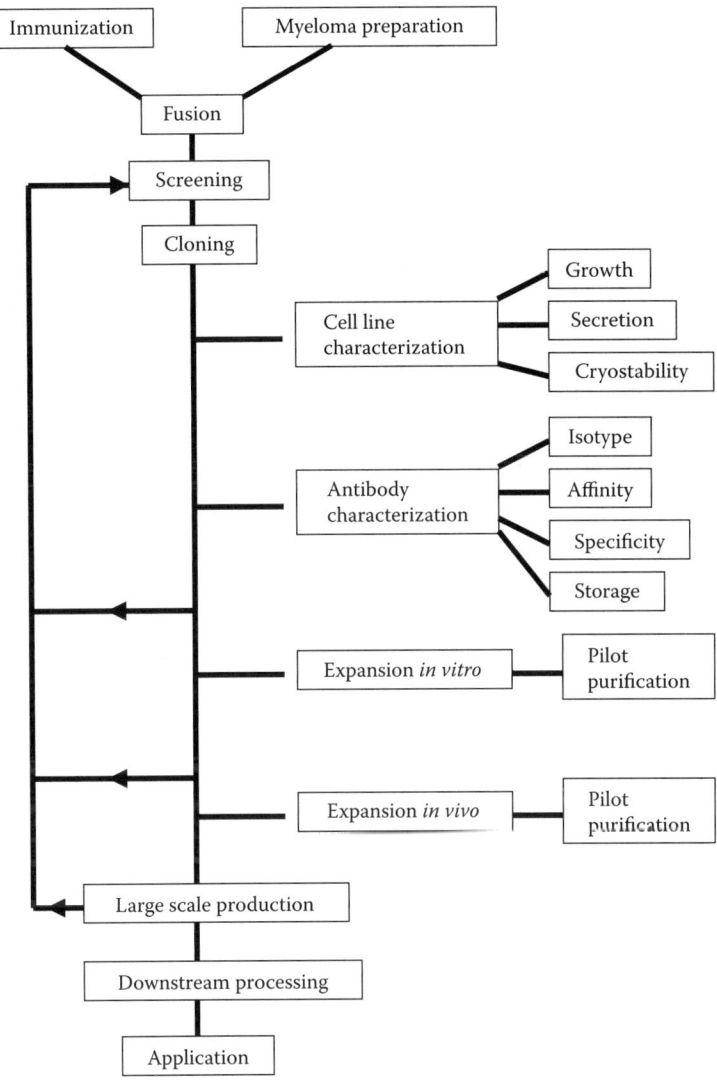

**FIGURE 4.2** Monoclonal antibody production protocol.

response of the donor is allowed to wane to a constant low level between boosts. The animal is then sacrificed and the spleen removed aseptically. Spleen cells are relatively rich in B lymphocytes; these are collected from the disrupted organ prior to fusion and separated as far as possible from the contaminating red blood cells, platelets, and other tissue structures, by buoyant density centrifugation.

The immunization schedule used will vary with the antigen and the characteristics of the antibody required. For a novel antigen, it is common to employ previously reported procedures successfully used with similar antigens and to adapt these on the basis of results. The use of insufficient or excessive amounts of antigen can lead to immune tolerance. Excessive immunization can also block lymphocyte antigen receptors and produce immune paralysis. Overimmunization has also been implicated in the production of relatively low avidity antisera by selecting for low affinity B lymphocyte clones.

Apart from the spleen, other lymphoid tissues, such as tonsils and the mesenteric or popliteal lymph nodes, can be used as a source of lymphocytes. In the preparation of MABs of human or veterinary origins it is often not possible to obtain lymphoid tissue, and there have been many reports of the successful use of lymphocytes separated from peripheral blood. In some cases, for ethical or practical reasons, it is not possible to immunize the lymphocyte donor, as when human MABs are required, or acutely toxic antigens are used. Also, antigen is not always available in sufficient quantities to perform a successful immunization *in vivo*. In these circumstances, it may be possible to perform the boosting stage or, indeed, the entire immunization procedure on the lymphocytes *in vitro*.

It is apparent that mice do not respond well to certain antigens, which may be masked by immunodominant epitopes, and that certain therapeutic treatments require specific effector functions not exhibited by mouse antibodies. In order to address these issues, work began on developing human MABs for therapy. Initially, B lymphocytes were immortalized by infection with Epstein Barr virus (EBV), but the secretion and stability of such lines is often poor, and attempts were made to improve these features by fusing the transformed lines with mouse myeloma lines. EBV transformation is specific for primate B cells, although the use of SV40 virus was tried, unsuccessfully, with rabbit B cells. This approach, together with direct fusion between human B lymphocytes and rodent myelomas has been successful in isolating many lines secreting human MABs. In order to improve the relatively poor efficiency of these processes, compared to mouse x mouse fusions, several groups have used hybrid mouse x human or even mouse x primate fusion partners to replace rodent myelomas. This approach has been used with a great deal of success in veterinary species.

Although these approaches have provided methods of immortalizing suitable B lymphocytes, the limitation of the system is in obtaining suitably sensitized lymphocytes. Because of ethical considerations, it is generally not possible to immunize humans specifically, although lymphocytes can be collected from individuals who have been exposed to the antigen of interest because of infection, vaccination, or accident. Consequently, much effort has been expended, with limited success, on *in vitro* immunization of lymphocytes. Standardized methods have not yet been developed and the majority of MABs derived are IgMs. Several commercial operations are now focused on the use of transgenic mice in which mouse embryos are transfected with human antibody genes. On subsequent immunization, human antibodies are produced. Several human MABs produced with these techniques are currently in clinical trials.

To circumvent some of the limitations of direct immunization, phage display technology has been applied to the preparation of fully human MABs. Gene libraries of cDNA from nonimmune or immunized donor B lymphocytes are expressed in bacteriophages. The bacteriophages display functional antibody fragments and can be rapidly selected and isolated for subsequent engineering into human antibodies. The resulting product is fully human, and several antibodies are under clinical evaluation. The technology is now being used in other species.

## MYELOMAS

Antibody producing hybridomas can be generated either by transformation of the cell line or by fusion of the sensitized B lymphocytes to plasmacytoma or myeloma cell lines. After the fusion, the cell population consists of hybridomas together with the unfused B cells and myeloma cells. The B cells die off rapidly; however, it is necessary to introduce selective mechanism to separate myeloma cells from hybridoma cells. In order to do this, the myeloma lines used are selected for a deficiency of hypoxanthine-guanine-phosphoribosyl transferase (HGPRT), a critical enzyme in the salvage pathway for DNA synthesis. Selection for cells deficient in the enzyme is effected by the addition of 8-azaguanine or 6-thioguanine to the culture medium. These purines are converted through the salvage pathway to cytotoxic nucleotides so that only mutants deficient in these pathways survive.

Fusion with the B cells compensates for this deficiency. When fused and unfused cells are incubated in the presence of the folic acid antagonist aminopterin, the *de novo* synthesis of purines and pyrimidines for DNA is blocked. Cells deficient in HGPRT die, whereas hybrid cells are able to bypass aminopterin blockage by metabolism of hypoxanthine and thymidine added to the medium. In the generation of mouse hybridomas, an number of myelomas deficient in HGPRT are available, all originating from MOPC 21, a spontaneous myeloma from the BALB/c mouse strain.

The first mouse myeloma used in the preparation of hybridomas was P3/X63/Ag8 (X63) by Köhler and Milstein in 1975. This line secretes an endogenous IgG1 with kappa light chains, possibly resulting in the production of irrelevant or chimeric antibodies by the resulting hybridomas. The myeloma X63 has been cloned, and nonsecreting myelomas were selected. One of the resulting lines in common use is P3/NS1/1.Ag4.1 (NS1), which produces, but does not secrete, kappa light chains. There are also several commonly used lines that do not produce any endogenous immunoglobulin components. These include P3/X63/Ag8.653 (653) and NS0/U, SP2/0-Ag14 (SP2) and F0. Successful fusions have been reported with all these lines, NS1 and SP2 perhaps being the most popular.

Rat plasmacytomas are available, derived from the ileocecal lymph nodes of the LOU/C strain. The first rat-rat hybridoma was described by Galfré and coworkers in 1979 using the aminopterin sensitive line R210Y3.Ag1.2.3 (Y3), which secretes immunoglobulin light chains. Further nonproducing rat myelomas YB2/0 and IR983F have since been derived.

A rabbit plasmacytoma line has been produced, after intensive studies of transgenic animals carrying c-myc and v-abl oncogenes, and has been used in the production of rabbit x rabbit hybridomas. Several chicken MABs have been produced using a thymidine kinase sensitized chicken B cell line established by *in vivo* transformation using avian reticuloendotheliosis virus and subsequent treatment with the chemical mutagen ethyl methylsulfonate. For other species the practical and ethical obstacles to generating fusion partners in this way may be too great. More hope may lie in the isolation of suitable naturally occurring lymphoblastoid tumors for sensitization. Such lines have been reported in humans, horses, cattle, cats, and pigs.

Extensive efforts have been made to generate a suitable cell line for the production of human hybridomas. Plasmacytomas are the most differentiated of lymphoid malignancies and are some of the most difficult human cell lines to establish in continuous culture. However, several plasmacytoma- or lymphoblastoid-derived lines of human origin have been used in the preparation of human–human hybridomas. Olsson and Kaplan first used a cell line of plasmacytoma origin, SK0-007, in 1980. The line still secretes both heavy and light immunoglobulin chains. Lymphoblastoid cells are more readily maintained in culture, but generally exhibit a lower rate of immunoglobulin production than plasmacytomas. They are also infected with Epstein-Barr virus. The more widely applied lymphoblastoid-derived lines include UC729-6, GM1500-6TG-2, and LICR-LON-Hmy2.

Karpas and coworkers in 2001 reported the development of a fully human myeloma cell line that was sensitized to hypoxanthine-aminopterin-thymidine (HAT) and selected for tolerance to polyethylene glycol (PEG). This myeloma, Karpas 707, secretes light chains, and has been used to derive fully human hybridomas. The line has been successfully fused with a range of cell types, including EBV-transformed B cells, and lymphocytes from tonsil tissue and peripheral blood. Although this approach has the potential to produce human MABs in pathological conditions, including autoimmunity and infection, as well as identifying specific and effective tumor antigens, the practical and ethical issues of immunizing human subjects mean that this line is unlikely to replace developments in humanization and *in vitro* antibody technology.

## Cell Fusion

In MAB technology, the objective of the fusion process is to produce hybrid cells that incorporate the immortal characteristics of the myeloma cell with the antibody secreting properties of the antigen-sensitized lymphocytes.

Spontaneous fusion of cultured cells occurs only rarely. However, the rate at which it happens can be markedly increased by the addition of certain viruses or chemical fusogens to the culture. Sendai virus, as used in early somatic cell fusions, has a lipoprotein envelope similar in structure to the animal cell membrane. It has been suggested that a glycoprotein in the envelope promotes cell fusion by an as yet unexplained mechanism.

Of the chemical fusogens, by far the most widely used is PEG. PEG is available with aggregate molecular weights of between 200 and 20,000. A wide range of molecular weights, concentrations, and conditions has been successfully used to induce cell fusion. PEG is toxic to cells at high concentrations, a property that is more pronounced with the lower molecular weight polymers. The higher molecular weight polymers tend towards greater viscosity and higher melting points. Most cell fusions, therefore, use PEG with molecular weights of between 600 and 6000. PEG is a highly hydrophilic molecule that acts by complexing water molecules in the medium, removing the hydration shell from the cells and forcing them into intimate contact. Cell membranes adhere because of hydrophobic interactions of membrane phospholipids and a small proportion of them break down and reform so that multinucleate cells are formed. In most of these, the nuclei fuse to produce hybrid cells containing chromosomes from both parental cells.

Because of the toxicity of PEG at high concentrations, it is necessary to balance the exposure with the increase in fusion efficiency of longer durations. Most reported methods use 1 or 2 minute incubation of the cells with a 50% solution of PEG in the medium followed by a gradual dilution and washing of the cells in fresh medium.

There have been a number of reports of the use of electrofusion to produce antibody-secreting hybridomas. Electrofusion is an elegant technique of great potential that has so far failed to find widespread application in MAB technology, probably due to the comparatively high cost and complexity of determining optimal conditions, relative to traditional PEG fusion. In this technique, low concentrations of myeloma cells and sensitized lymphocytes are subjected to an alternating electric field that induces dielectrophoresis in which cells become polarized and adhere, positive to negative poles, in long chains attached to the electrodes. Fusion is induced by a field pulse, a high voltage DC pulse lasting for microseconds, that causes reversible breakdown of the area of membrane contact; a membrane pore develops and cytoplasmic continuity occurs.

## Screening Assays

Any fusion will produce a high proportion of clones that do not secrete antibodies of interest. Any cloning of cell cultures will also result in a proportion of previously secreting lines ceasing production. Therefore, it is vital that there are suitable methods available to determine which cells are, and which are not, producing antibodies. Inability to discriminate between the two will result in the nonsecreting lines overgrowing those carrying the metabolic burden of production.

Even a simple fusion can generate a very large number of clones. Without a rapid and reliable screening assay, the level of work required to maintain and process lines can quickly become excessive. The final application of the antibody is also important in the selection of the screening method to be used. For example, if the MAB is to be used as an assay reagent, then it is best screened in that system; if an immmunocytochemical reagent is required, then the antibody is best screened against the tissue of interest. Careful thought should be given to the screening assay so as to eliminate possible cross-reactants and formatting problems. For most purposes, the enzyme-linked immunosorbent assay (ELISA) or dot-blot formats are convenient.

Assay of cell culture supernatants is best performed soon after fusion rather than later when a large number of cells have been grown. Early identification of wells containing secreting lines can allow early cloning and avoid the problem of overgrowth by nonsecretor lines. This requirement imposes a need for screening assays that are as sensitive as possible.

## Cloning

The full potential of the monoclonal antibody technique can be realized only if the cell culture producing the antibody is truly monoclonal, consisting solely of cells derived from a single progenitor. Cloning is necessary in the postfusion stage of hybridoma generation and later in the existence of the cell line when nonsecreting or class switch variants can arise. In the early postfusion stage, fused cells undergo a random sorting and loss of chromosomes. It is necessary to isolate those viable cells still containing functional genes coding for immunoglobulin synthesis. A significant proportion (as much as 50%) of a hybridoma's synthetic apparatus and a corresponding proportion of its metabolic energy can be devoted to immunoglobulin synthesis. Cells that lose the chromosomes responsible for antibody production inevitably have a greater proportion of energy available for growth and division. Consequently, nonsecreting lines may overgrow the desired hybridoma cells.

Two methods are commonly used to derive monoclonal populations from heterogeneous cell mixtures. The first is cloning in soft agar. The heterogeneous population is dispersed in warm, molten agar and poured into petri dishes. As the cells begin to divide, individual colonies can be picked up and transferred to tissue culture plates for expansion and screening. The second method, cloning by limiting dilution, is less labor intensive and more widely used. A heterogeneous cell population is diluted and aliquoted into 96-well tissue culture plates, generally in the presence of feeder cells to provide the growth factors necessary at low cell densities, with the object of including only a single hybridoma cell in each well. Obviously, even at the correct dilution, some wells will contain no cells, while others will contain two, three, or more cells. The cell distribution can be theoretically modeled according to the Poisson distribution and this enables the probability of monoclonality in individual wells to be defined during repetitive subclonings. Routine visual inspection of the distribution and morphology of cell colonies in the tissue culture plates is likely to be of more immediate practical assistance in deciding whether or not a culture is truly monoclonal. After plating, clones are allowed to grow from 7 to 10 days before visual inspection for monoclonality and screening for the antibody of interest.

Cloning can also be performed using a FACS. Although a powerful technique, it is available to only a limited number of laboratories because of the cost of the equipment. Antigen coupled to a fluorescent marker is incubated with the cells. Cells with surface immunoglobulins of relevant specificity can then be isolated and sorted into individual tissue culture wells.

Although it may be apparent visually (soft agar) or statistically (limiting dilution) that a particular antibody secreting colony is monoclonal, this must be verified objectively by analysis of the antibody produced. No single method can demonstrate monoclonality unequivocally. Uniformity of isotype, affinity, and target epitope can indicate whether the antibody is monoclonal. Isolectric focussing can be used to demonstrate the homogeneity of an antibody preparation, although microheterogeneity in terms of glycosylation and deamination may result in multiple banding even if the antibody is indeed monoclonal.

## CELL LINE CHARACTERIZATION

Once a hybridoma has been cloned, it is necessary to characterize the line to determine the conditions for optimal growth and secretion. Regular monitoring of these parameters enables the detection of changes in the cells and their performance, permitting rectification of any reversions that may occur.

If a hybridoma is to be successfully exploited, it must be stable during freezing and thawing to enable the line to be stored long term in liquid nitrogen. After cryostability has been established, growth rate and antibody secretion rates can be examined using defined media. Most murine hybridomas have a doubling time of 12 to 36 hours and secrete 5 to 100 µg of antibody per ml of medium. Lines should then be karyotyped to determine the chromosome complement; this is especially important with interspecific hybridomas. If the hybridoma is to be grown on a large scale, it is useful to optimize the medium composition and to determine the purification profiles so that departures from the normal can be detected.

## ANTIBODY CHARACTERIZATION

The final application of the antibody must be borne in mind when deciding the extent of characterization. Initially, the antibody must be tested to establish whether binding occurs with the immunogen, with and without any carrier molecules used in the immunization. This test should be carried out with reference to the intended application to control for bridge binding. For instance, a reagent intended for use in a capture ELISA should be tested when coated onto the assay solid phase. If the antibody is intended for *in vivo* immunoneutralization, it should be tested initially in a liquid phase assay. If the final use is to be immunocytochemical, then the testing should be conducted on tissue sections.

When the reactivity of the antibody with the antigen of choice has been established, it should be checked for cross reactivity with potential interfering substances. Again, this check should use, as far as is possible, the intended final format, since specificity will often vary with different situations. For example, a high affinity antibody to estradiol that cross reacts with other estrogens is best suited to immunoneutralization *in vivo*, while a direct assay reagent requires low cross-reactivity with other circulating steroids, but is less dependent on high affinity, since estradiol levels are quite high.

Once the specificity of the antibody has been established, it can be further characterized with respect to it isotype. This may be relevant to its application, but is also important if class switch variants are to be detected.

The determination of antibody affinity (in $M^{-1}$) is useful for comparison with other reported antibodies. However, the important test of affinity is how well the antibody performs. If more than one antibody has been prepared, a direct comparison of antigen binding curves will indicate the most avid preparation.

# GENETIC ENGINEERING

## RECOMBINANT ANTIBODIES—HUMANIZATION AND DEIMMUNIZATION

*A detailed discussion of genetic engineering techniques is beyond the scope of this chapter; however, genetic manipulation has been used to modify the characteristics of traditionally isolated MABs and to isolate novel binding moieties, or to combine them with new effector functions.*

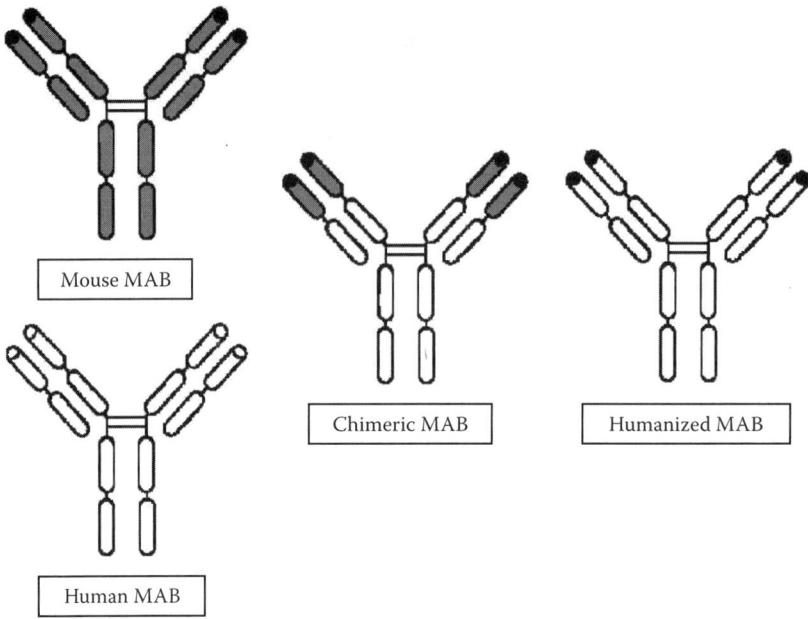

**FIGURE 4.3** Antibody types.

The HAMA response of patients to murine MABs is primarily a reaction to the Fc component of the antibody. Early protein engineers fused murine exons coding antibody Fv with human Fc exons to generate chimeric genes. The resulting chimeric MABs preserve the binding characteristics of the mouse antibody but have the effector functions of the human antibody and a reduced potential for antiglobulin reaction. This technique is now well established and several chimeric MABs have been approved by the FDA for sale (see Table 4.1). This approach can be taken further to produce "humanized" MABs by grafting the CDRs of the murine MAB onto a human antibody framework (Figure 4.3). A number of humanized MABs are currently available or in clinical trials.

Where antigens are poorly immunogenic or toxic a technique known as phage display is of particular use (Figure 4.4). This procedure relies upon the fact that the lymphocyte pool of any individual contains an enormous library of genetic material with the potential to produce over $10^{12}$ different antibody sequences. The lymphocyte pool can be prepared from randomly selected donors, which will tend to produce a naïve repertoire of IgM antibodies, or from donors with particular diseases or histories of antigen exposure, which will generate IgG libraries. The libraries are produced using PCR amplification of B cell DNA. Alternatively, libraries can be generated *in vitro* using "randomized wobble" primers; this provides the equivalent somatic hypermutation during the immune response and may produce antibody specificities or affinities that would not occur *in vivo*. To screen the libraries for antibodies of interest the DNA is transfected into bacteria. Following infection with bacteriophages the cell produces many new filamentous phages, each expressing

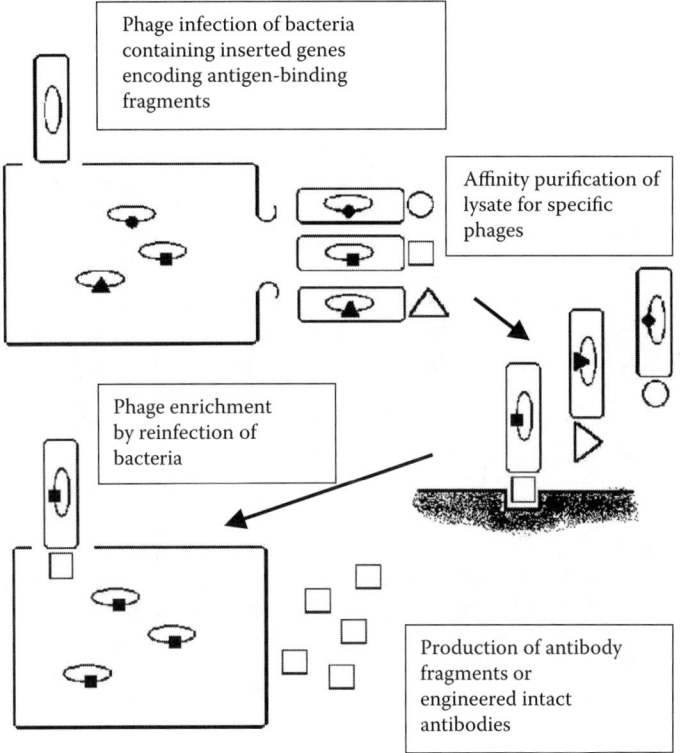

**FIGURE 4.4** Phage display.

different antibody fragments. By screening against solid phase antigen, bacterio-phages carrying the genes of interest can be isolated and produced in bacterial culture or inserted into mammalian cells for expansion.

## ANTIBODY FRAGMENTS AND CONSTRUCTS

In nature, mammalian antibodies occur in five distinct classes: IgG, IgA, IgM, IgD, and IgE. These differ in structure, size, amino acid composition, charge, and carbo-hydrate components. The basic structure of each of the classes of immunoglobulins consists of two identical polypeptide chains linked by disulfide bonds to two identical heavy chains. Differences between classes and subclasses are determined by the makeup of the respective heavy chains. IgG is the major serum immunoglobulin and occurs as a single molecule; IgA also occurs as a single molecule but also polymer-izes, primarily as a dimer and also associates with a separate protein when secreted. IgM occurs in the serum as a pentamer, with monomers linked by disulfide bonds and the inclusion of an additional polypeptide component, the J-chain. IgD and IgE occur primarily as membrane-bound monomers on B-cells, or basophils and mast cells, respectively.

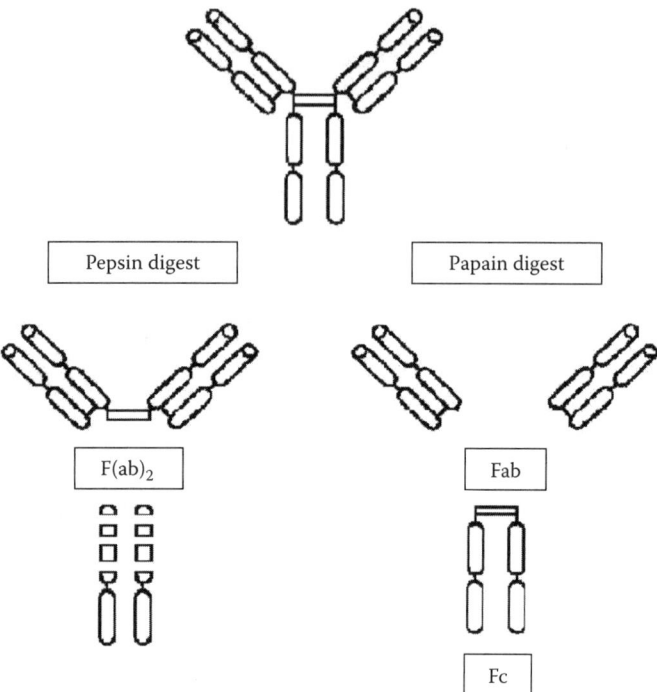

**FIGURE 4.5** Enzyme cleavage of antibody structure.

Each Y-shaped monomer consists of two identical antigen binding sites located on the short arms, and a constant region including a hinge region and variable amounts of carbohydrate residues. Experiments with enzymatic digests revealed that monomers can be digested to separate components with binding characteristics of the intact molecule but without the effector functions (Figure 4.5). The use of pepsin cleaves the immunoglobulin molecule beneath the disulphide hinge region and generates a divalent $F(ab)_2$ fragment. If papain is used proteolysis produces two identical monovalent Fab fragments. Fab fragments are often used in immunoassay in order to reduce background from nonspecific adsorbtion of the constant regions. They are also used in therapeutic applications where interaction of the intact antibody with cell-bound Fc receptors would neutralize the desired effect or localize targeted antibodies away from the desired location. Labeled Fab fragments are also smaller, by a factor of 3 or more, than intact antibodies and can therefore be more readily distributed through tissues—an advantage in therapy and imaging as well as immunocytochemistry.

Recombinant technology can be used to produce novel antibody constructs. The potential to produce antibody fragments consisting only of variable regions, or in extreme case of single chain constructs on which the hypervariable regions are assembled has been explored. Protein engineering can also be used to produce constructs of binding moieties linked with novel effector regions such as markers or toxins or to produce reagents with multiple specificities. By replacing the constant region of the engineered antibody with one or more peptides which bind to particular effector ligands the action of the antibody *in vivo* can be precisely controlled. By inserting T-cell epitopes into the constant region antibodies can be targeted to particular antigen presenting cells and used in vaccination applications.

## Transgenic Mice

One successful approach to circumventing the ethical issues of immunizing humans with a range of antigens while avoiding the incompatibility of murine antibodies with patients has been the development of transgenic mice carrying human genes for immunoglobulins. Companies using this technology include Medarex, Inc. (Princeton, NJ), Abgenix, Inc. (Fremont, CA), and XTL Biopharmaceuticals (Rehovat, Israel). Mice immunized with a range of immunogens respond by producing human immunoglobulins that can then be immortalized using standard murine techniques.

# PRODUCTION METHODS

## In Vivo Production

It is possible to produce antibody from an established hybridoma line by propagating the cells in the peritonea of mice or rats. The injection of mineral oil such as pristane (tetramethylpentadecane) into the peritoneal cavity makes rodents susceptible to the development of plasmacytomas. If hybridoma cells are injected into pristane-primed rodents of the same strain as the myeloma (generally BALB/c mice), the cell line proliferates and the secreted antibody accumulates in the ascitic fluid. Ten to 50 mL of fluid can be collected containing up to several mg/mL of antibody. This technique has been widely used in research applications and, in some countries, in commercial production. There are, however, ethical and legal, as well as practical, problems with the continued use of ascites. The inherent variability of the animals can also result in some lack of consistency; some lines produce solid, rather than diffuse, tumors, and some produce none at all or kill the host. The poorly defined and highly variable nature of ascitic fluid also makes subsequent purification of the antibody difficult. Contaminants can include irrelevant mouse antibodies as well as damaging proteases and nucleases. Monoclonal antibodies produced in this way are generally considered to be unsuitable for therapeutic applications because of the possibility of viral contamination. Increasing restrictions on the use of animals in biological research, allied with the real alternative of *in vitro* production for MABs, have already resulted in restrictions on the use of ascites in Europe and this is certain to happen elsewhere before long.

## Mammalian Cell Culture

Most MABs are produced initially in cell culture dishes and flasks in generalized mammalian cell culture media, such as Dulbecco's MEM or RPMI 1640, supplemented with serum to provide critical growth factors. Cultures are carried out in batches and are not mixed. Several different approaches have been used to increase production to allow evaluation and eventually production of suitable MABs. "Static" cultures can be expanded using multiple flasks or tray systems such as Nunc's "Cell Factory." Advantages include simple extrapolation of conditions defined in early flask cultures, disadvantages include the cost of disposable items, incubator space, and the manipulation of large volumes of media.

Hybridoma cells are generally anchorage independent and so can be grown in suspension culture. Mixing the cultures improves gas and nutrient exchange and greater cell densities and product yields can be achieved compared with static cultures. Culture volumes of up to 50,000 liters can be achieved in a single vessel. Mixing can be accomplished with paddles or stirrers or by airlift devices. All have the potential to damage cells with shear forces or frothing. Media costs at these scales are very important and the protective and nutrient nature of added serum has to be offset against the cost, availability of suitable quantities and quality, and frothing. Serum-free media formulations are available but costs are also high. Suspension culture of mammalian cells for commercial production also suffers from high capital costs and the batch sizes mean that any contamination event can be hugely expensive.

Static and suspension systems are generally operated on a batch basis. Cells can also be grown in perfusion systems that can operate on a continuous or semi-continuous basis. Cells are immobilized on a matrix or within a semipermeable membrane and media is continuously circulated over the cells. When the cells are retained behind a semipermeable membrane, generally with a molecular weight cutoff of about 10,000 Daltons, the serum can be retained on the cell side of the membrane while the antibody, and any waste products cross into the circulating media and can be removed without disturbing the cells. Cell densities approaching those found in tissues can be achieved using these methods. There are savings in media supplements as well as downstream costs. Validation of the process is critical to ensure that changes do not occur in the product as the cell population ages.

Mammalian cell culture uses well understood technology and the patterns of glycosylation and assembly of the antibody is likely to be preserved during production. The limitations to the process and the long culture periods limit scale-up to about 500 kg/year for any product. There are currently only ten large-scale plants in operation with the capacity to produce industrial quantities of MABs to FDA-certified standards. A bottleneck in production capacity has been identified for some years but the capital investment necessary to build the 25 or so plants required to service the 100 MABs estimated to be approved within the next 10 years has not been forthcoming. Cost-effective alternatives to traditional cell culture such as the use of transgenic animals may fill the gap. Transgenic approaches, such as the harvest of MABs from milk, promise costs of less than a third of culture methods with reduced capital costs. The major hurdle is the demonstration of bioequivalence and purity comparable to MABs generated from mammalian cell culture. Until the first products using these new technologies have been approved by the FDA many companies are unwilling to commit their candidate therapies and would rather compete for the limited available bioreactor capacity.

## BACTERIAL CELL CULTURE

Bacterial culture is used for many commodity products and the technology is well understood. Production of mammalian MABs from transfected bacterial hosts has the potential for multitonne yields with high recovery. There are questions over the glycosylation of the product as patterns resulting from changes to culture method or medium can alter the properties of the antibody and jeopardize its regulatory approval. This method may be best suited to production of antibody fragments, rather than complete immunoglobulins. Since many anticancer MABs do not function fully without Fc receptor binding and other effector functions, glycosylation is very important. Increasing the expression of galactosyl transferase in CHO cells producing MABs has been shown to increase the degree of glycosylation 20-fold and could enhance the efficiency of the antibody in mediating antibody-dependent cellular-cytotoxicity.

## TRANSGENIC ANIMALS

By incorporating the selected genetic material into animal embryos MABs can be recovered from the milk of cattle and goats or from the eggs of chickens. Several advantages have been claimed for chickens, glycosylation is similar to that in human immunoglobulins, there have been no prion diseases identified in chickens, and the species barrier between birds and humans may limit the potential for zoonotic infections. Glycosylation is correct and production costs are currently comparable with mammalian cell culture although capital costs could be much lower and operational costs may be as little as one-third. Expression levels of 2 to 6 g/L for immunoglobulins in the milk of transgenic goats have been reported. Scale-up is achieved by increasing the herd or flock size. Cloning techniques have been developed by the Roslin Institute (Edinburgh, Scotland), PPL Therapeutics (Edinburgh, Scotland), and the University of Tennessee for sheep, cows, and pigs. For MABs produced initially in cell culture there must be an extensive phase of validation to demonstrate identity. Several companies are manufacturing recombinant therapeutic proteins but no transgenically produced material has yet been approved by the FDA.

Transgenic technologies may provide relief for capacity bottlenecks, especially for products required in large quantities. For products required at more than 150 Kg/year transgenic animals may provide the most efficient route. The biopharmaceutical industry is generally cautious of committing to this route for MAB production until it has been proven—a catch-22 situation. Major concerns include the contamination with adventitious agents, viruses and prions; the demonstration of bioequivalence and glycosylation and any postproduction modification; finally, the use of animals may involve social and political acceptance and it is anticipated that that new regulatory measures will be introduced.

## TRANSGENIC PLANTS

Several companies have developed technologies to use plants for the production of animal proteins. Tobacco and maize are the favored crops. Expression cassettes with the required gene(s) and an associated marker are inserted and screened plants expressing suitable levels of the target protein are multiplied and grown up. The estimated time from transfection to production may be a short as 10 months. Solvent extraction and purification can yield up to 1 mg/gm plant material. Set up and harvest costs are modest as standard agricultural methods are used. The absence of plant pathogens able to cross to humans, together with the ease and speed of scale up make this an attractive method for bulk production. Products requiring full mammalian-pattern glycosylation may not be suitable candidates. Monsanto (St. Louis, MO), through its Integrated Protein Technology unit is the first company to produce a parenteral product for clinical trials from transgenic corn. Biolex (Pittsboro, NC) has produced a number of "plantibodies" using transfected tobacco plants and have examined their potential as topical antiviral agents.

## PHARMACEUTICAL, REGULATORY, AND COMMERCIAL ASPECTS

### PHARMACEUTICAL REGULATION

In the United States, FDA clearance is required before any therapeutic, imaging, or diagnostic reagent can be marketed. The FDA, under the auspices of the Center for Drug Evaluation and Research and the Center for Biologics Evaluation and Research, have issued guidance for the producers of MABs. The rapid development of the associated technologies means that it is very important to be able to justify all decisions made in processing and purification. Special attention is paid to monitoring for viruses and prions from supplementing sera as well as from the lymphocyte donor and originating cell lines. It is important to be aware of current FDA guidelines on exposure of production cell lines and parental lines to bovine sera before commencing any developments.

## FURTHER READING AND WEBSITES

Antibody Resource, http://www.antibodyresource.com/educational.html

Campbell, A.M. (1984). *Monoclonal Antibody Technology.* Elsevier, Amsterdam.

Ezzell, C. (2001). Magic bullets fly again. *Scientific American,* October, 28–35.

FDA. (2001). *Guidance for Industry Monoclonal Antibodies Used in Drug Manufacturing.* Center for Biologics Evaluation and Research, Rockville, MD.

FDA. (1997). *Points to Consider in the Manufacture and Testing of Monoclonal Antibody Products for Human Use.* Center for Biologics Evaluation and Research, Rockville, MD.

Goding, J.W. (1996). *Monoclonal Antibodies: Principles and Practice.* Academic Press, London

King, D.J. (1998). *Applications and Engineering of Monoclonal Antibodies.* Taylor & Francis, London.

Zola, H. (2000). *Monoclonal Antibodies: Preparation and Use of Monoclonal Antibodies and Engineered Derivatives.* Springer-Verlag, New York.

# 5 Proteomics: A New Emerging Area of Biotechnology

*N. O. Sahin, Ph.D.*

## CONTENTS

## INTRODUCTION

Since the discovery of the DNA double helix by Watson and Crick in 1953, the field of genetics has been explored extensively. Until 1990, molecular and cell biologists studied the structure of individual genes and proteins using various techniques such as polymerase chain reaction (PCR), blotting (Northern, Southern, Western, Slot), sodium dodecyl sulfate-polyacrylamide gel electrophoresis (SDS-PAGE), acrylamide and agarose gel electrophoresis. On the October 1, 1990, the genome sequencing project to form a database of human genes (Human Genome Project, or HGP) was initiated. The completion of the project in 2000 provided a significant tool for researchers to search entire genomes for specific nucleic acids or protein sequences. A parallel development in bioinformatics enabled extraction of information on human genomes from these genomic databases. Unfortunately, information gathered from RNA in genomic databases was not sufficient to predict the regulatory status of corresponding proteins at cell levels. In addition, analytical methods for genomics are not suitable for analysis of proteins. Now, scientists are turning to decoding the network of proteins in human cells and tissues. The new project is called the Human Proteome Project because it is similar to the genome project. The aim of the project is to investigate the interactions among protein molecules, which will provide information in order to understand how these interactions become dysfunctional in disease states and to find and catalog the most vulnerable proteins for drug and diagnostic kit development.

The terms *proteomics* and *proteome* were coined by Wilkins et al. in 1994 to describe the entire collection of proteins encoded by genomes in the human organism. Proteomics differs from protein chemistry at this point since it focuses on multiprotein systems rather than individual proteins and uses partial sequence analysis with the aid of databases. Proteins are the bricks and mortar of the cells, carrying out most of the cellular functions essential for survival of the organism. In the cell, proteins

- Provide structure
- Produce energy
- Allow communication, movement, and reproduction
- Provide a functional framework

Proteins exhibit dynamic properties whereas genes are static. Unlike genomic structures proteins are much more complex and, hence, difficult to study for the following reasons.

1. Proteins cannot be amplified like DNA. It is therefore difficult to detect less abundant sequences.
2. Secondary and tertiary structures of proteins are maintained during analysis.
3. Proteins are susceptible to the denaturing effect of enzymes, light, heat, metals, and so on.
4. Analysis of some proteins is difficult due to poor solubility.
5. Genes are not expressed in all the cells with similar characteristics. Some genes (those encoding for enzymes of basic cellular functions) are expressed in all cells while others are coded for functions of specific cells. Hence, genomes are not differentiated in an organism, but most proteins are.
6. Most proteins exist in several modified forms influencing the molecular structure, function, localization, and turnover.
7. Genes specify the type of the amino acid forming a certain protein, but it is not easy to identify the structure and the function of such protein based on its amino acid sequence.
8. The shape of all genes are linear whereas proteins vary in shape upon folding.
9. The number of genes are approximately 40,000, but a human organism consists of hundreds of thousands of distinct proteins.
10. As an outcome of the HGP, the old dogma claiming that one gene makes one protein is no longer valid. In fact, one gene can encode a variety of proteins (Figure 5.1).

Despite the fact that it is difficult to identify a completely new tool, analytical techniques may have the capacity to help us for this purpose over the next couple of years.

## WHAT IS A PROTEOME?

The term *proteome* defines a complete protein entity encoded by a specific gene of an organism or cell. Proteome is susceptible to any change in environment unlike the genome. Proteome plays a key role in intracellular signaling pathways of the immune system and intercellular metabolism as being the interface between the cell and the environment. A specific gene produces a protein molecule following a complex pathway (Figure 5.2). The amino acid sequence of this molecule can easily be predicted from the nucleotide sequence of the gene. Subsequently, this protein molecule undergoes further modifications at the post-translational stage, resulting in the formation of a protein with different biological activities at the cellular level.

Approximately 200 different types of post-translational modifications (folding, oxidation of cysteine, thiols, carboxylation of glutamate) have been described. Modified proteins are then delivered to specific locations in cells to function. Post-translational modifications significantly influence the process of degradation. For example, some proteins undergo degradation following conjugation with ubiquitin.

Proteins have a modular structure comprised of motifs and domains. Short peptide sequences bring specificity for certain modifications while in most cases

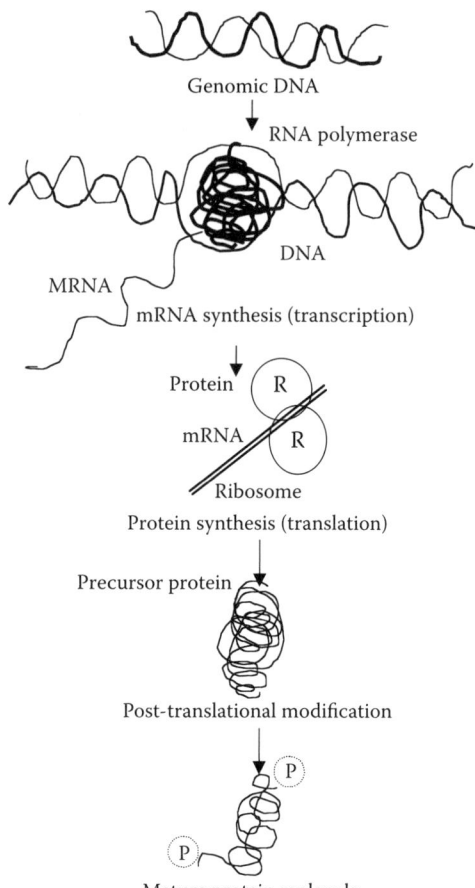

**FIGURE 5.1** Protein synthesis.

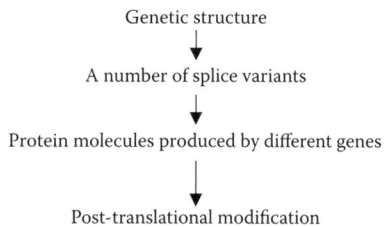

**FIGURE 5.2** Production of a protein molecule following a complex pathway.

longer ones form domains reflecting bulk physical properties as a result of formation of helices in the secondary structure.

The translation of the peptide sequence to functions in a protein molecule can be expressed as modular units (motifs and domains) that confer similar properties or functions in a variety of proteins. Based on characteristics of modular units, proteomes can be classified into families of proteins possessing related functions. For instance, some proteins play a key role in intercellular signaling pathways and some are structural, while others participate in metabolism. Up to 40% of proteins are encoded by the genome and carry out unknown functions, which still need to be discovered.

The classification of proteomes revealed the occurrence of "core proteome," indicating the complexity of an organism or a genome and varying number of paralogs among organisms. What makes the human organism complex is not the size of the human genome, but the diversity of human proteomes.

Another important issue is what can be used to predict change in protein levels. In a review of the literature, it is apparent from the measurements in yeast and mouse liver that there is yet no strong correlation between mRNA level and that of proteins, especially in the case of poorly expressed genes. Therefore, conducting gene expression measurements may not be sufficient to infer protein expression. Thus, the analytical methods used in proteomics must provide detection of proteomes present in multiple-modified forms at relatively low levels.

Sequences of proteome can be predicted by analytical methods and these will be covered in later parts of this chapter and can be used to define biological activity of proteins.

## TECHNOLOGIES FOR PROTEOMICS

The main goal of the proteomic research is to find the distinction between quantitative regulation and structural proteomics. Today, the core technology of proteomics is 2DE (two-dimensional electrophoresis) coupled with MS (mass spectrometry). It offers the most widely accepted way of gathering qualitative and quantitative protein behavioral data in cells, tissues, and fluids to form proteomic databases.

Protein expression and function under various physiological conditions can be investigated by employing 2DE techniques. Quantitative regulation of proteomics allows us to monitor quantitative changes in protein expression within a cell or tissue under different circumstances, particularly in a disease state.

The first important step of proteomics research is "separation of proteins." 2DE research is the most effective way of separating proteins as it will make it possible to identify diseases-specific proteins, drug targets, indicators of drug efficacy and toxicity. Again, separation of post-translationally modified protein from the parent one is usually achieved by 2DE. Several thousands of different proteins can be separated from each other in one gel by 2-dimensional polyacrylamide gel electrophoresis.

Basicly, the protein separation is performed depending on the charge (isoelectric point) or mass of protein molecules. The former is called "isoelectric focusing" in which proteins migrate in a pH gradient toward the pI where they possess no charge. The most commonly used method of this kind is IPG-Dalt (immobilized pH gradient associated with 2DE-PAGE). IPG-Dalt has higher resolution, improved reproducibility, and higher loading capacity. Separation is conducted by size in a vertical direction and by the isoelectric point in horizontal direction.

Traditional gel electrophoresis methods were not sufficient enough to separate complex protein mixtures and possessed such problems as insufficient sensitivities and limited linear dynamic ranges of silver- and Coomasie-brillant blue staining. This problem can be overcome by labeling with fluorescent dyes in 2DE as it will increase sensitivity and widen dynamic separation range. Gels used in 2DE are also selective, especially for membrane proteins and other poorly soluble ones. Unlike silver staining in traditional methods, 2DE does not require subsequent analysis and, hence, is preferred to apply along with mass spectrometry.

Traditional methods of electrophoresis were time consuming as only a few gels could be analyzed in a day and required 1 to 8 hours for manual editing per gel. On the contrary, 2DE does not have limitations when 200 to 400 gels per week are required to be analyzed.

Apart from 2DE, microarray techniques are still being developed for protein separation. This promising new technique is used for protein purification, expression, or protein interaction profiling. The principal behind the technique is the tendency of antibodies, receptors, ligands, nucleic acids, carbohydrates, or chromatographic surfaces to bind to proteins. Following the capture step, the array is washed to minimize nonspecific binding. Subsequently, it is subjected to laser light for short intervals. Protein molecules leaving the array surface are analyzed by laser desorption/ionization time-of-flight mass spectrometry.

The second step in proteomic research is to identify the separated proteins. This can be achieved by MS. Here, proteins are differentiated based on their mass-to-charge ratio (m/z). At first, the protein molecule is ionized. The resultant ion is propelled into a mass analyzer by charge repulsion in an electric field. Ions are then resolved according to their m/z ratio. Information is collected by a detector and transferred to a computer for analysis. The most commonly used ionization methods are

- Matrix-assisted laser desorption/ionization (MALDI)
- Electrospray ionization (ESI)

The data obtained in MS after the proteolytic digestion of gel-separated proteins into peptides and mass analysis of the peptides will enable the researcher to search protein and nucleotide sequences, which can then be checked against their theoretical fingerprints in databases.

Separated and identified protein undergoes post-translational analysis that leads to information on activity, stability, and turnover. Post-translational modifications, such as phosphorylation and glycosylation, are analyzed by use of glycosylases and phosphatase.

Crystal structure of a protein molecule can also be determined by x-ray crystallography. Purified protein is crystallized either by batch methods or vapor diffusion. X-rays are directed at a crystal of protein. The rays are scattered depending on the electron densities in different positions of a protein. Images are translated onto electron density maps and then analyzed computationally to construct a model of the protein. It is especially important for structure-based drug designs.

Another method of determining the secondary and tertiary structure of a protein is NMR (nuclear magnetic resonance) spectroscopy. NMR spectroscopy reveals detailed information on specific sites of molecules without having to solve their entire structure.

Determining the activity of the protein molecule under investigation is most often achieved by screening protein or peptide libraries. These libraries are formed with the aid of viral surfaces. Another way to define activity of a protein molecule is COLT (cloning of ligand targets). In COLT, small peptide sequences bind to larger domain units within a protein molecule, leading to discovery of new domains and protein molecules.

All information obtained using the above techniques are stored in databases that hold information to analyze biological, biochemical, and biophysical data on proteins. The field of science based on such databases is called *bioinformatics*. A simpler way of defining bioinformatics is that it is like a huge computational expert system on proteins. Fast developing IT (information technology) have made significant contributions to bioinformatics. Using data stored in a protein database, a new structure or a therapeutic agent may of then be discovered. The techniques outlined above will be explained in detail later in this chapter.

## PROTEIN IDENTIFICATION

The basic identification process is analysis of the sequence or mass of six amino acids unique in the proteome of an organism, then to match it in a database. Put another way, protein identification is achieved converting proteins to peptides, determining the sequence of peptides, and then, matching with corresponding proteins from matching sequences in a database (Figure 5.3).

It may be assumed that a protein mixture consists of protein molecules with different molecular weights, solubilities, and modifications. Analysis of the mixtures begins with separation of the proteins from each other and then, cleavage of these proteins to peptides by digestion as mass spectrometry cannot be performed using intact proteins. The resultant peptides are analyzed employing one of the following techniques

- Matrix-assisted laser desorption ionization time of flight (MALDI-TOFF): measures the mass of the peptides
- Electrospray ionization (ESI)-tandem MS: determines the sequence of peptides

Finally, collected data are matched with those in a software-assisted database to identify peptide and peptide sequences.

The best analysis can be achieved when the complexity of a protein mixture is reduced. This will increase the capacity of the MS instrument for analysis. Considering those with approximately 6,000,000 tryptic peptides, it becomes highly challenging unless mixtures are separated to their components (single proteins or peptides) prior to analysis.

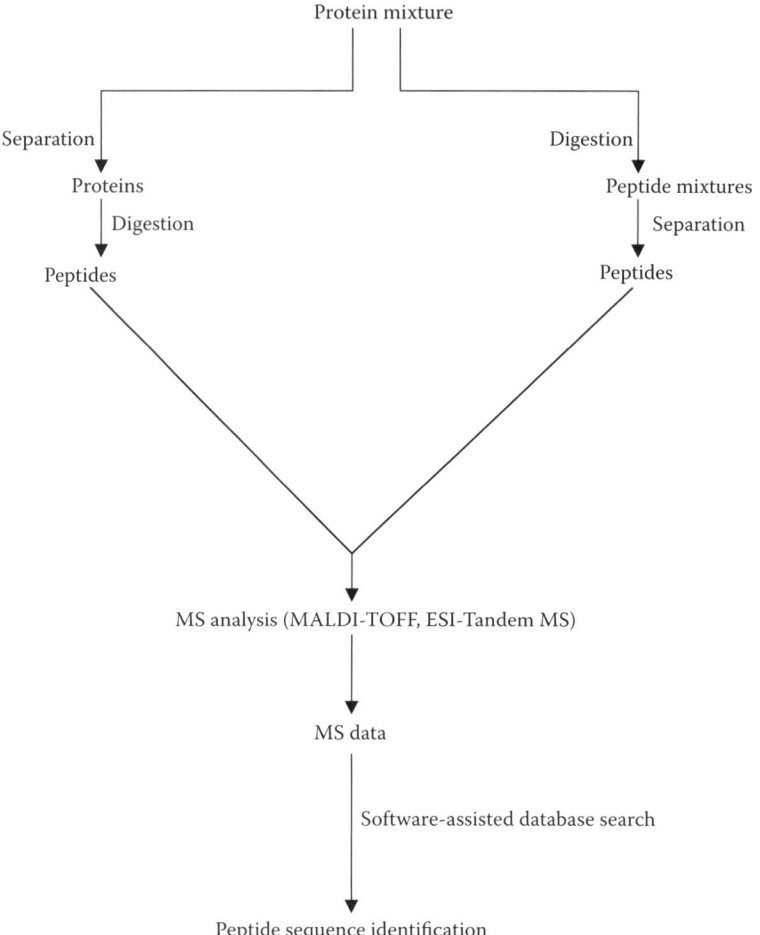

**FIGURE 5.3** Schematic presentation of basic proteomic analysis for identification.

The biological sample in consideration for analysis is first pulverized; then, it is homogenized, sonicated, or disrupted to form a mixture containing cells and subcellular components in a buffer system. Proteins are extracted from this mixture using these substances

- Detergents: SDS, CHAPS
- Reductants: DTT, thiourea
- Denaturing agents: urea, acids
- Enzymes: DNAse, RNAse

The aim is to extract protein molecules as pure as possible. Detergents generally help membrane proteins to dissolve and separate from lipids. Reductants are used to reduce disulfide bonds or prevent oxidation. Denaturing agents alter ionic strength and pH of the solution and then destroy protein–protein interactions, disrupting secondary and tertiary structures. Digestion is achieved by enzymes. Protease inhibitors are often used to prevent proteolytic degradation. Conducting an extraction process, one must consider the possibility of experiencing the interference of these substances during the analysis. Some protease inhibitors and some detergents may also interfere with digestion.

Extracted proteins are separated employing one of the following techniques

- 1D-SDS-PAGE (1-dimensional sodium dodecyl sulfate-polyacrylamide gel electrophoresis)
- 2D-SDS-PAGE (2-dimensional sodium dodecyl sulfate-polyacrylamide gel electrophoresis)
- IEF (isoelectric focusing)
- HPLC (high performance liquid chromatography)
- Size exclusion chromatography
- Ion exchange chromatography
- Affinity chromatography

1D-SDS-PAGE allows separation into smaller fractions whereas 2D-SDS-PAGE leads to separation into many fractions. Resultant fractions are used for digestion or further separation prior to MS analysis.

These techniques are the most widely used ones and are described in detail below.

## 1D-SDS-PAGE

This technique is based on the complexity of the system that is not favored in proteome analysis. It is preferred in the separation of some proteins in biological fluids (e.g., cerebrospinal fluids) and protein–protein interactions where relatively few proteins are involved.

Steps of the procedure are as follows

1. Preparation of a loading buffer containing a thiol reductant (e.g., DTT) and SDS
2. Dissolving protein in the loading buffer
3. Binding of SDS to protein to form a protein-SDS complex
4. Applying to the gel
5. Applying high electric voltage to the ends of the gel
6. Migration of the protein-SDS complex
7. Formation of bands on the gel in order of molecular weight

One of the important parameters to be considered in such separation is the degree of crosslinking of polyacrylamide. While choosing the optimum extent of crosslinking, the molecular weight of the protein to be separated should be taken into account. For example, the better way of separating a mixture of proteins with low molecular weight involves use of a highly crosslinked gel. In some cases, gradient gels were produced in which the degree of crosslinking increased gradually from the top to the bottom of the gel to separate proteins with a broad range of molecular weights. 1D-SDS-PAGE usually gives a single, distinct band for proteins with diverse molecular forms. Better resolution is achieved with 2D-SDS-PAGE because it separates the sample into multiple spots on the same band based on isoelectric points. This is due to post-translational modifications not affecting the formation of the SDS-protein complex.

## 2-DE

Proteome analysis begins with separation, visualization, and subsequent analysis of complex mixtures containing several thousands of proteins isolated from cells or tissues. Although sophisticated methods using chip-based technologies, microarrays, affinity tags, and large-scale yeast two-hybrid screening have been developed recently, 2-dimensional polyacrylamide gel electrophoresis still remains the most favorable technique for separation of protein mixtures because it generates reproducible high resolution protein separation. 2-DE separation is conducted based on the electrical charge and molecular weight of the proteins. Using staining procedures, separated protein molecules are visualized according to their size, brightness, and location on the gel. The method used should be objective, reproducible, sensitive, and compatible with computer-aided methods to process images, provide large amounts of data, and allow quantitative analysis.

### Historical Development of 2-DE

In 1956, Smithies and Poulik first used 2-DE combining paper and starch gel electrophoresis to separate serum proteins. Nearly 20 years later, polyacrylamide was applied as a support medium. Charge-based protein separation followed as isoelectric focusing (IEF), applied to SDS-PAGE. Later, urea and nonionic detergents were used in IEF-2DE. The most significant achievement was the separation of proteins from *E. coli*.

### Steps of 2-DE

2-DE technique can be applied with the following steps

1. Preparation of the sample
2. Solubilization
3. Reduction
4. IPG-IEF
5. Equilibration
6. SDS-PAGE

*Preparation of the Samples for 2-DE*

The diverse nature of proteins leads to the problem of applying a unique, single method in preparing samples for 2-DE analysis. The most important point to be taken into consideration should be minimizing protein modifications while preparing samples. The method with minimum modification should be chosen, otherwise artifactual spots may form on the gel and mislead the operator.

Serum, plasma, urine, cerebrospinal fluid (CSF), and aqueous extracts of cells and tissues are encountered as soluble samples and often require no pretreatment. They can be directly analyzed by 2-DE following a solubilization step with a suitable buffer (e.g., mostly phosphate buffered saline, or PBS).

Liquid samples with low protein concentrations or large amounts of salt should be desalted and concentrated prior to 2-DE. Desalination can be achieved by dialysis or liquid chromatography. The desalted sample is then lyophilized, subject to dialysis against polyethylene glycol or precipitated with acetone to remove interfering compounds.

A solubilization process is also applied to solid tissue samples. The sample is broken up in the frozen state with the aid of liquid nitrogen (i.e., cryosolubilization). Resultant specimens are then crushed between cooled blocks simply using a pestle and a mortar under liquid nitrogen or homogenized in a buffer using a rotating blade-type homogenizer. The major problem with preparation of tissue samples is to obtain an homogenous cell content. The problem has been substantially overcome using laser capture microdissection (LCM) by Banks et al. (2000). They applied the technique to obtain dissection of epithelial tissue from a sample of normal human cervix. In this technique, a transfer film was placed on a tissue section. A laser beam was directed onto the selected area of film under an inverted microscope. This part of the film bonds to the cells of the tissue on contact. Subsequently, this film with the cells was removed and used for further analysis by 2-DE.

The best method of preparing samples from the cells of an *in vitro* culture medium or the circulating cells as erythrocytes or lymphocytes is by centrifugation, followed by washing up and solubilizing in PBS. If the cells are grown on solid substrates in culture medium, at first, the medium should be removed prior to washing up process. This will minimize the salts interfering with the IEF. Later, cells are lysed directly by adding small amount of PBS.

Highly viscous cell samples containing large amounts of nucleic acids should be treated with a protease-free DNAse + RNAse mixture.

In case protein mixtures are needed to be fractionated, samples are subjected to subcellular fractionation, electrophoresis in the liquid phase, adsorption chromatography, and selective precipitation or sequential extraction based on solubility prior to fractionation.

## Solubilization

In order to avoid misleading spots on the 2-DE profile and to remove salts, lipids, polysaccharides, or nucleic acids interfering with separation, samples should be solubilized. Solubilization procedure involves disruption of all noncovalently bond protein complexes into a solution of polypeptides. It is the most critical step of 2-DE.

The most favorable solubilization method is the one described by O'Farell in 1975. O'Farell used a mixture of 9.5 M urea, 4% w/v of NP-40 (a nonionic detergent), 1% w/v of DTT (a reducing agent), and 2% w/v of SCA (an ampholytic synthetic carrier) to obtain appropriate pH ranges for separation. However, this method is not valid for membrane proteins. Four percent w/v of CHAPS (3[(cholamidopropyl)dimethylammonio]-1-propane sulfonate) can be used in a mixture with 2 M thiourea and 8 M urea to solubilize membrane proteins. Although SDS is a suitable detergent for solubilization of membrane proteins, it cannot be used as main solubilizing agent on IEF gels due to its anionic character. SDS is often used in presolubilization procedures. Then, SDS is removed from the sample and replaced with the solubilizing detergent, which is nonionic or zwitterionic in nature. Consequently, the sample becomes solubilized and ready for further analysis.

Beside selection of the best solubilizing agent, the most important parameter is the ratio of protein to detergent. The amount of SDS is also important as an excess amount may cause interference with IEF.

Solubilized samples should not contain nucleic acid as it increases the viscosity of the sample, leads to the formation of complexes with the protein under investigation, and may cause misleading migration on the gel. This problem may be overcome by degrading nucleic acid with the aid of protease free endonucleases.

### Reduction of Proteins

This step in 2-DE involves reduction of disulfide bonds in the protein samples. DTT or β-mercaptoethanol are the most widely used reducing agents. However, these thiol-containing agents cause significant problems during IEF procedures. They are charged molecules and therefore, they migrate out of the IEF gel. This leads to reoxidation of the protein sample causing loss of solubility. Thus, noncharged reducing agents (e.g., tributyl phosphine: TBP*) have been preferred recently.

### IPG-IEF

At first, IPG gel is prepared according to the method summarized below:

1. Preparation of the slab gel: IPG and SDS gels are prepared at a desired pH.
2. Casting: The slab gel is cast on a polymer film (e.g., GelBond PAG film) and polymerized. Resultant polymerized gel is washed up with deionized water, left in 2% w/v of glycerol, and dried at ambient temperature. Finally, the gel is covered with a plastic film. The shelf life of IPG gel is approximately 1 year at 20°C.
3. Preparation of strips: IPG gels are cut off to obtain the strips of desired length (e.g., 3–5 mm in width) for 2-DE. The width of strips is important to obtain separation with high resolution. For instance, the strips of 18–24 cm provide high resolution, but those with 4–11 cm are suitable only for rapid screening. There are also commercially available strips in different widths.
4. Rehydration strips: The strips are rehydrated in a reswelling cup containing the mixture of 2 M thiourea, 5 M urea, 0.5–4% w/v detergents (e.g., Triton X-100 or CHAPS), 15 mM reducing agent (e.g., DTT or TBP), and 0.5% w/v of SCA. They are then left for equilibration.

   Rehydrated strips are then placed onto a custom-made cooling plate or a commercially available strip tray of a horizontal-type electrophoresis apparatus. The strip tray is fitted with bars holding the electrodes. The bars fitted with sample cup onto the tray allow easy application of the

---

* TBP is a noncharged reducing agent which does not migrate during IEF and therefore, improves protein solubility (especially in the case of wool proteins).

samples onto the gel. In order to protect the gel from air, silicon oil is filled into the tray.

The major problem is that horizontal streaking is observed at the basic end of the protein profiles when basic IPG is applied for the first dimension of 2-DE. It may be prevented by using an extra electrode strip soaked in DTT and placed onto the IPG strip along with the cathodic electrode strip or replacing DTT with TBP.

5. Application of the samples: Silicon rubber frames or special sample cups are used for sample application. Sample cups are usually placed at the anodic or cathodic end of the strips.

6. Optimum conditions for electrophoresis: The voltage of 150 V is applied for 30 minutes initially to obtain maximum sample entry, and increased gradually until it reaches 3500 V. The optimum time for application depends on the following factors: (a) type of sample, (b) amount of protein in the sample, (c) length of the IPG strips, (d) pH gradient.

The optimum temperature is 20°C. Below this temperature, urea is crystallized. The best value for the gel length is 18–24 cm. The pH gradient is 1–9.

The major problem of protein precipitation is experienced particularly in the application of high protein loadings (>1 mg). To avoid this, two solutions are suggested.

(a) Reswelling IPG strips in the solution containing the protein under investigation: In the case of protein loadings above 10 mg, loss of proteins with high molecular weight occurs during this treatment. This problem is also experienced with membrane proteins.

(b) Use of a special instrument (e.g., IPGphor, IPGphaser, Protean IEF cell) usually comprised of a strip holder allowing rehydration of IPG strips and, in some cases, separate sample loading. The apparatus can run with an 8000 V power supply. The maximum applicable current is 0.05 mA per strip. The reswelling process is performed at 30 V for 12–16 hours. In case of reswelling high molecular weight proteins, lower voltage is applied.

Initial analysis of an unknown sample is usually done with a linear pH gradient of 3.5–10. However, it must be kept in mind that many proteins have pI values in the range of 4–7 and thus, the operation at this pH gradient of 3.5–10 may cause loss of resolution. This problem may be resolved by using a nonlinear pH gradient. Commercially available IPG strips are of pH 4–7 providing better separation. These type of gel strips allow analytical (sample loading of 50–100 microgram) or micropreparative (sample loading <1 mg) analysis.

Alkaline proteins with similar pI values (e.g., ribosomal and nuclear proteins with pIs in the range of 10.5–11.8) can be differentiated by narrow range (pH 10–12 or 9–12) IPG gels. Substitution of DMAA (N,N-dimethylacrylamide) and addition of propan-2-ol to the rehydration solution provide high reproducibility and suppresses the reverse electroendosmotic flow. To obtain better resolution for this type of sample, sample cups should be loaded at the anode side and the IEF should be conducted under silicon oil.

Complex proteins of eukaryotic cells are best separated after sample prefractionation or using multiple, overlapping narrow range IPG gels of 1–5 pH units (e.g., zoom gels, composite gels, subproteomics). IPGs running at pH 4–7 give rise to better resolution along with better rehydration, sample cup loading, and allow micropreparative separation.

The time required to achieve best quality and reproducible IEF patterns at a steady state is called optimum focusing time. Streaking upon horizontal or vertical application is caused by keeping optimum focusing time too short. As a result, electroendosmosis (active transport of water) occurs, leading to excess water exudation on the surface of the IPG gel, distortion of protein patterns, and loss of proteins. Selection of the optimum focusing time should be accomplished taking the following parameters into consideration: (a) content of protein sample, (b) degree of protein loading, (c) optimum pH range, and (d) length of IPG gel strips.

7. *Equilibration step:* Following IPG-IEF dimension, a second dimension of separation can be applied. However, gels are often equilibrated prior to second dimension analysis in order to allow separated proteins to interact with SDS. This interaction will provide migration during SDS-PAGE analysis. Equilibration can be achieved by incubating the strips for 15 minutes in 50 mM Tris buffer of pH 8.8 in the presence of SDS, DTT, urea, and glycerol. Urea and glycerol prevent transfer of reduced protein from the first to second dimension (electroendosmosis). After incubation, free DTT is alkylated with iodoacetamide to avoid migration of DTT to the second dimension. An alternative way to avoid this last step is to use TBP in place of DTT. Equilibrated strips are drained on a filter paper for 1 minute to remove excess liquid. In many cases, strips are immediately used for the second dimension. Otherwise, they are stored between two sheets of plastic film at 80°C.

## Application of the Second Dimension

The second dimension of 2-DE is SDS-PAGE. Laemmli buffer systems and IPG gel strips are usually preferred. The strips are applied onto the surface of horizontal or vertical SDS-PAGE gels.

Beside custom-made slab gels, commercially available ones (e.g., ExcelGel) can also be used for horizontal applications. Six percent stacking gel and 12% homogenous or gradient resolving gel must be used. The plastic support holding the gel in the system prevents variations in gel size during the staining process. As the gel thickness is much lower than that of vertical application, higher voltages can be applied to reduce the running time. This will lead to reduced protein diffusion. Also, spots are sharper for the same reason.

Maximum reproducibility of 2-DE protein patterns in large-scale proteome analysis can be achieved by simultaneous electrophoresis of SDS-PAGE. Amersham Pharmacia Biotech (Piscataway, NJ) firm developed a multiple vertical second-dimension

SDS-PAGE system called DALT in 1975 to provide this benefit. Investigator and Protean II are the new versions of such systems.

In contrast to horizontal systems, it is not essential to use stacking gels in a vertical system. In such a system, protein zones in the strips are concentrated and IEF gel with low polyacrylamide concentration behaves like a stacking gel.

### Resolution of 2-DE Gels

The separation length is the most significant factor influencing the resolving capacity of he gel. Application of IPG IEF gel of 18 cm and second-dimension SDS-PAGE of 20 cm long allows resolution of a complex mixture with approximately 200 proteins. Rapid screening can be achieved by mini-gel formats. However, only a few hundred proteins can be separated by such system. So, the amount of the protein separated depends upon the size of the gel.

### Reproducibility of Protein Profiles Obtained by 2-DE

The reproducible high-resolution separation of protein mixtures is the main purpose of proteome analysis. O'Farell's classic tube gel technique has limited reproducibility. It is often difficult to compare the protein profiles obtained using O'Farell's method in different laboratories. In some cases, the data obtained even in the same laboratory by different operators are not comparable.

Recent developments in 2-DE techniques and apparatus (e.g., ISO-DALT, Investigator) have overcome this problem. Today, samples obtained from different sources of cells or tissues can be analyzed with high reproducibility and the data from different laboratories are highly comparable.

### Stains and Dyes of 2-DE

Staining in proteome analysis is important to visualize proteins in gels. Properties of an ideal 2-DE stain are summarized in Table 5.1.

1. *Bromophenol dye:* The first dye used for protein visualization was an organic dye solution developed by Durum in 1950. Durum used bromophenol dye for staining on filter paper.
2. *Amido black dye:* Two years later, Grassman and Hanning developed another organic dye to be used on filter paper after electrophoresis: amido black stain. It has moderate sensitivity. Today, amido black dye is used for colorimetric determination of electroblotted proteins on PVDF (polyvinylidene difluoride) and nitrocellulose membranes.
3. *Coomasie blue dye:* Ten years later, Commasie blue dye, a commonly used staining organic dye was developed by Fazakas de St. Groth et al. and gained increasing popularity in both qualitative and quantitative protein analysis using polyacrylamide gel electrophoresis. Over the years, approximately 600 different versions of Coomasie blue dye have been utilized to improve linearity of staining and sensitivity of protein analysis on electrophoresis. Although it is sensitive and capable of detecting

**TABLE 5.1**
**Properties of an Ideal 2-DE Stain**

| Property | Explanation |
| --- | --- |
| Safety | No stain should contain a hazardous material such as radioisotopes, cyanide, and cacodylic acid. |
| Sensitivity | The broad linear dynamic range is required. Detection sensitivity should be at subnanogram levels. |
| Simplicity | Simple incubation of gel or blot is essential, endpoint stains are preferred. |
| Specificity | Stains should be protein specific and be able to differentiate different classes of proteins. |
| Speed | Rapid detection and longer incubation periods are desired. |
| Stability | Both the staining solutions and the stained gels should be chemically stable at room temperature for days to months. Stains possessing long shelf life are usually preferred. |
| Compatibility | A stain should be easily used to most electrophoretic or separation techniques without causing irreversible modifications in protein molecules. |

30–100 ng of proteins, this is considerably less than that obtained using silver staining and fluorescence staining. Coomasie blue dye can also be used for detection of electroblotted proteins on nitrocellulose membranes with a very high background. However, in many instances, it interferes with subsequent immunodetection procedure. The most popular use of Coomasie blue dye is background-free detection of proteins in PAGE. Colloidal variation of the dye is preferred for this purpose because unlike free dye, colloidal dye is excluded during the equilibrium of staining process, preventing background staining. However, its detection capacity is limited to 8–10 ng of proteins. Selection of the most suitable solvent for preparation of Coomasie blue dye is important regarding subsequent protein modification analysis by mass spectrometry. Use of trichloroacetic acid or alcohol (e.g., ethanol) causes irreversible esterification of glutamic acid-side chain carboxyl groups, which in turn leads to misinterpretation of peptide mapping data from MS.

4. *Silver stain:* Apart from Commasie blue, silver staining is also widely used in protein analysis. Although it was first used for electrophoresis in 1972, its background goes back to the 1800s when it had been used in photography and histology. Unlike Coomasie blue, silver staining was initially devised for agarose electrophoresis to detect both DNA and proteins with high sensitivity. In 1979, Switzer et al. successfully utilized silver staining in polyacrylamide gel electrophoresis. Improved methodology of silver staining allows protein analysis at nanogram range. Over the years, more than 100 different silver staining techniques have been developed. However, detection of glycoproteins using silver staining is a problem as they stain poorly. This may be overcome by using a prestaining procedure with cationic dyes prior to silver staining. Another disadvantage of this dye is

that it interacts with cysteine residues and causes alkylation of α- and ε-amino groups of proteins due to the use of glutaraldehyde and formaldehyde during the staining procedure. Modifications brought by glutaraldehyde and formaldehyde are not desired in analysis where Edman-based sequencing is utilized as they reduce the sensitivity and uniformity and result in low sequence coverage of stained proteins. In analysis where MS is used for further detection, these modifications are not the major problems since they can be removed from the stain. The most widely used forms of silver staining are:

(a) *Alkaline/silver diamine* involves formation of soluble silver diamine complex in the presence of ammonium hydroxide followed by visualization of proteins by reduction of free silver ions to elemental metallic silver with formaldehyde in acidic environment. The best results can be obtained with the use of piperazine for crosslinking. This form of silver staining is not preferred for IPG, analysis of acidic proteins, and the resolution of low molecular weight proteins. The most favorable use of this stain is for analysis of basic proteins. It is not a suitable method for quantitative analysis due to low reproducibility and nonstochiometric properties.

(b) *Acidic/silver nitrate* is comprised of gel impregnation with silver ions at acidic pH, subsequent reduction of silver ions at alkaline pH in the presence of formaldehyde. It is a photographic-based method. Acidic/silver nitrate is often used for analysis of acidic proteins and has high sensitivity. Although it has limited image development time, this form of silver staining can also be used for quantitative analysis of proteins.

5. *Reverse stains:* In 1974, Wallace et al. developed reverse staining just to improve recovery of proteins for subsequent microchemical characterization. However, reverse stains are not suitable for quantitative analysis and staining of transfer membranes. The procedure is based on detection of proteins as transfer zones on a black background and used for visualization of proteins, passive elution of intact proteins from gels, and analysis by MS. Over the years, several different forms of procedures have been developed for reverse staining (Table 5.2).

The most widely used methods of reverse staining uses potassium chloride, copper chloride, and zinc chloride procedures. Zinc chloride staining is the most sensitive among all three procedures.

Zinc-imidazole staining is another popular method of reverse staining and involves separation of proteins tightly bound to zinc ion following the precipitation of free ions and the formation of clear zones on a semiopaque background. Detection capacity of zinc-imidazole staining is limited by 10–20 ng/band for proteins (except separation of BSA: 40–80 ng/band), 1–10 ng/band for lipopolysaccharides, and 10–100 ng/band for oligonucleotides of 28 bp–1.3 kbp. Proteins separated by this method are compatible with Edman-based sequencing following electroblotting, in-gel tryptic digestion conducted for detection of peptide sequences by MALDI-TOFF mass spectrometry.

## TABLE 5.2
### Historical Development of Reverse Staining

| Year | Investigator | Parameter or Metals for Staining |
|------|--------------|----------------------------------|
| 1974 | Wallace et al. | Low temperature |
| 1977 | Takagi et al. | Cationic detergent |
| 1979 | Higgins and Dahmus | Sodium acetate |
| 1980 | Hager and Burgess | Potassium chloride |
| 1984 | Merril et al. | Silver |
| 1985 | Casero et al. | Colloidal gold |
| 1987 | Lee et al. | Copper chloride |
| 1988 | Dzandu et al. | Zinc chloride, cobalt acetate, or nickel chloride |
| 1990 | Berube | Complexation of a planar dye (rhodamine, eosin or fluorescein) with a metal salt (potassium dichromate, barium chromate, magnesium chloride, or calcium chloride) |
| 1992 | Ortiz et al. | Zinc-imidazole |
| 1996 | Candiano et al. | Methyl trichloroacetate |

6. *Stains in the form of colloidal dispersions:* The need for highly sensitive detection methods for proteins after blotting (e.g., electroblotting, dot-blotting, slot blotting) applied to nitrocellulose or PVDF membranes drove scientists' attention to develop new stains. It was discovered that the stains in the form of colloidal dispersions are suitable for such applications. The most commonly used colloidal stains are

   (a) *India ink stain* is the first colloidal dispersion stain used for detection of electroblotted proteins. It is simple and economic. General formula of India ink contains acidic (sulfonates or sulfonamides) or basic dyes, fine dispersions of carbon, copper phthalocyanide, glycerol, surfactants, antioxidants, and viscous agents. It is compatible with direct, colorimetric or autoradiographic immunostaining of proteins. However, duration of staining may cause interference in binding steps with antibody at immunodetection. This may be resolved by combining chemoluminescence with India ink staining. Other disadvantages of this type of staining are high protein to protein variability, incorporation of detergents (e.g., Tween 20) removing proteins from transfer membranes, and the need for in-house optimization of the method.

   (b) *Colloidal silver stains* method was developed by Janssen Pharmaceuticals (Titusville, NJ), but never commercialized.

   (c) *Colloidal iron stains* method was first developed under the name of FerriDye. However, the production of FerriDye was later discontinued as it was highly toxic due to its cacodylic acid content.

   (d) *Colloidal gold stain* possesses similar sensitivity as silver staining for transfer membranes in PAGE. Detection sensitivity is superior to other colloidal stains. However, it requires pretreatment of the membrane with a blocking agent (e.g., Tween 20) and subsequent incubation in a detergent-stabilized colloidal solution for 2–18 hours. Also, detection capacity of the method is very limited (1–20 ng). Colloidal gold staining is compatible with traditional colorimetric blotting procedures and chemoluminescent immunodetection, but incompatible with protein microchemical methods, MALDI-TOFF and ESI type mass spectrometry applications.

7. *Fluorophore stains:* Fluorescent detection of proteins following electrophoresis during large-scale proteome analysis is achieved by fluorophore stains. Their detection sensitivity is superior to colorimetric stains. The most commonly used fluorescent stains are

   (a) *Covalently bound electrophoresis*: Proteins are covalently bound to the stains, altering the mobility of labeled proteins and causing rapid decay of the protein and limited sensitivity that depends upon the availability of the functional groups for modification by fluorophore. The detection capacity of direct fluorescence staining method is 5–10 ng/protein.

   (b) *Noncovalently bound fluorophores*: The most commonly used stains in this category are ANS (1-anilino-8-naphtalene sulfonate), bis-ANS, Nile red, SYPRO orange, and red dyes. They do not cause fluorescent in aqueous solutions whereas they fluoresce in nonpolar solvents. Among them, SYPRO dyes are the most sensitive (Table 5.3) and the only commercially available ones. They provide a simple, single-step staining procedure for protein detection in SDS-PAGE.

## TABLE 5.3
### Noncovalently Bound Fluorescent Stains

| Stain | Sensitivity | Limitations | Application |
|---|---|---|---|
| ANS | Less than Coomasie blue staining | Low sensitivity | SDS-PAGE for proteins |
| Bis-ANS | 100 ng/protein | Low sensitivity | SDS-PAGE for proteins |
| Nile red dye | 20–100 ng/protein | Low sensitivity and stability of stains, susceptibility to precipitation in aqueous solutions, difficult handling and image analysis | Edman sequencing and immunodetection |
| SYPRO dyes (red and orange) | 2–10 ng/protein | — | Edman-based protein sequencing, immunostaining, SDS-capillary gel electrophoresis for protein quantification at picomolar range |

8. *Metal chelate stains:* This class of stains has been developed to be used along with modern technologies of proteome analysis. The major advantages of this class of stains are

(a) Compatibility with common microchemical characterization procedures

(b) No need to contain harmful chemicals such as glutaraldehyde, formaldehyde, or Tween 20

(c) Easy use for proteomic platforms including automated gel stainers, image analysis and protein digestion work stations, robotic spot excise apparatus, and mass spectrometers

(d) Better sensitivity than silver staining
   The most commonly used metal chelate stains are

(a) *Colorimetric stains* are the earliest stains of this class. Detection limit is 600 ng/protein. Incorporation of radioactive $^{59}$Fe into bathophenantroline disulfonate complex improved the sensitivity to 10–25 ng/protein, which is almost similar to that of Coomasie blue dye. The major problem is radioactivity. Staining on nitrocellulose transfer membranes can also be achieved with the same detection limit. This class of stains are compatible with Edman-based sequencing, mass spectrometry, and immunoblotting.

(b) *Luminescent stains:* The principle behind luminescent staining is that the measurement of light emission is more sensitive than the absorbance when metal chelate stains include certain transition metals as europium, terbium, and ruthenium. Examples for these types of metal stains are SYPRO rose protein blot stain (bathophenanthroline disulfonate/europium), SYPRO ruby protein blot stain. The latest stain is a ruthenium-based metal chelate stain and is more sensitive than silver stains. One of the most significant benefits of this dye is that staining time is not critical as it is in an endpoint stain. It provides one-step, low background staining in PAGE. Its performance is superior to those of silver and Coomasie blue dyes. Detection capacity (2–5 ng/protein) is similar to that of colloidal gold and is time efficient (2–4 hours of staining) regarding electroblotting. It is preferred to colloidal gold as it does not interfere with MS or immunodetection procedures.

## Image Analysis in 2-DE

Image analysis of the gel is performed with the following steps

1. Data acquisition is actually a scanning process that converts the analog gel image into a digital image with the aid of computer-based techniques.

2. Image processing is essential to detect the exact location of spots on the gel and to determine the shape and intensity of the spots as it is related to the abundance of the proteins.

3. Gel matching is actually a comparative analysis of identical spots in a series of gels. It is performed under various experimental conditions. Gel matching

is important to analyze the changes of proteins. Matching is done at either pixel level or spot level. In the former, only the pixels of two gel images are matched. In the latter, only the proteins of two or more gels are matched. Local or global polynomial transformations are used for mapping.

4. Analysis of data is performed to provide information on quality assurance and qualitative and quantitative changes in protein expression. The number of spots is tabulated against numbers of gels to determine the changes in protein expression. Qualitative difference among gel groups are usually determined using "on/off" spots, which are the indicators of alterations under different experimental conditions. Uni- and multivariate statistical approaches are used for the analysis of alterations. Quantitative analysis is performed based on finding the changes in the intensity of the spots. Initially, intensity values should be normalized according to the average intensity of the gel image. Alterations in intensities can be compared using t-test and Mann-Whitney test.

5. Presentation and subsequent interpretation of data require each spot to have corresponding molecular weight and isoelectric points. Molecular weight to isoelectric point of the protein sample under investigation is compared to that of marker proteins (e.g., bovine serum albumin [BSA], actin) with known values. The variations in spot intensities are schematized using graphs or charts and further analyzed by computer-based statistical programs (e.g., SAS). Information on protein name and molecular weight/isoelectric point values are obtained and interpreted in this step.

6. Forming databases is essential for storage of the data. During the gel analysis, these data can be used as a reference standard. Some gel analysis systems have their own internal databases including useful descriptive information on protein spots. However, these databases are not shareware. Nowadays, most proteome investigators use Internet-based databases offering even some free software packages useful for the construction of 2-DE databases. These products allow convertion of TIFF images of gels into the specific image format and can be provided from the ExPASy molecular biology server at *http://www.expassy.ch/ch2d/tiffmel.html*. Comparison of spots can also be made on the Internet, checking the slopes and lengths of edges between the points of the reference image and of sample image by computational geometry. This type of matching is performed using CAROL software system at *http://gelmatching.inf.fu-berlin.de*.

## Drawbacks of 2-DE/SDS-PAGE

Despite the superiority of the 2-DE technique, it has several problems.

- Reproducibility can be a significant problem when images of the stained gels of two samples are compared by means of 2-DE/SDS-PAGE
- Incompatibility of some proteins (large, hydrophobic proteins) with the first-dimension IEF step
- Relatively narrow detection range for stained gels

2-DE seems to be the best analysis technique for highly expressed, abundant, long-lived proteins. However, most of the biologically significant proteins are expressed at low levels and are rapidly turned over. Thus, other analytical approaches are often preferred for the analysis of such proteins.

## ISOELECTRIC FOCUSING (IEF)

Apart from the IEF applied in the first step of 2-DE/SDS-PAGE, preparative IEF is also used for protein separation. Separation procedure is carried out on IPG strips, in a tube gel, or a solution. Among all three, the separation in solution is most often preferred. Polycarboxylic acid compounds, which are soluble ampholytes in nature, are used to generate stable pH gradient when voltage is applied across the focusing gel. Subsequently, the protein sample is added. Proteins in mixture are separated based on their isoelectric points (pI) following the voltage application. With the commercially available IEF apparatus, proteins can be separated into 12–20 fractions. Then, the ampholytes can be removed by dialysis or gel filtration.

The advantages of IEF are

- Recovery ratio for proteins is significantly high: 85–90%
- This technique presents diversity in physical properties of intact proteins
- Large sample capacity (mg to g)
- Easy and simple running

## HIGH PERFORMANCE LIQUID CHROMATOGRAPHY

High performance liquid chromatography (HPLC) is often used for protein purification. It is used in the initial step of fractionation. In this manner, HPLC is as useful as preparative IEF and presents diversity of separation modes. In this technique, chromatographic separations can be performed using

- Ion exchange chromatography: charge-based separation
- Size exclusion chromatography: size/molecular weight-based separation
- Affinity chromatography: interactions with specific functional groups drive separation
- Reverse phase chromatography (RPC): hydrophobicity-based separation

In the case of peptide separation by HPLC, separation modes are combined in series. This approach is called tandem LC. For instance, ion exchange associated with RP is used for peptide separation. Multidimensional protein identification technique (MudPIT) involving use of microcapillary columns (SCX cationic column and RP column) linked in series and eluted into MS is preferred for separation of complex peptide mixtures (Figure 5.4).

Tandem LC is superior to 2-DE for two reasons.

1. Only the proteins with visualized spots on 2-DE gel can be analyzed by MS. If no spot is detected on the gel, no data can be obtained from subsequent MS analysis. Tandem LC does not present such problem.

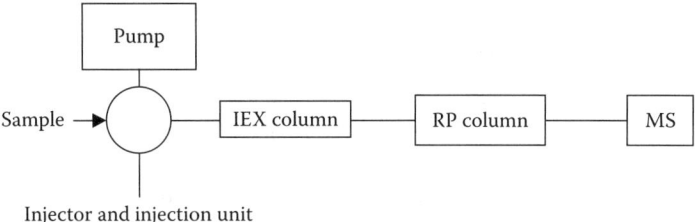

**FIGURE 5.4** Schematic representation of Mud-PIT.

2. Analysis of very dilute samples with small amounts of material will present a problem of having fractional loss of less abundant peptides due to high interaction with surfaces and other processing components. This problem is experienced when 2-DE analysis is utilized.

## CAPILLARY ELECTROPHORESIS

Operational procedure of capillary electrophoresis (CE) is similar to IEF. Proteins are separated in an electrical field, migrating until they reach the point where they carry zero charge. Analysis is carried out in a microcapillary tube which provides high resolution. CE can be coupled directly to a MS instrument. The superiority of CE to other analytical techniques is its high resolution. However, it is not widely used for proteomic analysis as there is no commercially available and reliable CE-MS instrument.

## PROTEIN DIGESTION

Particularly, proteins of complete cell or tissue extracts and heterogenous protein mixtures are usually analyzed after digestion to peptides. There are several reasons behind this approach.

1. MS instruments used for the analysis of separated proteins run for peptides with fewer errors.
2. Because the greater the mass of the protein, the greater the possibility of obtaining inaccurate results.
3. It is often difficult to perform MS on very large and hydrophobic proteins.
4. Sensitivity of mass measurements of peptides is superior to that of proteins.
5. Currently available MS instruments are much more suitable for peptide analysis and the data obtained can be directly used for comparison with protein sequences derived from proteome databases.

Keeping all these in mind, proteins are cleaved prior to MS analysis in many cases. Digestion should be performed at certain specific amino residues to yield peptides of optimal length for MS analysis. For an ideal digestion, peptide fragments of between 6 and 20 amino acids are used. Fragments shorter than 6 amino acids are not sufficient to produce unique sequences to be used for matching with databases. Peptides longer than 20 amino acids are difficult to analyze by MS. Thus, the optimum length of a peptide fragment should be between 6 and 20 amino acids. The digestion process is achieved by use of enzymes, which should be stable, well-characterized, possess high degree of specificity, and be available in sufficient quantity and high purity.

The most common approach for digestion is "in-gel" digestion. It is used for cleavage of proteins separated by 1-DE or 2-DE. This technique involves cutting out the bands from the gel, destaining, and treatment with proteases (e.g., trypsin). Digestion occurs in the gel matrix. Resultant peptides are eluted by washing and then, subjected to MS analysis. The choice of gel-staining technique is the critical issue regarding the success of in-gel digestion. Aldehyde type fixatives or long treatments with acids should be avoided for staining. Use of highly crosslinked gels will retard penetration of digestive enzymes. Again, residuals of SDS-PAGE should be removed completely to avoid their inhibitory effect on proteases.

"On-blot" digestion is another method used for proteins blotted onto nitrocellulose or PVDF membranes. Procedure is similar to "in-gel" digestion with the exception of elution from the membrane surface.

Different enzymes are used for digestion. *Proteases* are the most widely used group of enzymes. The most commonly used proteases are

1. *Trypsin* is the most frequently used serine protease. Generally, porcine or bovine pancrease is used as the source of pure trypsin. Trypsin usually digests proteins at their lysine and arginine residues. The superiority of trypsin is that it displays good activity both in solution and in-gel digestion protocols.

2. *Glu-C*, so-called V8-proteases, is an endoprotease digesting proteins at carboxyl side of glutamate residues in the buffer solutions (ammonium acetate or bicarbonate buffers). It is used only in in-gel digestions. Glu-C possesses higher specificity than trypsin, allowing analysis of proteins with high content of lysine and arginine.

3. *Nonspecific proteases* such as subtilysin, pepsin, proteinase K, or pronase are also used in proteomics. Upon cleavage, they produce multiple overlapping peptides due to lack of specificity. This may be beneficial in obtaining sequence data over a high percentage of each protein sample analyzed. Precaution must be taken by keeping cleavage period short to prevent undesired long digestion.

4. *Cyanogenbromide (CNBr)* is the most widely used chemical digestion agent. It cleaves proteins at methionine residues. Although CNBr offers a high degree of specificity, it also yields large fragments which are not useful for sequencing data by MS.

## MASS SPECTROMETRY (MS) FOR PROTEOMICS

Separated, fractionated, and digested protein samples are analyzed by means of mass spectrometry. A spectrometer is composed of three main parts

- The ion producing source
- A mass analyzer: converts components of a mixture into ions based on their mass/charge ratio (m/z ratio)
- A detector to detect the resolved ions

The data obtained by MS is analyzed automatically by a data analyzer and retrieved for interpretation either manually or with the aid of a computerized system. MS instruments for proteomics should possess the following characteristics

- Generation of good data on peptide masses, describing peptide fragmentation
- High sensitivity: the instrument should be capable of generating data on femtimole ($10^{-15}$ mole) quantities of peptides or less
- Sufficient resolution for distinguishing ions with very similar m/z ratios
- Capable of distinguishing ions with different m/z ions at a minimum of 1 Da
- Sufficient mass accuracy: this is important particularly in the case of identification of peptides by comparison with real database values.

The most frequently used instruments of MS-based proteome analysis are as follows.

## MALDI-TOFF

MALDI-TOFF stands for *matrix-assisted laser desorption ionization time of flight*. MALDI refers to the source of ionization whereas TOF indicates type of the mass analyzer.

The procedure involves mixture of sample protein with a chemical matrix which is a small organic molecule containing a chromophore. The matrix absorbs light at certain wavelengths. The most well-known matrix forming compounds are sinapinic acid and α-cyano-4-hydroxycinnamic acid. This mixture is then applied on a small plate and left to evaporate the solvent, forming a crystal lattice. Peptide to be analyzed is captured inside this lattice. It is then placed in the source which is equipped with a laser beam. The laser beam is fired onto the target and the chemical matrix absorbs the beam leading to the excitation. The excess energy is delivered to the peptides under investigation and ejected into the gas phase. This ionization process produces positive ions for peptides and proteins. Upon accepting a proton from the matrix, each peptide becomes singly charged. The resultant ions are then extracted and transferred into the TOF mass analyzer.

The mechanism of TOF analyzer is based on the time required for ions to pass from one end to the other end of the analyzer tube. Ions passed through the tube strike the detector. The speed of this passage is dependent upon m/z ratio of the ions. There is a positive correlation between two parameters. Ions with greater m/z ratio pass through the tube faster. TOF analyzer operates in a linear mode, indicating the direct transfer of ions from the source to the tube and then to the detector. However, it has a significant drawback in that there is poor resolution. In MS, resolution refers to the capability of distinguishing ions with slightly different m/z ratios. This may rise from the variations in velocities of the ions with the same m/z ratios. This problem may be overcome by introducing either a reflectron or pulsed-laser ionization with delayed extraction. The reflectron allows the ions with the same m/z ratios to reach the detector at the same time. The pulsed-laser ionization with delayed extraction (PLIDE) causes a delay between the laser pulse for ionization and the transfer of ions to the tube. In this case, ions with the same m/z ratios will start running down the tube together and reach the detector at the same time. This dramatically affects the resolution. If an instrument permits analysis of peptide ions distinguished with 0.001 amu of m/z ratios, the resolution is considered as ideal.

Superiorities of MALDI-TOFF can be summarized as

- An easy, simple, user-friendly, and robust method as it does not have any HPLC instrument interface
- Very suitable for walk-in routine use by different users in a shared proteomics facility
- Compatible with most robotic sample preparation devices, reducing labor and increasing speed and reproducibility
- High accuracy and resolution of TOFF
- Very sensitive and generates quality MS data on low femtimole quantities

Problems associated with the use of MALDI-TOFF analyzers are as follows:

- In general, these instruments are capable of measuring peptide masses. However, this kind of information has limited use in proteome analysis.
- The quality of the sample influences the quality of the data obtained by MALDI-TOFF. Contaminations inhibit ionization. As no in-line HPLC system is included in the system, contaminants cannot be removed from the sample. Fortunately, some new clean up tools (e.g., ZipTips) for removal of contaminants have been developed recently (Figure 5.5).

### ESI Tandem MS

ESI tandem MS stands for *electrospray ionization mass spectrometry performed in multistage*. This technique is conducted based on the production of multiply charged ions from proteins and peptides. In this technique, ionization procedure is carried out within the instrument. Three types of mass analyzers are used individually or in combination.

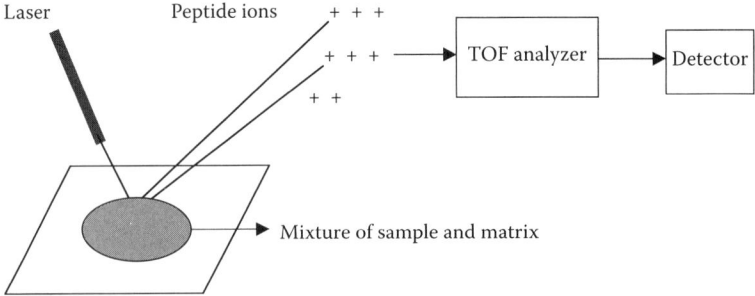

**FIGURE 5.5** Schematic representation of a simple MALDI-TOFF analyzer.

**TABLE 5.4**
**Optimum pH Values of Functional Groups for Ionization**

| Functional Groups | Optimum pH for Ionization |
|---|---|
| Carboxylic acid | > 5.0 |
| N-terminal amines | < 7.0 |
| Nitrogen of histidine | < 7.0 |
| Nitrogen of lysine | < 8.5 |

- Quadrupole analyzer
- Ion-trap analyzer
- TOF analyzer

The first two have more limited mass range compared to TOFF.

In ESI-MS, samples are introduced in the form of aqueous solutions since peptides (particularly, the ones derived by tryptic digestion) exist as singly or multiply charged ions in solutions. The critical parameter is the pH of the solution as ionization occurs depending on the pH of the media. Some of the functional groups are listed in Table 5.4 with their optimum pH values for ionization. Peptides ions with positive charge can undergo fragmentation. Therefore, ESI analyses are conducted in positive ion mode for acidic samples.

At equilibrium, intact proteins in solutions exist in multiply charged states since they have many proton accepting sites. ESI mass spectrum of the multicharged protein is converted to the actual protein mass by a specific software.

Peptides carry single or charged ions based on their size, molecular weight, and the number of amino acid residues. In many cases, double-charged ions of peptides are predominant in mass spectrum.

The sample to be analyzed is introduced to the ESI source by means of a flow stream from an HPLC instrument. The sample flows through a stainless-steel needle and then, sprays out in the form of a mist whose droplets hold peptide ions and mobile phase of HPLC. Peptide ions are separated from the mobile phase and subsequently, transferred into a mass analyzer either by a heated capillary or a curtain of nitrogen gas. Desolvation process can be carried out by a vacuum system.

## TECHNIQUES USED FOR STRUCTURAL PROTEOMICS

Structural proteomics aims the determination of three-dimensional protein structures in order to better understand the relationship between protein sequence, structure, and function. NMR and x-ray crystallography have been significant methods and indispensable tools to determine the structure of macromolecules, especially proteins. Many biotechnology companies have been using these two techniques for enlightening protein structure (Table 5.5).

**TABLE 5.5**
**Biotechnology Companies Using NMR and X-ray for Proteomic Studies**

| Company | Technologies Used for Proteomics |
|---|---|
| Vertex Pharmaceuticals | NMR, x-ray, computational chemistry, chemogenomics |
| RiboTargets* | NMR, x-ray, RiboDock software |
| GeneFormatics | NMR, x-ray, Fuzzy Functional Form Technology, bioassay development |
| TRIAD Therapeutics | Integrated Object-oriented PharmacoEngineering (IOPE) |
| Integrative Proteomics | NMR, x-ray, high throughput protein production |
| Metabometrix* | NMR metabonomics |
| Novaspin Biotech* | NMR, software development, biochemistry |

*Academic spin-out company.

To obtain optimal results, protein should possess minimum 95% purity. Thus, the molecule under investigation should be purified by gel or column separation, dialysis, differential centrifugation, salting out, or HPLC prior to structural analysis.

## X-Ray Crystallography

X-ray crystallography is used to determine the tertiary structure of a protein. Prior to structure analysis, optimum crystallization conditions must be tailored for each individual protein molecule. Employing these conditions, crystals in sufficient size and quality are obtained. They are mounted and snap frozen by cryogenic techniques in order to prevent formation of ice-lattice. Then, x-ray beams are directed at this crystalline sample. The x-ray beams scatter based on the electron densities in different proportions of a protein molecule. The crystal must remain supercooled during data collection in order to minimize radiation damage and backscatter. Images are translated into electron density maps and recorded on an x-ray film. The diffraction pattern is composed of spots on the film, and the crystal structure can be determined from the positions and intensities of the diffraction. In the final step, maps are superimposed either manually or with software and then the construction of the model is completed. It is a powerful technique, revealing very precise and critical structural data useful for structure-based drug design (e.g., the HIV-protease inhibitors amprenavir and nelfinavir). However, it is time consuming, expensive, and requires specific training and equipment.

Mainly two companies, Syrrx and Structural Genomix (SGX), both in San Diego, CA, have been using x-ray crystallography for protein discovery process. Syrrx has been implementing robotic techniques, which were developed using the ones in the automotive industry. All the procedures starting from purification to crystallization have been done robotically on an assembly line. Structural Genomix also has been using a similar methodology. All the information generated is stored in a software called "Protein Data Bank" (PDB).

Much information about flexibility of protein structure has also come from x-ray crystallography data. It is either in the form of atomic mean-square displacements

(AMSDs) or B factors. AMSD profile has been developed by Hale et al. (1988) It is essentially determined by spatial variations in local packing density of a protein molecule. With the statistical mechanics and generic features of atomic distributions in proteins, it is possible to predict an inverse correlation between the AMSD and the contact density. In other words, the number of noncovalent neighbor atoms within a local region of approximately $1.5 \text{ nm}^3$ volume can easily be determined by AMSD. This local density model is then compared to a set of high-quality crystal structures of 38 nonhomologous proteins in PDB. The result gives accurate and reproducible peaks in the AMSD profile and this profile helps to capture minor features, such as the periodic AMSD variation within $\alpha$ helices. The models developed using AMSDs suggest that this technique provides a quantitative link between flexibility and packing density.

### Drawbacks of X-Ray Crystallography

Despite the enlightening information on protein structure, the production of crystals for x-ray studies can sometimes cause structural anomalies. They might mask native architectural features. Also, it is well known that membrane proteins are not readily amenable to existing crystallization methods. Although most of the information on protein structures that are now available through the Protein Data Bank have mainly been gathered using x-ray crystallography, few of these structures are intact membrane-associated peptides and proteins. Thus, advanced technologies need to be generated to obtain structural and functional information for membrane proteins by x-ray crystallography.

### Nuclear Magnetic Resonance

Although x-ray crystallography is advantageous since it defines ligand-binding sites with more certainty, NMR measures proteins in their native state. It has a distinct advantage over x-ray crystallography. Precise crystallization, which is often difficult, is not necessary for conducting structural analysis by NMR. Furthermore, NMR is increasingly being recognized as a valuable tool, not only in three-dimensional structure determination, but also for the screening process. Proteins with large molecular weight (up to 30 kDa) can be analyzed. The most significant advantages of NMR spectroscopy are

- It reveals details about specific sites of protein molecules without a need to solve the whole structure.
- It is sensitive to motions of most chemical events which in turn provides direct and indirect examination of motions within micro-time scale (milliseconds to nanoseconds, respectively).

The advanced NMR spectrometers are generally computer-controlled radio stations with antennas placed in the core of magnets. Nuclear dipoles in the protein sample align in the magnetic field and absorb energy. With the aid of this energy gain, it flips from one orientation to another, then readmits this energy. The alignment occurs depending on the strength of the magnetic field. The stronger the dipole, the greater the energy of alignment. Radio waves are directed and pulsed to the sample by a computer. Some of the pulse is absorbed by the sample and after a while, radio signals are readmitted. These signals are amplified by a receiver and stored in a software. The energy radiated by the molecules with different chemical structure will be different. This allows scientists to interpret NMR data using software routines.

The hydrogen nucleus has been widely used in NMR analysis as it is the most highly abundant organic element and possesses one of the strongest nuclear dipoles in nature. Using pulse sequence with hydrogen nucleus, the chemical bond connectivity, spatial orientation. and distance geometry of proteins can be mapped. Combining this information with those obtained from molecular modeling, the structure of many proteins can be enlightened.

Surprisingly, the advances in NMR technologies are usually used by smaller biotechnology companies who make deals directly with the NMR technology providers in a customized offering. For instance, both GeneFormatics and Integrated Proteomics have recently announced partnerships with Bruker BioSpin, Bruker Daltonics, and Bruker AXS. These partnerships are actually two-sided. Another example is RiboTargets and De Novo Pharmaceuticals' deal with Sun Microsystems.

## PHAGE DISPLAY TECHNIQUE FOR FUNCTIONAL PROTEOMICS

Advances in genomics led to the formation of databases of short DNA sequences, so-called expressed sequence tags (ESTs), in 1991. ESTs represent most of the expressed human genome in a fragmented state. However, interpretating such data is much more difficult than generating it. In fact, proteomics require understanding of the expression patterns of proteins, which will give key information on their functions under normal conditions and/or disease states. Therefore, scientists have directed their interest toward characterizing and cataloging the proteome.

In the conventional methods of proteomics, 2-DE and MS use such approaches working backward from the fractionated proteins to their corresponding DNA sequences. Unfortunately, these approaches represent a significant bias toward the abundant proteins, and functionally important proteins of low abundance are often not detected by this way.

This problem can be overcome by a targeted approach involving generations of protein-specific probes based on DNA sequence data. These probes can be used to identify individual proteins and to investigate their functions at cell or tissue levels. This solution helps scientists to obtain useful, comprehensive and easily interpretable data. It is faster and easier than the conventional approach. These targeted molecular probes are formed using recombinant human antibodies by means of phage display technique.

Antibodies raised in animals have been used for quantitative detection of proteins in tissues or extracts for many decades. They are useful, specific, and sensitive tools to localize proteins at cell levels in histological materials, to recognize, and to track target molecules. The use of recombinant human monoclonal antibodies has been a great interest of scientists as they are much more advantageous to traditional monoclonal or polyclonal antibodies. Some of the main reasons behind it are that traditional antibodies require animal use, and are produced in relatively longer time. Also, it should be kept in mind that regulations for animal usage are becoming more and more stringent. In contrast, recombinant antibodies can be produced without the use of an animal and the process is not time-consuming. They can be generated against an enormous number of antigens (see Chapter 4). In addition, increasing the affinity and specificity of recombinant antibodies is very simple and relatively fast. Thus, large numbers of different antibodies against a specific antigen can be generated in one selection procedure. Therefore, researchers aimed to develop faster and easier ways to improve the target-recognition qualities of antibodies. With all these in mind, phage display was developed in 1990 as a new method that enables researchers to quickly evaluate a huge range of potentially useful antibodies and then produce large quantities of the selected ones. It allows us to produce recombinant human monoclonal antibody fragments against a large range of different antigens (e.g., RNA, various protein antigens, peptides, proteoglycans, etc.) in order to identify functions of individual proteins. Identification of the function of a protein is achieved by preventing the interaction of the target protein of the antibody under investigation with its natural ligand.

Phage display is an important tool in proteomics. Proteins can be engineered for specific properties and selectivity, employing phage display methods. A variety of display approaches can be used for the engineering of optimized human antibodies, as well as protein ligands, for such diverse applications as protein arrays, separations, and drug development. The utilization of phage display in screening for novel high-affinity ligands and their receptors has been crucial in proteomics because it allows us target essential components and pathways within many different diseases, including cancer, AIDS, cardiovascular disease, and autoimmune disorders. Although phage display technology is still at early stages of development, it has already been used for the production of human antibodies potentially applicable in therapy.

In this technique, synthetic antibodies that have all the target-recognition qualities of natural ones are produced and selected to form phages. Production of the antibodies is accomplished by using the same genes that code for the target-recognition or variable region in natural antibodies from mammalian systems.

Phages are genetically engineered antibodies. A phage is comprised of the displayed antibody's phenotype coupled to its genotype. In this structure, a particular antibody is fused to a protein on the phage's coat and the gene encoding the displayed antibody is contained inside the phage particle. This allows the DNA to code for the selected antibody which will be available for future use. Selection is made using biopanning procedure (Figure 5.6). These antibody covered phage populations form the phage library in each of which billions of different antibodies exist.

As the part of this approach, the procedure of cloning immunoglobulins (i.e., repertoire cloning) is carried out entirely under *in vitro* conditions and enables the production of any kind of antibody without involving an antibody-producing animal. At first, the total messenger RNA (mRNA) of human B lymphocytes is extracted from peripheral blood (Figure 5.7). Corresponding DNA chains are deducted from this mRNA and then, synthesized using the enzyme called reverse transcriptase. Resultant cDNA (complementary DNA) chains are subject to amplification by PCR (polymerase chain reaction). Subsequently, the genes are multiplied by PCR to enable the selection of immunoglobulin fragments specific for a given antigen. These antibody fragments should possess such a format that is a protein composed of the

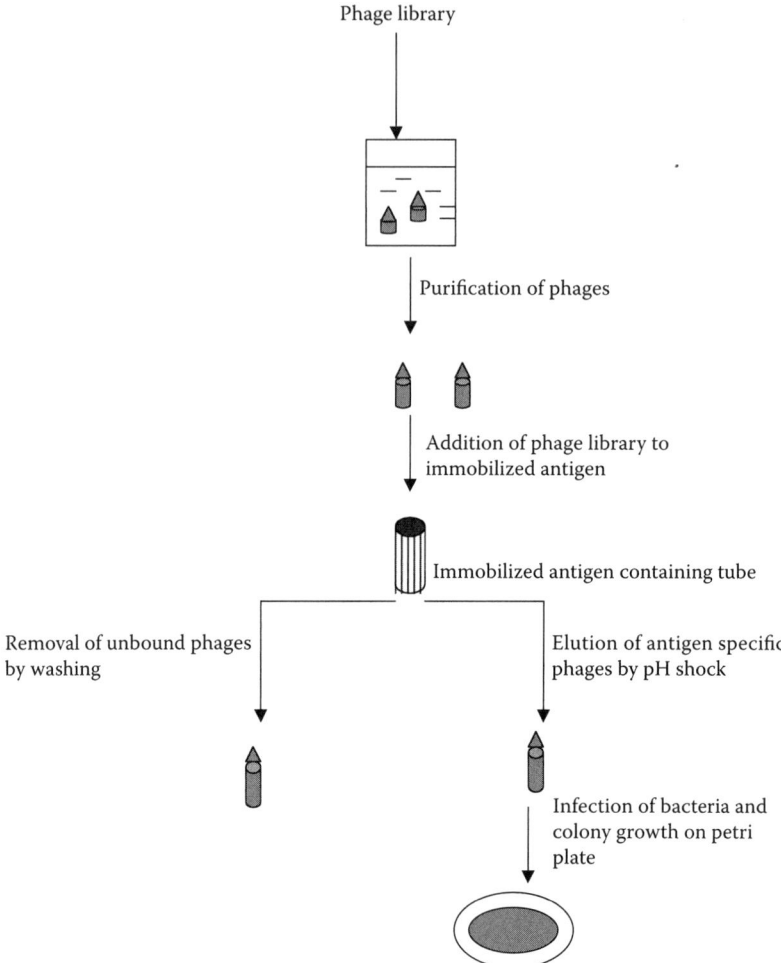

**FIGURE 5.6** Schematic representation of panning.

variable regions of the IgG heavy and light chains (VH and VL) connected by a flexible peptide linker. This format is usually symbolized as scFv (single chain Fv) (Figure 5.8). The resultant products of cloning are introduced in phagemid vectors which are easier to manipulate and have high transformation efficiencies.

These phagemid vectors (e.g., *Escherichia coli*, yeast, cells) are expressed in the appropriate bacteria and produce corresponding proteins in large amounts when cultured in a fermenter. The products are human antibodies' fragments which are very similar to natural antibodies in structure. The last step is the expression of a scFv immunoglobulin molecule on the surface of the filamentous bacteriophage as fusions to N-terminus of the minor phage coat protein (g3p). These bacteriophages represent a mixture of immunoglobulins with all specificities included in the repertoire, from which antigen specific immunoglobulins are selected.

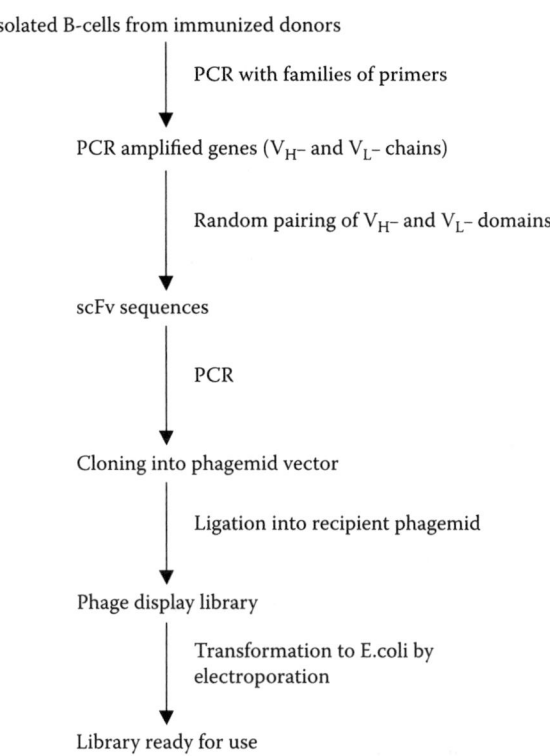

**FIGURE 5.7** Schematic representation of the construction of a phage library.

Specificity and affinity are the important features for a phage library. The larger the library, the greater the possibility of possessing high affinity with desired specificity. The best part of this technique is that human antibody libraries readily yield clones able to recognize any human protein under investigation. The largest library in existence contains $10^{11}$ different sequences. Clearly, the best advantageous aspect of phage display technique is that it functions as if it is an *in vitro* immune system. Therefore, it is the most efficient way of producing antibodies for medical purposes.

Phage particles are often used as reagents in medical diagnostic techniques such as ELISA, immunocytochemistry, or flow cytometry. The most significant benefit arising from the application of phage display technique in diagnosis is its high speed for screening databases without using a pure protein reagent.

ProxiMol® and ProAb® (Cambridge Antibody Technology, UK) are patented techniques using phage display approach. In ProAb, peptide sequences are defined for an open reading frame, then corresponding antigens are produced using an automated synthesizer. Automatization provides a significant benefit such that the generation of antibodies never becomes the rate limiting step. The best part of this technique is that it does not require complete DNA sequence or protein antigen expression. It generates desired antibody probes directly from information stored in

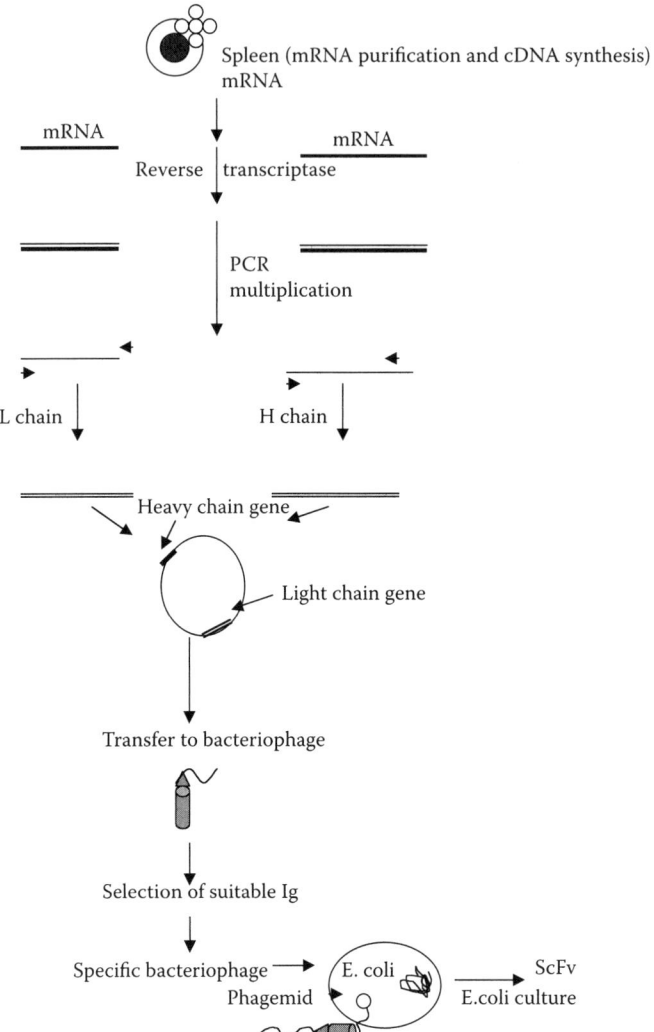

**FIGURE 5.8** Construction of a ScFv.

DNA databases (ESTs) (Figure 5.9). The driving force behind the development of ProAb was the need to correlate gene expression to disease indications in high throughput. The aim is to translate the data in ESTs into synthetic peptide antigens. These antigens are usually used for selection of phage antibodies. Thus, ProAb forms a direct relation between ESTs and disease state.

ESTs contain the richest data on DNA sequence. Interesting segments of DNA are selected from ESTs and then, translated into proteins by ProAb. Generally, a known protein with interesting biological activity and known structure is used to search for family members. At first, peptides (e.g., 15-residue antigenic peptides)

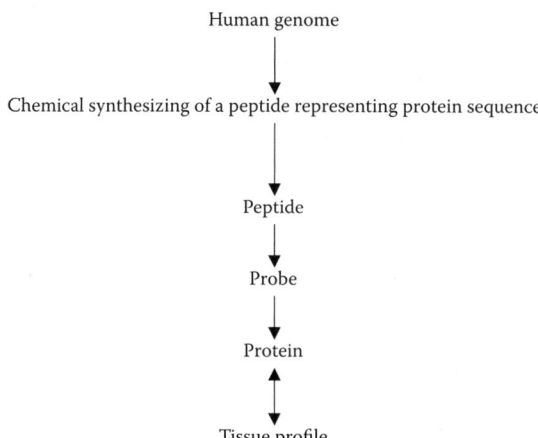

**FIGURE 5.9** Schematic representation of ProAb technique.

are selected from ESTs by means of bioinformatics, then downloaded to an automated solid-phase peptide synthesizer (e.g., advanced chem tech synthesizer). They are synthesized chemically with an N-terminal cysteine residue or N-terminal biotinylation to represent the protein sequence under investigation. These peptides should posses some important features such as having a native-like structure, being located in a solvent-accessible region of the native protein, having high affinity to specific antibody. They are deprotected, cleaved from the beads and the reagents are removed by precipitation. Cleaned peptide subnatants are redissolved and tested for quality control by reversed phase liquid chromatography and electrospray mass spectrometry. Optimum purity should be a minimum 85%. Resultant peptides are used as antigens for the selection of phage antibodies. Selected antibodies can be used as detection agents in immunocytochemistry. In most cases, BSA (bovine serum albumin) is used as the carrier protein following activation procedure by bifunctional succinimide-maleimide reagent. Efficiency of coupling (peptide-protein) is tested by using SDS-PAGE and MALDI mass spectrometry. The optimum coupling ratio is 20:25 peptides per protein molecule. These peptide-protein conjugates are coated onto multiwell (96 or 384 wells) microtiter plates. In most companies, microtiter plate reactions are carried out by generic robots. The selection of specific antibody is performed based on affinity capture: "panning." The procedure is simple and requires a small volume of liquid (e.g., 100 µl). As in DNA encoding the antibody is covered by a phage particle which displays the scFv (single chain antibody fragment of IgG) specificity, the protein and its gene are coselected. For this purpose, coated conjugate is incubated with the target; the surface is washed; specific phages are infected into *E. coli*; the bacterial culture is spread onto a culture plate to produce bacterial colonies. Each colony produces a monoclonal antibody. Subsequently, a colony picking process is performed for ELISA positive clones using bioinformatics. Generally, 12 antibody clones are picked per peptide sequence and archived as glycerol stocks at −70°C. This enables

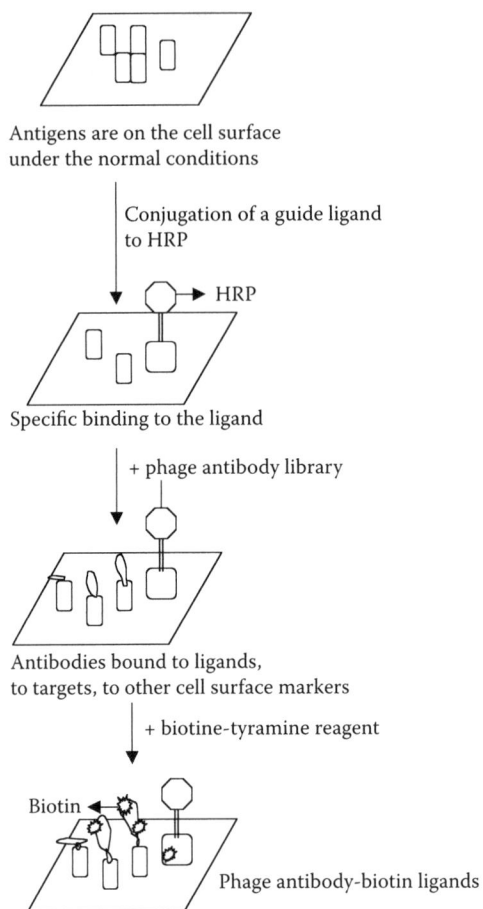

Antigens are on the cell surface
under the normal conditions

Conjugation of a guide ligand
to HRP

HRP

Specific binding to the ligand

+ phage antibody library

Antibodies bound to ligands,
to targets, to other cell surface markers

+ biotine-tyramine reagent

Biotin

Phage antibody-biotin ligands

**FIGURE 5.10** Schematic representation of ProxiMol technique.

us to gain understanding of the distribution and abundance of the genes in both normal and disease states.

This technique (ProxiMol) was developed by Osburn et al. in 1998 With ProxiMol, antibodies are recognized by a guide molecule (Figure 5.10). The guide molecule used in this technique can be a carbohydrate ligand or a conjugated fatty acid, with affinity for either the target protein or the near neighboring molecule. It also possesses the ability to couple with HRP (horseradish peroxide) directly or indirectly and retains sufficient biological activity to remain associated with the target. In this technique, the guide molecule and phages from antibody library are added together to the intact cells; it is incubated; unbound phage is removed by a washing off procedure; HRP-conjugate is added and then incubated. Again, the unbound phage is washed off. Biotin-tyramine and hydrogen peroxide are added, incubated, and washed. TEA (triethylamine) is added for elution of bound biotinylated phages. These phage antibodies are recovered using streptavidin-magnetic beads. Biotinylation is carried out in close proximity due to the short half-life of the biotin tyramine free radicals. Phage antibodies identified by the ProAb technique can be used as the source of antigen for the ProxiMol technique. Screening of the ProxiMol output can be carried out either on tissue sections or cells by means of immunochemistry, flow cytometry, or ELISA.

In immunocytochemistry, signals are usually amplified by means of catalyzed reporter deposition. In ProxiMol, catalyzed reporter deposition has been modified to permit the isolation of phage antibodies binding to near neighbor biotinylated guide molecule. Deposition of biotin molecule can be detected by electron microscopy.

Phages specific for the guide molecule's binding partner and its neighbor can easily be recovered using the ProxiMol technique. This technique does not require purification or cloning of a protein from its native state. In order to target new epitopes, repertoire of phage antibodies can be expanded using this technique.

For target selection, adhesion molecules are expressed on the surface of endot-helial cells. For this purpose, sialyl Lewis X is used as the guide molecule in the case of expression of P- and E-selectins for TNFα–activated HUVEC (human umbil-ical vein endothelial cells). Thus, targeting membrane-bound antigens *in situ* is also possible with this technique.

The most significant disadvantage of this technique is that the use of a receptor ligand limits the isolation of antibodies that do not bind at the ligand binding sites and neutralize the receptor-ligand interaction.

In conclusion, the ProxiMol technique may enable us to form the maps of multiprotein complexes as it provides both targeting specific cell surface receptors and other proteins associated with this receptor. No matter what technique is used to form phage libraries, it is not possible to interpret the data manually; it is accomplished by using bioinformatics.

Immunochemistry is the most important medical diagnostic technique where phage particles find optimum applicability. However, it has not been used universally because of the difficulty of operating the technique high throughput as it requires analysis of nearly 1000 samples per month. It requires a link with clinical centers having diverse tissue banks in order to get easy and fast access to tissues which should be preserved promptly postmortem. Optimal processing should be done to preserve antigens by means of cryosections of tissue samples and mild fixation with cold-acetone. This will increase the speed and facilitates cross-comparisons. Thus, it is well known that forming tissue banks is not easy and cost-effective.

Phage antibodies are directly used to label tissue sections. Interpretation of the data is performed by a team of experts including histochemists and must be subjected to cross-examination. Images are captured using microscopes with a digital camera attachment and archived in databases for further analysis of pathologists.

## Fields of Application for Phage Display Technique

Phage display techniques can be used in any field where molecular biological approaches find application. It can be successfully used for epitope mapping, vaccine development, identification of proteinkinase substrates and nonpeptid ligands, selec-tion of functional interactions of hormones, enzymes, enzyme inhibitors, mapping of protein-protein bonds, identification of ligands of cell surface receptors, and *in vivo* screening for diagnosis.

Antibody produced by this technique is used for many different purposes in medicine and pharmacy. As the size of a scFv molecule produced by phage display (27 kDa) is smaller than that of a complete antibody (~150 kDa), it is used for diagnosis and therapy. Monoclonal antibodies produced by conventional approaches for tumor detection, slowly diffuse into the tissues and are removed from circulation due to their large size. It seems that scFv will soon replace them.

Small antibodies in the form of scFv can easily pass over the cell boundaries and this is the main reason behind their use for diagnosis and therapy. Also, rapid removal of these small structures from serum and tissues presents a significant advantage. Removal of digoxin from the circulation is achieved by this approach. Now, it is possible to produce protector antibodies against HIV and hepatitis B is possible using phage display technique. Another field of application for phage display techniques is the synthesis and the improvement of immunogenicity of synthetic peptide vaccines.

## BIOINFORMATICS

With the development of advanced automated DNA sequencers and the collaborative efforts of the Human Genome Project (HGP), the amount of information on gene sequencing, proteins, gene and protein expressions has increased enormously. Human genome sequence offers wealth of useful knowledge to scientists. It should be possible to categorize, store, and execute the data on sequence information generated by genomics and proteomics for interpretation when needed. Popular sequence databases (e.g., GenBank, EMBL) have been expanding with new sets of data. This enormous information has necessitated the development of a new approach. It is believed that this new approach will provide two significant implications: (a) use of data management and data analysis tools to deal with all this information, and (b) enlightening complex interactions. In the 1990s, information technology (IT) was applied to biology to produce the field called *bioinformatics*.

Proteins interact with each other in many different ways. These interactions may be structural, evolutionary, functional, sequence based, and metabolical. Life depends on such biomolecular interactions. Among all these interactions, structural ones are the simplest and easiest to investigate because they are the most definite. Therefore, the main goals of bioinformatics are to create and to maintain the databases of the biological information which may lead to better understanding of these interactions.

Both the nucleic acid sequences and the protein sequences derived from the biological information are collected in most such databases. Large amounts of data in these databases need to be sorted, stored, retrieved, and analyzed. Selection of subsets of data for particular analysis should also be done. IT providers designed such a data warehouse and developed an interface that provides an important benefit to researchers by making it easy to access the existing information and also to submit new entries (i.e., datamining) (Table 5.6). Middlewares and structured query language (SQL) softwares were developed for this purposes. The former one is used

**TABLE 5.6**
**Diagram of Information Chain in Bioinformatics**

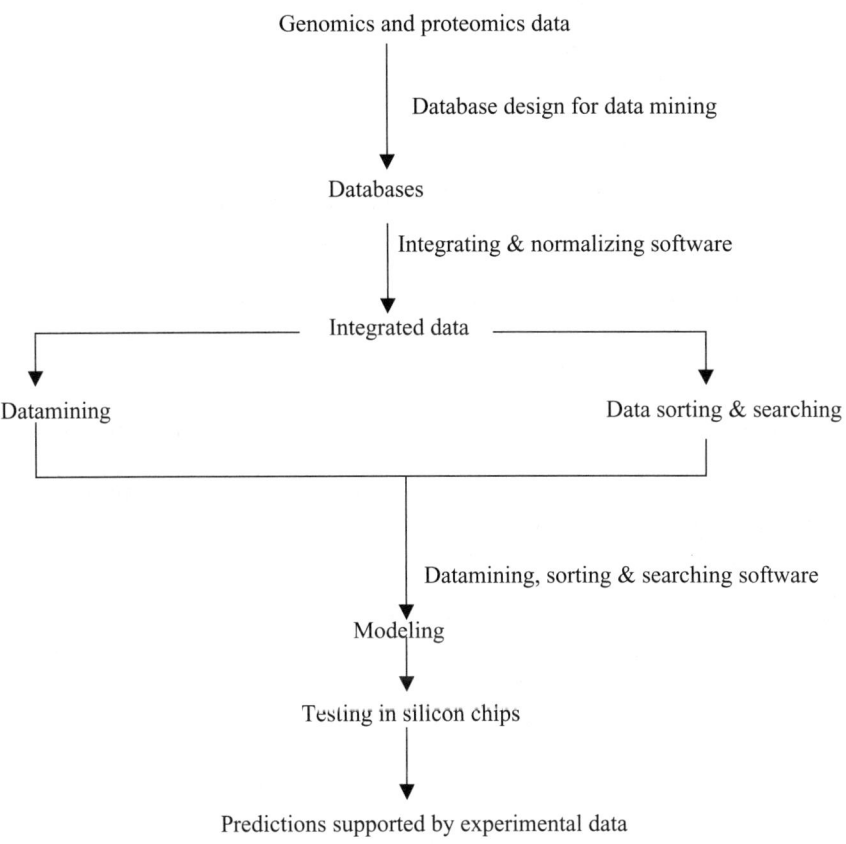

to normalize the data and to eliminate duplication. The latter one allows specific information searches (e.g., a specific gene containing a specific sequence). Data mining for unknown relationships is a process of identifying valid, novel, potentially useful, and clear patterns in data and is conducted through building algorithms and looking for correlations and patterns in data. Statistical methods, pattern recognition, neural networks, fuzzy logic, cluster analysis, case-based reasoning (CBR) are some of the major datamining methodologies. Outcome of data mining can be used to produce models of protein interactions, disease pathways, regulatory cascades. Subsequently, these models are used to make predictions. In the later stages of bioinformatics, *in silico* testing is carried out to predict the behaviors-based models created by datamining. For example, a new set of potential pharmaceutical active substances can be subjected to *in silico* testing and the ones with less potential are discarded without conducting any sets of experiments.

IT providers and biologists work together to form an approach to analyze the data on sequence information stored in databases and they called this approach *computational biology.* The particular objectives of computational biology (CB) are

- To find the genes of interest in the stored DNA sequences
- To develop new methods to predict the structure or function of newly discovered proteins
- To classify and find the category of protein sequences under investigation into families of related sequences
- To develop new protein models
- To generate phylogenetic trees by aligning similar proteins
- To examine evolutionary relationships among proteins

The DNA sequences in nature encode proteins with very specific functions. In CB, algorithms are used to enlighten the 3-D structure of these functional proteins. This requires the combinatorial work of physics, chemistry, and biology. The information obtained from the analysis of other proteins with similar amino acid sequences form the basis of this work. This allows the use DNA sequences to model protein structure.

Most biological databases are composed of the sequences of various nucleotides (e.g., guanine, adenine, thymine, cytosine, and uracil) and/or amino acids (e.g., threonine, serine). Each sequence of amino acids represents a particular protein. Before the development of bioinformatics, sequences were represented using single letter designations as suggested by the International Union of Pure and Applied Chemistry. This was to reduce the space needed for storage of information and to increase processing speed for analysis. With the implication of information technologies to biology, this problem was overcome. The data on sequences have been stored and organized into databases and then analyzed rapidly by means of computational biology. Everyday mathematicians and/or IT scientists develop new and sophisticated algorithms for allowing sequences to be readily compared using probability theories. These comparisons become the basis for determining gene function, developing phylogenetic relationships, and simulating protein models. With the invention of the Internet, nearly everyone has access to this information and the tools necessary to analyze it.

Outcome of the methods used in bioinformatics allow scientists to build a global protein structural interaction map. The first developed map is called PSIMAP (Protein Structural Interactome Map). It has low resolution and allows production of a draft map for very large-scale protein interaction study. Protein maps reveal that protein structures have distinct preferences for their interacting partners and the interactions are not random. Some proteins have only one interaction partner whereas some have more. Some protein groups function as separately while others work within larger complexes. Also, many proteins possess homointeraction.

The collecting, organizing, and indexing of sequence information into a database provides the scientist with a wealth of information on human genome and proteome. What makes this database so useful and powerful is its analysis, which may lead to information indicating that the sequence of DNA in question does not always constitute only one gene; it may contain several genes.

As all genes share common elements, it has been possible to construct consensus sequences, which may be the best representatives for a given class of organisms. Common genetic elements include promoters, enhancers, polyadenylation signal sequences, and protein binding sites' and they have also been further characterized into subunits. This common elemental structure of genes makes it possible to use mathematical algorithms for the analysis of sequence data. Software programs are developed to find genes with desired characteristics. In these programs, probability formulae are used to determine whether two genetic sequences are statistically similar or not. This allows pattern recognition. Then, the data tables containing information on common sequences for various genetic elements are formed. As different taxonomic classes of organisms possess different genetic sequences, these taxonomic differences are also included in an analysis to speed up processing and to minimize error. Analysis of the data is carried out following the instructions provided with the software defining how to apply algorithms. The acceptable degree of similarity and the criteria to include the entire sequences and/or fragments in the analysis are given by these instructions. It is best to enable users to adjust these variables themselves.

Protein modeling is challenging work and comprised of the following steps.

1. *Finding the place of transcription*: In this step, a genetic locus is located for transcription. It provides the information on start and end-codons for translation.

2. *Location of translation*: Although it is known that the start-codon in mRNA is usually AUG, the reading frame of the sequence must also be taken into consideration. In a given DNA sequence, only six reading frames are possible. Three frames on each strand must be considered. As transcription takes place away from the promoters, the definitive location of this element can decrease the number of frames to three. Finding the location of the appropriate start-codon will include a frame in which they are not apparent abrupt stop-codons. Information on the molecular weight of the protein will be helpful to complete the analysis. Incorrect reading frames usually show themselves with relatively short peptide sequences. Although it is easy to carry out such process in the case of bacteria, the scientists working with eukaryotes face a new obstacle, namely, introns.

3. *Detection of intron/exon splice sites:* Introns in eukaryotes cause discontinuation of the reading frame. If the analysis is not focused on a cDNA sequence, these introns must be spliced out and the exons joined to form the sequence that actually codes for the protein. Intron/exon splice sites can be predicted based on their common features. Most introns begin with the nucleotides GT and end with the nucleotides AG. There is a branch sequence near the downstream end of each intron involved in the splicing event.

4. *Prediction of 3-D structure of the protein under investigation:* After completing the analysis of primary structure, modeling the 3-D structure of the protein is carried out using a wide range of data and CPU-intensive computer analysis. In most cases, it is only possible to obtain a rough model of the protein. This may not be the key to predict the actual structure as several

conformations of the protein may exist. Patterns are compared to known homologues whose conformation is more secure. X-ray diffractional analysis is conducted and the data obtained may be useful for the protein of interest. Then, biophysical data (physical forces and energy states for bonds) and analyses of an amino acid sequence can be used to predict how it will fold in space. All of this information is computed to determine the most probable locations and bonds (bond energies, angles, etc.) of the atoms of the protein in space. Finally, this data is plotted using appropriate graphical software to depict the most suitable 3-D model of the protein under investigation.

As seen in the last step, protein models are often developed based on graphs. Graphs may be used to nondirectional molecular interactions and associations between two or more molecules. However, graphs may rapidly grow in complexity and overwhelm even the most powerful supercomputers today.

Many databases already store data on various kinds of pathways of biological systems and molecular interactions. Some of them are

- For metabolic pathways: KEGG, EcoCyc, BRENDA, WIT2
- For signaling patways: CSNDB, AFCS, SPAD, BRITE, TransPath
- For gene regulatory networks: TRRD, TransPath
- For protein interactions: BIND, DIP, MIPS, GeneNet

Data is stored by various ways in these databases. Some are stored in flat files form, some relational, and others object oriented. The most important disadvantage of bioinformatics is the lack of standardization preventing widespread commercialization of new tools. This leads to a challenge to represent, analyze, and model molecular interactions in systems biology for determination of the way in which proteins function at cell level. This requires development of new representations and algorithms. Some modeling tools have been used to overcome problems of modeling in engineering and information technologies may be useful to solve this obstacle. To discuss the problems and the possible ways of solving them, BioPathways Consortium (BPC) was established in June 2000. The main objectives of the consortium are to synthesize and disseminate information from different sources, to identify scientific and IT-related problems, to form a set of standards, and to promote university–industry collaboration.

Beside company-owned private databases, there are also public databases that anyone can have access to. Use of private databases requires a fee. Some of them do not allow public access regardless of the payment of a fee. Most of the accessible databases are Web-based. Sometimes databases have different levels for access. Generally, the basic access is free or requires a minimal fee. Higher fees are asked for the access to higher levels which contain the more difficult to obtain data or complex search and algorithms for datamining. Software to access information from these databases is also public software (i.e., shareware) or the commercially available ones. There are also boutique-type companies offering expertise on specific databases and applications. The most commercially available software has the ability to analyze clusters, annotate genes based on known functions, and display image of gene positions and different profiles of the same gene, and at the same time, quantify expression levels, analyze data, search for regulatory sequences, and identify the linked genes. It can also perform data extraction for a specific gene.

As the 2D-PAGE is the core separation technology with high resolution, many scientists working in the field of proteomics have been using 2D protein databases. Implication of this technique to systematically analyze human proteome requires a reproducible 2D-gel system along with a computer-assisted technology to scan the gels, make synthetic images, assign numbers to individual spots and match them. At first, a digital form of the gel image is prepared as a standard. Scanners (laser type, array type, or rotating type), television cameras, and multiwire chambers are used as tools for this purpose. Subsequently, the 2D-protein gel image is analyzed and processed using various software programs such as PDQUEST (BioRad), MEL-ANIE 3 (Swiss Institute of Bioinformatics), ProXPRESS (Perkin-Elmer), PHORETIX (Nonlinear Dynamics Ltd). Synthetic images can be stored on a disk for further editing. The editing process is not time efficient and requires expertise. Misleading scratches are erased, streaks are canceled, closely packed spots are combined, restored, and added to the gel during the editing process. The standard image is then matched automatically by the computer to other standard gels. Proteins are matched based on their gel position. After preparing a standard map of a given protein sample, related qualitative and quantitative information can be entered to establish a reference database. Then, categories are created on the basis of physical, chemical, biochemical, physiological, genetic, structural, and biological properties. Unfortunately, there is no software program available in the market to provide full automatic analysis of 2D gel images. There are data warehouses or galleries that can be used for the comparison of protein expression profiles between human tissues and cells, on the Internet (e.g., ZOO-PLANT gallery of Gromov et al. 2002).

In conclusion, the protein mapping by bioinformatics approach offers an effective route to drug discovery by pinpointing signaling pathways and components that are deregulated in specific disease states. The most important challenge to be addressed before applying proteome projects to the treatment of diseases is to identify and functionally characterize targets that lie in the pathway of diseases. This is of great importance to the pharmaceutical industry from the drug screening perspective. A recently developed technique called laser capture microdissection (LCM) may be useful for this matter as it provides rapid and reliable extraction of cells from a specific tissue section, and recovery of a protein suitable for 2D-PAGE analysis and MS. Thus, LCM coupled 2D-PAGE approach may have great impact on analysis of proteins in healthy and disease states.

Much of the basic research in bioinformatics applications for proteomics originated in universities. Although most of the universities do not have specific departments of bioinformatics, they have biology, bioengineering, computer engineering, chemistry, pharmacy, and information management departments working on projects directly or indirectly on data for bioinformatics. There are also many collaborative ongoing projects at public research institutions. Since bioinformatics requires a multidisciplinary approach (Figure 5.11), the scientists with different backgrounds and skills gather for such projects and generally, the universities become the start point. Every day, a new academic spin-off company is established, bringing a particular

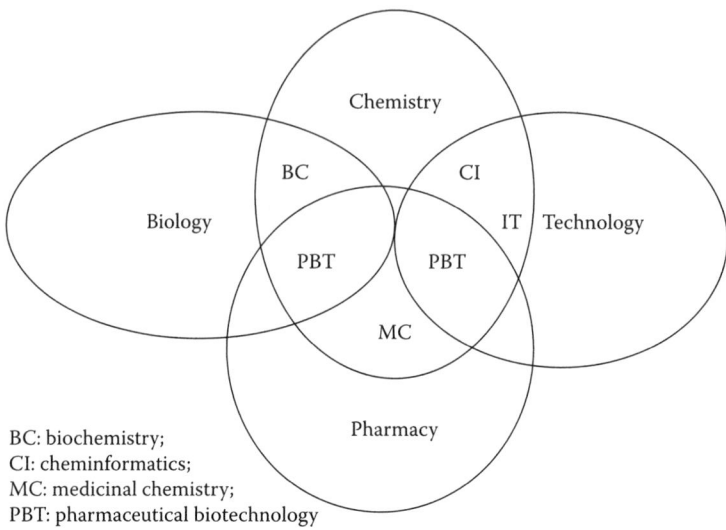

BC: biochemistry;
CI: cheminformatics;
MC: medicinal chemistry;
PBT: pharmaceutical biotechnology

**FIGURE 5.11** Bioinformatics as a multidisciplinary approach.

combination of skills that are required for bioinformatics (Table 5.7). Molecular Mining Co., Open Text Co., and Nanodesign, Inc. are just some of them. The latest one is a drug discovery company that designs virtual molecule libraries focused on a targeted therapeutic activity. Moreover, university–industry collaborations exist. Some of the bioinformatics companies are also listed in Table 5.7. Many new companies develop new or improved datamining and data management software. Collaborations in the field of bioinformatics are usually formed among software developer companies, academic researchers, and traditional pharmaceutical or biotechnological companies. Apart from pharmaceutical companies, hardware and software companies in Silicon Valley also have a great interest in bioinformatics. For instance, IBM made a $100 million investment in biotechnology. Also, Oracle, Hitachi, and Myriad invested in a collaborative project. Unfortunately, most of the small bioinformatics companies may not survive in the competitive marketplace. Presumably, in the next 5 or 10 years, parallel to the advent of bioinformatics, a new class of professionals will emerge, the *bioinformaticists*. They will be different from the biologists with some IT knowledge or some IT experts with some biology knowledge. Bioinformaticists will be expert on developing the software and using it as a tool after training in future bioinformatics departments of universities.

Other anticipated trends include

- Software will become more user friendly over time
- *In silico* testing is expected to blossom in the future
- The accuracy and the ability of DNA and protein microarray technologies to detect small differences will get better
- Computing speeds and storage capabilities will improve

**TABLE 5.7**
**Bioinformatics Companies**

| Company | Collaborator | Projects |
|---|---|---|
| Molecular Mining Co. | Queens University | Development of software to analyze and visualize gene expression data. |
| | | Building predictive models for *in silico* testing of drug candidates. |
| Nanodesign Inc. | University of Guelph | Software development to provide pharmaceutical candidate generation and optimization. |
| | | Design of molecular libraries for targeted therapeutic activity. |
| Open Text Co. | Waterloo University | Searching and managing text-based biological information. |
| Curagen | — | Identification and development of protein and antibody drugs. |
| | | Discovery of small molecule drug targets. |
| AxCell Biosciences (a subsidiary of Cytogen Co.) | — | Intercellular protein interactions $\rightarrow$ a complete characterization of one protein domain family. |
| Informax Inc. | — | Integrated software solutions for project management and analysis of available proteomic information. |
| Accelrys, Inc. | Oxford University | Software combining bioinformatics and cheminformatics. |
| Double Twist Inc. | Sun Microsystems Oracle | Software and Web-based tools for analyzing, annotating, managing, and mining information from a large number of biological and chemical sources. |

- Data mining tools will become more powerful and accurate
- The possibility of bench testing for screening potential pharmaceutical active substances for confirmation of the outcomes of silicon tests. This may shorten the time frame to reach clinical trials for potential drug candidates and allow a more systematic drug development process

In summary, the information collected by the use of bioinformatics will lead to understanding of complex regulatory and disease pathways and will be helpful to design new treatment strategies.

## DNA AND PROTEIN MICROARRAY TECHNOLOGIES

It is well known that the genome of an organism and its products (i.e., RNA and proteins) function in a complicated and well-harmonized way that is believed to be the key to the mystery of life. However, conventional molecular biology methods generally suggest an outcome that the throughput is very limited, and to determine the entire picture of the function of a gene is really difficult and challenging.

Multiple material transfer steps in genomics and proteomics (e.g., pipetting, chromatography, elution from gels) present a significant problem: "loss of peptides." This can be overcome with miniaturization (micro- or nanoscale analysis). Since miniature-scale technology offers high-sensitivity analytical work, during the last decade, scientists developed a new technology called *DNA microarray*. It has attracted tremendous interests with molecular biologists. It is basically a new protein separation technology and holds promise to monitor the whole genome on a single chip. This could allow researchers a better picture of the interactions among thousands of genes simultaneously and use this information for further protein analysis. Microarrays allow a very large number of experiments to be performed and analyzed automatically. Common characteristics of the chips used in microarray technology are that they are usually made of silicon and very similar to those used in microcircuits. Microarray technologies facilitate incorporation of electronic controls and of detectors into the devices. These devices allow us to conduct simultaneous processing and analysis of a number of samples at the same time. This approach leads to high throughput. Highly parallel implication of protein digestion, separation, and MS analysis can significantly increase the speed of proteome analyses. Microscale MS sources are used to coupled microscale peptide separation devices to MS analyzers. MALDI and ESI are used as ionization methods in prototypes for such sources and are highly sensitive.

A wide range of terminology has been used to describe this technology in the literature (e.g., biochip, DNA chip, DNA microarray, gene array). Some companies derived their own definition for their microarray technologies. For example, Affymetrix, Inc. (Santa Clara, CA) describes its own microarray technology as GeneChip®, which refers to its high-density, oligonucleotide-based DNA arrays. GeneChip is the most widely used terminology to describe DNA microarray technologies.

## DNA ARRAYS

Base pairing (i.e., A-T and G-C for DNA; A-U and G-C for RNA) or hybridization is the basic principle behind DNA microarray. An array is an orderly arrangement of samples that provides a medium to match known and unknown DNA samples based on base-pairing rules. Identification of the unknown samples is performed using automated tools. Common assay systems such as microplates or standard blotting membranes can be used for DNA array experiments. In these experiments, the samples can be deposited either manually or with robotics. In general, DNA arrays are described as *macro-* or *microarrays*. The only difference between them is the size of the sample spots. Macroarrays are comprised of minimum 300 microns of sample spots and can be easily imaged by existing gel and blot scanners. On the other hand, the size of the sample spots in microarray is maximum 200 microns in diameter. Microarrays contain thousands of spots and require specialized robotics and imaging equipment. The most significant drawback behind micrroarray technology is that these advanced instruments are not commercially available as a complete system.

DNA microarrays are produced using high-speed robotics on glass or nylon substrates to form probes with known identity. According to Phimister, "probe" stands for the tethered nucleic acid sample whose identity/abundance is being detected. These probes are usually used to determine complementary binding and allow scientists to conduct massive parallel gene expression and gene discovery studies. A DNA microarray experiment utilizing a single chip can provide information on thousands of genes simultaneously, which leads to an enormous increase in throughput.

Two different forms of the DNA microarray technology exist at present.

1. Probe cDNA which is 500~5,000 bases long, is immobilized onto a solid surface such as glass using robot spotting. Then, it is exposed to a set of probes either separately or in a mixture. This method was developed at Stanford University and is called "conventional DNA microarray."
2. An array of oligonucleotide which is composed of 20~80-mer oligos, or peptide nucleic acid probes, is synthesized either *in situ* (i.e., on-chip) or using conventional synthesis followed by on-chip immobilization. The resultant DNA array is then exposed to the labeled sample of DNA, hybridized, and the identity/abundance of complementary sequences determined. This method was developed at Affymetrix, Inc. and called "DNA chips." Today, oligonucleotide-based chips are manufactured by many companies using alternative *in situ* synthesis or depositioning technologies.

## Design of a Microarray System

A DNA microarray experiment can be designed and conducted with the following steps.

1. Determination of the DNA type to be used for probes in this experiment: Generally, cDNA/oligo with known identity is used.
2. Chip fabrication: Probes are placed on the chip by means of photolithography, pipetting, drop-touch, or piezoelectric (ink-jet).
3. Sample (target) preparation: The sample (RNA, cDNA) is fluorescently labeled.
4. Assay is carried out by means of hybridization, the addition of a long or short ligase or base, electricity, MS, electrophoresis, fluocytometry, or PCR.
5. Readout can be performed without using any probe (i.e., applying conductance, MS, electrophoresis) or by implication of fluorescence or electronic techniques.
6. Bioinformatics: Robotics control, image processing, data mining, and visualization are usually used for implementation of microarray experiments.

## Attachment of a Single DNA Molecule to a Silicon Surface

This is the most important step of microarray technology. Attachment of a single DNA molecule to a silicon surface can be done by either one of the following methods.

1. *Chemical attachment:* Chemicals usually specific to certain surfaces or sites are used for this purpose.

2. *Heat-aided attachment:* A more recent method that promises a more general and powerful method for depositing single DNA molecules onto specific locations of silicon chips. DNA arrays developed using this method are expected to provide a new approach for developing biosensors or bioelectronic circuitry. Shivashankar et al. were the first scientists to attach a single DNA molecule to a latex bead in water. Then, an "optical tweezer" (a focused laser beam) was used to trap the bead and hold the DNA molecule in place. Subsequently, AFM (atomic force microscope) tip was brought in contact with the bead. Meanwhile, the laser was left on to attach the tip of the probe to the bead. This method as given allows great manipulation flexibility, helps maintain the biological functionality of the DNA, and offers the possibility of studying DNA and protein interactions; therefore, it is more favored.

## How to Choose an Array

The choice of DNA array depends on cost, density, accuracy, and the type of DNA to be immobilized on the surface. The first criteria should be whether the chips contain immobilized cDNAs or shorter oligonucleotide sequences. The former must be spotted on the chips as complete molecules, but oligos can either be spotted or synthesized on the surface of a chip. The final criteria to select a chip should be whether the user makes or purchases the chip. Homemade systems offer limited number of spotting samples.

## PROTEIN ARRAYS

In proteomics, the corresponding match to DNA microarray is *protein array.* Although it sounds promising and easy to conduct, protein arrays present a major problem. DNA array technology is based on the hybridization of complementary sequences via Watson-Crick base pairing. However, hybridization does not take place between proteins and complementary sequences. Therefore, one-to-one matching between targets and probes that are the major requirement in DNA microarray, is not possible in proteomics. Protein arrays with different recognition elements are still being developed to be used for protein analysis by selective protein interactions.

Recognition elements can be nonselective or highly selective.

1. Nonselective recognition molecules
    (a) Ion exchange media: They bind proteins on the basis of charge under specific solution conditions
    (b) Immobilized metal affinity ligands: They recognize some protein functional groups such as phosphoserine, phosphothreonine residues
2. Highly selective molecules
    (a) Antibodies targeted to specific proteins
    (b) Nucleic acid aptamers: They are different oligonucleotide sequences for unique arrangements of hydrogen-binding donors and receptors in 3-D space. These sequences may specifically bind to specific protein structural motifs

The most well-known approach of using recognition elements in protein arrays was developed by Ciphergen Biosystems (Fremont, CA). This company offers a number of customized surface chips for protein capture to be used prior to MALDI-TOFF analysis.

The recognition elements can be used as:

1. Front end to capture proteins for MS (as used by Ciphergen)
2. Alternative approach to MS (non-MS detection): Arrays containing a number of different antibodies are used to capture a wide range of a collection of proteins targeted by antibodies. This approach allows researchers to conduct a screening process with high sensitivity and throughput for the specific proteins. The specificity and the affinity of antibodies, the influence of the antibody-attachment chemistry on antibody efficiency and the conditions under which antibody-protein target binding occurs, play the key role to achieve the goal in this approach.

Protein-specific arrays allow scientists to carry out parallel studies dealing with identification of protein–protein interactions and the influence of drugs, other chemical entities, and diseases on these interactions. Arrays printed on glass slides or multiwell plates are used to investigate protein–protein and protein–drug interactions. Subsequent analysis is conducted by mass spectrometry tools.

Various types of substances are bound to protein arrays. Some of them are antibodies, receptors, ligands, nucleic acids, carbohydrates, or chromatographic surfaces. Some of these surfaces possess a wide range of specificity and show nonselective binding to proteins (i.e., bind to entire proteome) whereas others possess selective binding characteristics. Nonselective binding is usually not favored and, therefore, the DNA array is washed to reduce this feature after the capture step. The retained proteins are uncoupled from the array surface upon application of laser light and then, analyzed utilizing laser disorption/ionization TOFF MS methods.

Some arrays used in proteomics contain antibodies covalently bound onto the array surface for immobilization. Then these antibodies capture corresponding antigens from a complex mixture. Afterwards, a series of analysis are carried out. For instance, bound receptors can reveal ligands. With this information in hand, binding domains for protein–protein interactions can be detected. The main problem in using microarray methods for proteomics is that protein molecules must show folding with the array in the correct conformation during the preparation and incubation. Otherwise, protein–protein interactions do not take place.

## APPLICATIONS OF DNA AND PROTEIN MICROARRAY TECHNOLOGY

DNA microarray technology can be applied for the following purposes.

- To identify sequence of a gene or a gene mutation
- To determine the expression level of genes
- To purify proteins
- To profile gene expression
- To profile protein interactions

**TABLE 5.8**
**Potential Applications of DNA and Protein Array Technologies**

| Field of Application | DNA Arrays | Protein Arrays |
|---|---|---|
| Diagnosis of diseases and/or pathological conditions | Current use is only for research | Significant use<br>No development yet |
| Pharmacogenomics | Current use is only for research | Significant use<br>No development yet |
| Screening | No development at present | Significant use<br>No development yet |
| Expression profiling | Significant use in research and drug discovery studies | Still in early stage of development |
| Toxicogenomics | Still in early stage of development | Significant use<br>No development yet |

DNA microarray technology has had a significant impact on genomics and proteomics study. It is widely believed that drug discovery and toxicological research will eventually benefit from the implication of DNA microarray technology. Also, DNA microarray devices called "microfluidics" were developed for diagnostic purposes near the end of the 1990s.

Design of DNA microarrays lead to significant improvement in the field of pharmacogenomics, which deals with the reasons behind individual differences in drug responses, toxicity, and multiple drug resistance and hence, aims to find correlations between therapeutic responses to drugs and the genetic profiles of patients.

Another important area for DNA chips is "toxicogenomics" where main goal is to find correlations between toxic responses to toxicants and changes in the genetic profiles of the objects exposed to such toxicants.

The major application areas of DNA and protein arrays are listed in Table 5.8. Although they offer promising applications in medicine and pharmacy, most of the fields of application are still being developed and some (diagnosis and pharmacogenomics) are available only for research. Among all of them, expression profiling has become the dominant application. Although the number of companies in this field is relatively small at present, it is estimated that the market share of expression profiling will be approx. US$434 million within a year as many companies have been entering the market. Currently, Affymetrix, Inc. is the leading company in this segment of the market.

Microarray technologies are expensive. Companies developed the most sophisticated and the most expensive systems. Recently, they have set up various marketing arrangements and collaborations between the industry and academia. For example, Affymetrics, Inc. developed a program called "AcademicAccess Program" which allows academicians to have access to Affymetrix's GeneChip technology with discount pricing. Companies also focus on developing cost-effective technologies with reasonable scale.

# PHARMACEUTICAL AND MEDICAL APPLICATIONS OF PROTEOMICS

Subsequent to the decoding of genomic information, the field of proteomics is booming since protein molecules, usually functional and structural units, represent essential targets for drug therapy. Techniques used in proteomics are complex and require state-of-the-art technologies to identify and determine interactions of protein molecules with themselves and other molecules (i.e., protein–protein, protein–drug interactions), to store and interpret proteome information to enlighten their structure and functions as may be useful for the development of new diagnostic kits and more efficient therapeutical approaches. It is believed that the databases created in proteomic studies will offer a global approach to investigate the chemical, physiological, biochemical, genetic, physical, architectural, biological properties, and function of proteome in health and disease.

Advances in complex proteome technologies accelerate site-specific drug development. Promising areas of research include: delineation of altered protein expression, not only at the whole-cell or tissue levels, but also in subcellular structures; in protein complexes and in biological fluids; the development of novel biomarkers for diagnosis and early detection of disease; and the identification of new targets for therapeutics; and the potential for accelerating drug discovery through more effective strategies to evaluate therapeutic effectiveness and toxicity.

Celis et al. (1994) have focused on establishing 2D proteomic databases useful for skin biology and in bladder cancer. They used noncultured cells such as keratinocytes (Celis et al., 1992) to study the former, and transitional and squamous cell carcinomas of the bladder (Celis et al. 1999) for the later. The human keratinocytes 2D-PAGE database is depicted at *http://biobase.dk/cgi-bin/celis*; it is the largest available database, listing 3629 cellular and 358 externalized polypeptides. The most significant benefit of this database is that it allows extensive search on proteomics with its large capacity. At present, approximately 100 information categories (e.g., protein name, localization, regulation, expression, post-translational modifications) are available in this database. It is possible to query the database searching by name, protein number, molecular weight, pI, organelle or cellular component, and to display its position on the 2-D image. This image can be compared to the master keratinocyte image. Nearly 100 polypeptides with known properties were stored as references. Again, the discovery of PCNA/cyclin, the first protein discovered by 2D gel technology, was carried out by Celis et al. It was found that PCNA/cyclin plays a significant role in DNA replication.

New precision technology, based on protein structures and function makes it possible for clinicians to detect cancer earlier than ever and provide individualized treatment. Following the discovery of new proteins and gene maps, cancer researchers believed that proteomics is a revolutionary approach to detect cancer and other major illnesses (e.g., HIV) during their early phases and to tailor individualized therapy.

Proteomics has the potential to revolutionize diagnosis and disease management. Profiling serum protein patterns by means of surface-enhanced laser desorption/ionisation time of flight (SELDI-TOFF) mass spectrometry is a novel approach to discover protein patterns that can be used to distinguish disease and disease-free states with high sensitivity and specificity. This method has shown great promise for early diagnosis of cancer (e.g., ovarian cancer).

The study of Cellis et al. (1999) on bladder cancer was performed with the aim of unraveling the molecular mechanism of underlying tumor progression. For this purpose, approximately 700 tumors have been analyzed by utilizing 2D gel technologies. Several biomarkers have been identified, assuming that they may be useful for classifying superficial lesions and determining the individuals at risk.

Moreover, Cellis et al. have been working on identifying protein markers that may be valuable for diagnosis and follow-up of bladder cancer patients. To achieve this goal, they have focused on the establishment of a comprehensive 2D gel database of urine proteins.

In 2001, both Lawrie et al. and Moskaluk coupled LCM (laser capture microdissection) with 2D gel proteomics to recover proteins from laser captured microdissected tissue in a form that can be used in 2D gel analysis and mass spectrometry. This may provide valuable information for protein profiling and databasing of human tissues in healthy and disease states.

Recombinant antibodies are used for analysis of a proteome of a cell. The quantitative detection can be performed using either Western blotting flow cytometry of labeled cells. Use of different antibodies, each one of which is directed toward a different cellular protein, can provide monitoring of protein levels. Microarray technology can be used to accelerate the process of monitoring. Facilitating such technology will offer parallel screening of selected antibodies against both diseased and nondiseased tissues. This will lead to the information on protein expression in disease states. The most significant benefit of microarray technology is that it enables the researcher to conduct an efficient proteomic study even when the sample size is limited in clinically derived tissues. Robotics and microfluidic networks have been used for this purpose along with microarray technologies (e.g., antibody array). This provides a better approach to localize antibodies and delivers small sample volumes to selected sites of the array.

Recently, polyclonal mouse antihuman IgG antibodies fluorescently labeled with Cy5 were used to detect the binding of human myeloma proteins to biotinylated monoclonal antibodies. Again, Cy5-labeled antibodies were used to determine bacterial, viral, and protein antigens bound to biotinylated IgGs.

Also, using DNA arrays for profiling gene expression has had a tremendous impact on pharmaceutical and medical research. Examples of disease-related applications include uncovering unsuspected associations between genes and specific clinical properties of disease. This will provide new molecular-based classifications of disease. As to cancer, most of the published data on tumor analysis by DNA microarrays have investigated pathologically homogeneous series of tumors for identification of clinically relevant subtypes (i.e., comparison of responders with nonresponders). These arrays have also examined pathologically distinct subtypes of tumors of the same lineage to identify molecular correlations or tumors of different lineages to identify molecular signatures for each lineage.

The best examples for the use of DNA microarray in cancer profiling are the studies on breast cancer to uncover new disease subtypes. Based on the data obtained in these studies, tumors leading to development of breast cancer have been classified into three distinct groups as a basal epithelial-like group, an ErbB2-overexpressing group, and a normal breast-like group. A later study on invasive breast cancers suggested a new classification for estrogen-receptor (ER)-negative breast cancer.

Although DNA microarray provides gene expression profiling in cancer, there is a significant need to identify many different features of proteins that are altered in the disease state. These include determination of their levels in biological samples and determination of their selective interactions with other biomolecules, such as other proteins, antibodies, or drugs. This led to the development of "protein chips" that can be used for assay of protein interactions. Protein chips or microarrays are being evaluated by scientists as a new way of tracking biological responses to therapy. Futhermore, development of tissue profiling at disease state, using protein microarrays is underway. Knezevic et al. studied protein expression in tissue derived from squamous cell carcinomas of the oral cavity through an antibody microarray. They used LCM for quantitative and qualitative analysis of protein expression patterns within epithelial cells. It can be correlated to the tumor progression in the oral cavity. Most of the proteins identified in both stromal cells surrounding and adjacent to regions of diseased epithelium were involved in signal transduction pathways. Therefore, they assumed that extensive molecular communications involving complex cellular signaling between epithelium and stroma play a key role in progression of the cancers in the oral cavity. Another important application of protein arrays is the identification of proteins which induce an antibody response in autoimmune disease (e.g., systemic lupus erythematosus).

Proteomics have been used to develop biomarkers. This is achieved by comparative analysis of protein expression in healthy and diseased tissues to identify expressed proteins to be used as new markers, by analysis of secreted proteins in cell lines and primary cultures, and by direct serum protein profiling. MALDI seems to be a good technique for direct protein analysis in biological fluids (e.g., the identification of the small proteins-defensin 1, 2 and 3 as related to the anti-HIV-1 activity of CD8 antiviral factor).

Increasing awareness of proteomics led proteomic researchers in academia gather to establish the Human Proteome Organization (*www.hupo.org*) with the leadership of Samir M. Hanash from the University of Michigan. The main goal of HUPO is to foster collaborative public proteome projects at the international level. These projects may offer useful outcome in the diagnosis, prognosis, and therapy of diseases. At first, member scientists agreed to concentrate mainly on the determination of proteins in blood serum, which may be useful for further studies on development of biomarkers.

Apart from academic organizations, the NCI (National Cancer Institute) and FDA (Food and Drug Administration) have initiated a collaboration to focus on using proteomic technologies to develop more targeted treatments and more reliable diagnostic kits for early detection of cancer. In mid-2001, researchers analyzed a series of tumor cells from different patients and came up with a roster of proteins present in cancer cells. These are distinct molecules not found in normal cells. In early 2002, scientists involved in this program reported that it is possible to compare the protein patterns of blood serums of ovarian cancer patients to those of the healthy volunteers. This could be a good step in development of biomarkers.

Standardized biomarkers for the detection of clinically significant immunological responses may be extremely valuable in immunotherapy. Most of the current assay techniques measure the frequency or function of antigen-specific T cells, or the titers of antibodies or immune complexes. Unfortunately, they posses inadequate sensitivity or too high a signal-to-noise ratio to reliably detect the low-frequency T-cell responses induced by cancer vaccines. Furthermore, these assay techniques indicate only one aspect of the immune response rather than the complete picture. On the contrary, proteomics offer a novel approach to immunological monitoring that complements existing immunological assays. Determination of the protein content of T cells responding to a vaccine or in the serum of vaccinated individuals may be helpful in developing quantitative methods to measure the magnitude of immunological responses. Yalin and Sahin have directed their recent study on the development of a vaccine for *H. pylori*–induced stomach cancer using phage display techniques to design recombinant antibodies and specific nonviral carriers (e.g., liposomes) to achieve immunotherapy.

Infectious diseases, still a leading cause of death worldwide, can also be cured by the aid of proteomics. The most significant obstacle in treatment of infectious disease is the development of drug resistance. This calls the need for developing effective new therapies. At this point, joint application of proteomics and microbiology may be valuable. Neidhardt characterized protein expression patterns in *E. coli* under different growth conditions. The identification of the complete sequence of a number of microbial genomes was helpful for identifying proteins encoded in these genomes using MS. The identification of new potential drug and vaccine targets against *Plasmodium falciparum*, which is the main cause of malaria, sets a good example for this kind of work. Proteomics can be used to enlighten the numerous significant aspects of microbial disease pathogenesis and treatment.

Beside medical applications, proteomics have enormous implications in the pharmaceutical industry. Most major pharmaceutical companies have implemented proteomics programs for drug discovery. The main reason behind this enormous interest is that most of the drugs target proteins. However, the pharmaceutical industry shows a cautious attitude as it is too early to reach a solid conclusion on the contributions of proteomics to drug development. It may be due to some uncertainty surrounding the adequacy and scalability of proteomics to meet the needs of the pharmaceutical industry. Nevertheless, it is expected that the use of proteomics may provide valuable information to be used in the progress of drug development, by identifying new targets and facilitating assessment of drug action and toxicity in preclinical and clinical phases.

Recent advances in genomic and proteomic research have exponentially increased the number of potential protein therapeutic molecules that may be used for unmet therapeutical needs. However, to fully understand the therapeutical potential of these substances, a parallel development should be made in protein delivery technologies. These new technologies presumably offer the ability to overcome biochemical and anatomical barriers to protein drug transport, without overcoming adverse events, to deliver the drug(s) at a favorable rate and duration, to protect therapeutic macromolecules from *in situ* or systemic degradation, as well as increase their therapeutic index by targeting the drug to a specific site of action.

Moreover, several studies in the literature illustrate the application of functional proteomics for identification of regulated targets. Identification of proteases that are most suitable for drug targeting, by means of an automated microtiter-plate assay, was modified for the detection of proteases in tissue samples: matrix metallopro-teases, cathepsins, and the cell serine proteases, tryptase, and chymase. Colorectal carcinoma biopsies representing primary tumor, adjacent normal colon and liver metastases were screened for protease activity. It was evident that matrix metallo-proteases were expressed at higher levels in the primary tumor than in adjacent normal tissue. On the other hand, the mast cell proteases were at very high levels in adjacent normal tissue. This type of activity-based screening studies provides a better way of selecting targets in the development of specific protease inhibitors. Protein chips can also be used for target identification. Proteomics also may provide increased efficiency of clinical trials through the availability of biologically relevant markers for drug efficacy and safety.

Toxicology is one of the most important applications of proteomics. 2D gel electrophoresis has been used for screening toxic agents and probing toxic mecha-nisms. Comparison of protein expression during a follow-up study may lead to identification of changes in biochemical pathways. After compiling a large number of proteomic libraries of the compounds with known toxicity, it may be possible to retrieve useful information to asses the toxicity of a novel compound before it enters clinical trials.

Toxicology-related proteomic projects involve identification of proprietary pro-teins associated with toxicological side effects. A few proteomics studies in toxicol-ogy have yet to be initialized. Wita-proteomics have focused on screening libraries on HEP G2 cells to identify novel proteins related to liver toxicity. In another ongoing study carried out by the Imperial College of London and GlaxoSmithKline, it was reported that by monitoring the proteins in urine of rats that had undergone glom-erular nephrotoxicity after an exposure to puramycin aminonucleoside, it was pos-sible to gain detailed understanding of the nature and progression of the protein-uria toxicity.

## CONCLUSIONS

The 1990s was devoted to the faster development of genomics. However, it seems that the first decade of the new millennium will become the "decade of proteomics." Recent advances in proteomic research have made it possible to generate quantitative data on protein expression that is compatible to those obtained at genomic level, with respect to sensitivity, specificity, and scale. Proteomics has begun yielding important findings that may advance major implications in the areas of pharmaceutics and medicine.

Ideal pharmaceuticals should be efficacious and target-specific. They should interact specifically with the target. One of the major requirements of drug devel-opment is the target validation. Drugs under investigation should inhibit the target without exerting any undesired effects. The most significant drawback of assays and studies in drug development and toxicology is that they depend on phenotypic observations. Molecular mechanisms of pharmacological and toxicological effects are not well understood. In order to gain more understanding at the molecular level, proteomics has emerged as a new technology. It provides deeper insight into the molecular processes of diseases as well. All this new information can improve decision making at critical stages in drug development. Analysis of proteomic data can enable us to detect and quantify hundreds of proteins in a single experiment and provides qualitative and quantitative assessments of changes in the protein synthesis between different tissues and cells with different phenotypes. Differences between healthy and diseased cells and/or tissues in the presence or in the absence of drug molecules can be detected. Also, response to different treatments or disease stages can be monitored, leading the study of the pharmacokinetics of the particular drug.

One of the important advances in this field is the ability to compile protein expression databases on cell types ranging from microbes to human cancer. It seems that proteomics will help to overcome many challenging obstacles in pharmaceutical biotechnology in the near future. Although 2D gel electrophoresis and MS are the driving forces behind this fast development, protein chips and antibody-based techniques will attract the interest of researchers the most. Instrumentation for MS analysis is evolving at an impressive rate. Nanosprays are underway. In addition to these developments, another noteworthy development is the application of MS as a "virtual imaging" approach for the analysis of protein distribution in cell and tissue samples. Now, it is possible to blot tissue slices onto a polymeric membrane, coat with a MALDI matrix, and use a series of MALDI analysis. Automation is also essential for large-scale analysis of proteomes.

Proteomics holds particular promise in the following fields

- Identification of disease-related markers (biosensors, diagnostic kits)
- Identification of proteins as potential candidates in the development of vaccine targets
- Identification of disease-related targets
- Evaluation of process and bioavailability of drugs (LADME)
- Evaluation of drug toxicity at tissue levels
- Validation of animal models
- Individualized drug design: pharmacoproteomics

Consequently, proteomics is a valuable tool for identification and validation of drug targets in early phases, the investigation of the mechanism of pharmacological drug activity and toxicity, and hence, individualized drug therapy. Use of biomarkers may lead to development of nonanimal models.

It seems that obstacles of proteomics can be overcome by using combined techniques. Thus, proteomics require a multidisciplinary approach for drug discovery—combining chemistry, biology, pharmacy, engineering, medicine, and information technology. More studies are still undergoing research as higher throughput and more predictable data must be obtained before drugs and diagnostics receive approval for clinical use. Unfortunately, it takes many years to develop safer, more efficient, and cost-effective drugs.

# REFERENCES AND FURTHER READING

Banks, R.E., Dunn M.J., Hochsstrasser D.F., et al. (2000). Proteomics, new perspectives, new biomedical opportunities. *Lancet,* 356, 1749–1756.

Benjamin, R.J., Cobbold, S.P., Clark, M.R., Waldmann, H. (1986). Tolerance to rat monoclonal antibodies, Implications for serotherapy. *J. Exp. Med.,* 163, 1539–1552.

Berman, J.E., Mellis, S.J., Pollock, R., et al. (1988). Content and organization of the human IgVH locus, definition of three new VH families and linkage to the Ig CH locus. *EMBO J.,* 7, 727–738.

Blackstock, W.P., Weir, M.P. (1999). Proteomics, quantitative and physical mapping of cellular proteins. *Trends. Biotech.,* 17, 121–127.

Boulianne, G.L., Hozumi, N., Schulman, M.J. (1984). Production of functional chimaeric mouse/human antibody region domains. *Nature,* 312, 643–646.

Bruggemann, M., Winter, G., Waldmann, H., Neuberger, M.S. (1989). The immunogenicity of chimeric antibodies. *J. Exp. Med.,* 170, 2153–2157.

Buluwela, L., Albertson, D.G., Sherrington, P., et al. (1988). The use of chromosomal trans-locations to study human immunoglobulin gene organization, mapping D-H segments within 35kb of the C-mu gene and identification of a new D-H locus. *EMBO J.,* 7, 2003–2010.

Burbaum, J., Tobal, G.M. (2002). Proteomics in drug discovery. *Curr. Opin. Biotechnol.,* 6, 427–433.

Canaan-Haden, L., Ayala, M., Fernandez-de-Cossio, M.E., Pedroso, I., Rodes, L., Govilondo, J.V. (1995). Purification and application a single-chain Fv antibody fragment specific to hepatitis B virus surface antigen. *Biotechniques,* 19(4), 606–608.

Carter, P., Presta, L., Gorman, C.M., et al. (1992). Humanization of an anti-p185Her2 antibody for human cancer-therapy. *Proc. Natl. Acad. Sci. USA,* 89, 4285–4289.

Cash, P. (2000a). Proteomics in medical microbiology. *Electrphoresis,* 21, 1187–1201.

Cash, P. (2002b). Proteomics, the protein revolution. *Biologist,* 40(2), 58–62.

Celis, J.E., Bravo, R., Larsen, P.M., Fey, S.J. (1984). Cyclin, a nuclear protein whose level correlates directly with the proliferative state of normal as well as transformed cells. *Leukemia Res.,* 8(2), 143–157.

Celis, J.E., Rasmusen, H.H., Madsen, P., et al. (1992). The human keratinocyte two-dimensional gel protein database, towards an integrated approach to the study of cell proliferation, differentiation, and skin diseases. *Electrophoresis,* 13(12), 893–959.

Celis, J.E., Celis, P., Ostegaard, M., et al. (1999). Proteomics and immunohistochemistry define some of the steps involved in the squamous differentiation of the bladder transitional epithelium, a novel strategy for identifying metaplastic lesions. *Cancer. Res.,* 59(12), 3003–3009.

Chen, P.P. (1990). Structural analysis of human developmentally regulated VH3 genes. *Scand. J. Immunol.,* 31, 257–267.

Chen, P.P., Liu M.F., Sinha S., et al. (1988). A 16/6 idiotype-positive anti-DNA antibody is encoded by a conserved VH gene with no somatic mutation. *Arthritis Rheum.,* 31, 1429–1431.

Chen, P.P., Siminovitch, K.A., Olsen, N.J., et al. (1989). A highly informative probe for two polymorphic VH gene regions that contain one or more autoantibody associated VH genes. *J. Clin. Invest.,* 84, 706–710.

Chothia, C., Lesk, A.M. (1987). Canonical structures for the hypervariable regions of immunoglobulins. *J. Biol. Chem.,* 196, 901–907.

Chothia, C., Lesk, A.M., Gherardi, E., et al. (1992). Structural repertoire of the human VH segments. *J. Mol. Biol.,* 227, 799–817.

Clark, M. (1995). General introduction. In: Birch J., Lennox, E.S., *Monoclonal Antibodies.* Hanser, Munich, Germany, 1–43.

Co, M.S., Deschamps, M., Whitley, R.J., et al. (1991). Humanized antibodies for antiviral therapy. *Proc. Natl. Acad. Sci. USA,* 88, 2869–2873.

Co, M.S., Avdalovic, N.M., Caron, P.C., et al. (1992). Chimeric and humanized antibodies with specificity for the CD33 antigen. *J. Immunol.,* 148, 1149–1154.

Cortese, J.D. (2000a). Array of options: Instrumentation to exploint the DNA microarray explosion. *Scientist,* 14(11), 26.

Cortese, J.D. (2000b). The array of today: Biomolecule arrays become the 21st century's test tube. *Scientist,* 14(17), 25.

Covacci, A., Kennedy, G.C., Cormack, B., Rappuoli, R., Falkow, S. (1997). From microbial genomics to meta-genomics. *Drug Dev. Res.,* 41, 180–192.

Davidson, H.W., Watts, C. (1989). Epitope directed processing of specific antigen by B-lymphocytes. *J. Cell. Biol.,* 109, 85–92.

Degen, W.G.J., Pieffers, M., Weli-Henriksson, E., van Venrooij, W.J., Raats J.M.H. (2000). Characterization of recombinant autoantibody fragments directed toward the autoantigenic U1-70K protein. *Eur. J. Immunol.,* 30, 3029–3038.

Dersimonian, H., Schwartz, R.S., Barrett, K.J., et al. (1987). Relationship of human variable region heavy chain germline genes to genes encoding anti-DNA autoantibodies. *J. Immunol.,* 139, 2496–2501.

DeRisi, J., Penland, L., Brown, P.O., et al. (1996). Use of a cDNA microarray to analyse gene expression patterns in human cancer. *Nat. Genet.,* 14(4), 457–460.

De Wildt, R.M.T., Finnern, R., Ouwehand, W.H., Griffiths, A.D., Van Venrooy, W.J., and Hoet, R.M.A. (1996). Characterization of human variable domain antibody fragments against the U1 RNA–associated A protein, selected from a synthetic and a patient derived combinatorial V gene library. *Eur. J. Immunol.,* 26, 629–639.

De Wildt, R.M.T., Steenbakkers, P.G., Pennings, A.H.M., van den Hoogen F.H.J., van Venrooij, W.J., Hoet, R.M.A. (1997a). A new method for analysis and production of monoclonal antibody fragments originating from single human B-cells. *J. Immunol. Methods.,* 207, 61–67.

De Wildt, R.M.T., Ruytenbeek, R., van Venrooij, W.J., Hoet. R.M.A. (1997b). Heavy chain CDR3 optimization of a germline encoded recombinant antibody fragment predisposed to bind to the U1A protein. *Protein Eng.,* 10, 835–843.

De Wildt, R.M.T., van den Hoogen, F.H.J., van Venrooij, W.J., Hoet, R.M.A. (1997c). Isolation and characterization of single anti U1A specific B cells from autoimmune patients. *Ann. N. Y. Acad. Sci.,* 815, 440–442.

De Wildt, R.M.T., Hoet, R.M.A., van venrooij, W.J., Tomlinson, I.M., Winter, G.. (1999a). Analysis of heavy and light chain pairings indicates that receptor editing shapes the human antibody repertoire. *J. Mol. Biol.,* 285, 895–901.

De Wildt, R.M., van Venrooij, W.J., Winter, G., Hoet, R.M., Tomlinson, I.M. (1999b). Somatic insertions and deletions shape the human antibody repertoire. *J. Mol. Biol.,* 294, 701–710.

De Wildt, R.M., Tomlinson, I.M., van Venrooij, W.J., Winter, G., Hoet, R.M.A. (2000a). Comparable heavy and light chain pairings in normal and systemic lupus erythematosus IgG(+) B cells. *Eur. J. Immunol.,* 30, 254–261.

De Wildt, R.M.T., Pruijn, G.J.M., Hoet, R.M.A., van Venrooij, W.J., Raats, J.M.H.. (2000b). Phage display as a tool to study the human autoantibody repertoire in systemic autoimmune diseases. In: Conrad, K. and Tan, E.M., eds., *Autoimmunity, Autoantigens, Autoantibodies,* 1, 306–326.

Dyer, M.J.S., Hale, G., Hayhoe, F.G.J., Waldmann, H. (1989). Effects of CAMPATH-1 antibodies in-vivo in patients with lymphoid malignancies, influence of antibody isotype. *Blood,* 73, 1431–1439.

Ekins, R., Chu, F.W. (1999). Microarrays, their origins and applications. *Trends Biotechnol.,* 17, 217–218.

Fritz, J., Baller, M.K., Lang, H.P., et al. (2000). Translating biomolecular recognition into nanomechanics. *Science,* 288(5464), 316–318.

Gerhold, D., Rushmore, T., Caskey, C.T. (1999). DNA chips, promising toys have become powerful tools. *Trends Biochem. Sci.,* 24(5), 168–73.

Gorman, S.D., Clark, M.R., Routledge, E.G., et al. (1991). Reshaping a therapeutic CD4 antibody. *Proc. Natl. Acad. Sci. USA,* 88, 4181–4185.

Gromov, P.S., Ostergaard, M., Gromova, I., Celis, J.E. (2002). Human proteomic databases, a powerful resource for functional genomics in health and disease. *Prog. Biophys. Mol. Biol.,* 80, 3–22.

Guillaume, T., Rubinstein, D.B., Young, F., et al. (1990). Individual VH genes detected with oligonucleotide probes from the complementarity-determining regions. *J. Immunol.,* 145, 1934–1945.

Hale, G., Dyer, M.J.S., Clark, M.R., et al. (1988). Remission induction in non-Hodgkin lymphoma with reshaped human monoclonal antibody CAMPATH-1H. *Lancet,* 2, 1394–1399.

Hanash, S. (2003). Disease proteomics. *Nature,* 422(6928), 226–232.

Harper, K., Ziegler, A. (1997). Applied antibody technology. *Trends Biotechnol.,* 15, 41–42.

Hoet, R.M.A., Raats, J.M.H., de Wildt, R.M.T., van Venrooij, W.J. (1995). Isolation and characterization of antibody fragments directed to disease markers isolated from patient derived combinatorial libraries. *Immunology,* 86(suppl 1), 96.

Hoet, R.M.A., van den Hoogen, F.H.J., Raats, J.M.H., Litjens, P.E.M.H., de Wildt, R.M.T., van Venrooij, W.J. (1996). Isolation and characterization of antibody fragments directed to disease markers isolated from patient derived combinatorial libraries (Abstract). Keynote Meeting, *Exploring and Exploiting Antibody and Ig Superfamily Combining Sites,* Taos, NM.

Hoet, R.M.A., Raats, J.M.H., de Wildt, R., et al. (1998). Human monoclonal autoantibody fragments from combinatorial antibody libraries directed to the U1snRNP associated U1C protein) epitope mapping, immunolocalization and V-gene usage. *Mol. Immunol.,* 35, 1045–1055.

Hoet, R.M., Pieffers, M., Stassen, M.H., et al. (1999). The importance of the light chain for the epitope specificity of human anti-U1 small nuclear RNA autoantibodies present in systemic lupus erythematosus patients. *J. Immunol.,* 163, 3304–3312.

Holt, L.J., Enever, C., de Wildt, R.M.T., Tomlinson, I.M. (2000). The use of recombinant antibodies in proteomics. *Curr. Opin. Biotechnol.,* 11, 445–449.

Hoogenboom, H.R., Marks, J.D., Griffiths, A.B., et al. (1992). Building antibodies from their genes. *Immunol. Rev.,* 130, 41–68.

Housman, D. and Ledley, F. (1998). Why pharmacogenomics? Why now? *Nat. Biotechnol.,* 16(6), 492–493.

Humphries, C.G., Shen, A., Kuziel, W.A., et al. (1988). A new human immunoglobulin VH family preferentially rearranged in immature B-cell tumours. *Nature,* 331, 446–449.

Huse, W.D., Sastry, L., Iverson, S.A., et al. (1989). Generation of a large combinatorial library of the immunoglobulin repertoire in phage lambda. *Science,* 246, 1275–1281.

Ideker, T., Thorsson, V., Ranish, J.A. (2001). Integrated genomic and proteomic analyses of a systematically perturbed metabolic network. *Science,* 292, 929–934.

Isaacs, J.D. (1990). The antiglobulin response to therapeutic antibodies. *Sem. Immunol.,* 2, 449–456.

Isaacs, J.D., Watts, R.A., Hazleman, B.L., et al. (1992). Humanized monoclonal antibody therapy for rheumatoid arthritis. *Lancet,* 340, 748–752.

Jenkins, R.E., Pennington, S.R. (2001). Arrays for protein expression profiling, towards a viable alternative to two-dimensional gel electrophoresis? *Proteomics,* 1(1), 13–29.

Johnson, M.T., Natali, A.M., Cann, H.M., et al. (1984). Polymorphisms of human variable heavy chain gene showing linkage with constant heavy chain genes. *Proc. Natl. Acad. Sci. USA,* 81, 7840–7844.

Jirholt, P., Ohlin, M., Borrebaeck, C.A.K., Söderlind, E. (1998). Exploiting sequence space: Shuffling in vivo rearranged CDR into a master framework. *Gene,* 215, 471–476.

Jirholt, P., Strandberg, L., Jansson, B., et al. (2001). A central core structure in an antibody variable domain determines antigen specificity. *Protein Eng.,* 4, 67–74.

Jones, P.T., Dear, P.H., Foote, J., et al. (1986). Replacing the complementarity–determining regions in a human antibody with those from a mouse. *Nature,* 321, 522–525.

Kabat, E.A., Wu, T.T., Perry, H.M., et al. (1991). Sequences of proteins of immunological interest. U.S. Department of Health and Human Services, U.S. Government Printing Office, Washington, DC.

Kane, M.D., Jatkoe, T.A., Stumpf, C.R., Lu, J., Thomas, J.D., Madore, S.J. (2000). Assessment of the sensitivity and specificity of oligonucleotide (50mer) microarrays. *Nucleic Acids Res.,* 28(22), 4552–7.

Kenten, J.H., Molgaard, H.V., Houghton, M., et al. (1982). Cloning and sequence determination of the gene for the human immunoglobulin e chain expressed in a myeloma cell line. *Proc. Natl. Acad. Sci. USA,* 79, 6661–6665.

Kettleborough, C.A., Saldanha, J., Heath, V.J., et al. (1991). Humanization of a mouse monoclonal-antibody by CDR-grafting—The importance of framework residues on loop conformation. *Protein Eng.,* 4, 773–783.

Khan, J., Saal, L.H., Bittner, M.L., Chen, Y., Trent, J.M., Meltzer, P.S. (1999). Expression profiling in cancer using cDNA microarrays. *Electrophoresis,* 20(2), 223–239.

Kishimoto, T., Okajima, H., Okumoto, T., et al. (1989). Nucleotide sequences of the cDNAs encoding the V regions of H and L chains of a human monoclonal antibody with broad reactivity to a malignant tumor. *Nucl. Acids Res.,* 17, 4385–4385.

Kodaira, M., Kinashi, T., Umemura, I., et al. (1986). Organization and evolution of variable region genes of the human immunoglobulin heavy chain. *J. Mol. Biol.,* 190, 529–541.

Köhler, G., Milstein, C. (1975). Continuous cultures of fused cells secreting antibody of predefined specificity. *Nature,* 256, 495–497.

Kricka, L. (1998). Revolution on a square centimeter. *Nat. Biotechnol.,* 16, 513.

Kueppers, R., Fischer, U., Rajewsky K., et al. (1992). Immunoglobulin heavy and light chain gene sequences of a human CD5 positive immunocytoma and sequences of four novel VHIII germline genes. *Immunol. Lett.,* 34, 57–62.

Lanzavechia, A. (1990). Receptor-mediated antigen uptake and its effect on antigen presentation to class-II restricted T-lymphocytes. *Ann. Rev. Immunol.,* 8, 773–793.

Lawrie, L.C., Curran, S., McLeod, H.L., Fothergill, J.E., Murray, G.I. (2001). Application of laser capture microdissection and proteomics in colon cancer. *Mol. Pathol.,* 54(4), 253–258.

Lee, K.H., Matsuda, F., Kinashi, T., et al. (1987). A novel family of variable region genes of the human immunoglobulin heavy chain. *J. Mol. Biol.,* 195, 761–768.

Liebler, D.C., Yates, J.R. (2002). *Introduction to Proteomics.* Humana Press, Totowa, NJ.

Lockhart, D.J., Winzeler, E.A. (2000). Genomics, gene expression and DNA arrays. *Nature,* 405(6788), 827–836.

van Loghem, E. (1986). Allotypic markers. *Monogr. Allergy,* 19, 40–51.

MacBeath, G., Schreiber, S.L. (2000). Printing proteins as microarrays for high-throughput function determination. *Science,* 289(5485), 1760–1763.

Marshall, A., Hodgson, J. (1998). DNA chips—An array of possibilities. *Nat. Biotechnol.,* 16(1), 27–31.

Marx, J.D., Hoogenboom, H.R., Bonnert, T.P., et al. (1991). By-passing immunization, human antibodies from V gene libraries displayed on phage. *J. Mol. Biol.,* 222, 581–597.

Marx, J. (2000). DNA arrays reveal cancer in its many forms. *Science,* 289, 1670–1672.

Mathieson, P.W., Cobbold, S.P., Hale, G., et al. (1990). Monoclonal antibody therapy in systemic vasculitis. *N. Engl. J. Med.,* 323, 250–254.

Matsuda, F., Shin, E.K., Hirabayashi, Y., et al. (1990). Organization of variable region segments of the human immunoglobulin heavy chain, duplication of the D5 cluster within the locus and interchromosomal translocation of variable region segments. *EMBO J.,* 9, 2501–2506.

Matthyssens, G., Rabbitts, T.H. (1980). Structure and multiplicity of genes for the human immunoglobulin heavy chain constant region. *Proc. Natl. Acad. Sci. USA,* 77, 6561–6565.

Middlemiss, N., Jonker, K., Jonker, A. (2001). Bioinformatics—Its role in advancing from gene to therapy. *Pharmaceut. Canada,* 2(4), 21–24.

Morrison, S.L., Johnson, M.J., Herzenberg, L.A., et al. (1984). Chimeric human antibody molecules, Mouse antigen–binding domains with human constant *Proc. Natl. Acad. Sci. USA,* 81, 6851–6855.

Moskaluk, C.A. (2001). Microdissection of histologic sections. Manual and laser capture microdissection techniques. In: Powel, S.M., ed., *Colorectal Cancer.* Humana Press, Totowa, NJ, 1–13.

Neuberger, M.S., Williams, G.T., Mitchell, E.B., et al. (1985). A hapten-specific chimeric immunoglobulin E antibody which exhibits human physiological effector function. *Nature,* 314, 268–271.

Needleman, S.B., Wunch, C.D. (1970). A general method applicable to the search for similarities in the amino acid sequence of two proteins. *J. Mol. Biol.,* 48, 443–453.

Neumann, E.K., Schachter, V. (2002). The informatics making sense of the genome, a progress report from the BioPathways Consortium. *Comp. Funct. Genom.,* 3, 115–118.

Ohlin, M., Jirholt, P., Thorsteinsdottir, H.B., Lantto, J., Lindroth, Y., Borrebaeck, C.A.K (1998). CDR–shuffling, targeting hyper-variable loops for library construction and selection. In: Talwar, G.P., Nath, I., Ganguly, N.K., Rao, K.V.S., eds., *Proceedings of the 10th International Congress of Immunology.* Monduzzi Editore, Bologna, Italy, 1525–1529.

Ohlin, M., Jirholt, P., Thorsteinsdottir, H.B., Borrebaeck, C.A.K. (2000). Understanding human immunoglobulin repertoires in vivo and evolving specificities in vitro. In: Zanetti, M. and Capra, J.D., eds. *The Antibodies,* vol. 6. Harwood Academic Publishers, Amsterdam, 81–104.

Ortho Multicentre Transplant Study Group. (1985). A randomized clinical trial of OKT3 monoclonal antibody for acute rejection of cadaveric renal transplants. *N. Engl. J. Med.,* 313, 337–342.

Padlan, E.A. (1991). A possible procedure for reducing the immunogenicity of antibody domains while preserving their ligand–binding properties. *Mol. Immunol.,* 28, 489–498.

Pascual, V., Capra, J.D. (1991). Human immunoglobulin heavy-chain variable region genes, organization, polymorphism, and expression. *Adv. Immunol.,* 49, 1–74.

Pennington, S.R., Dunn, M.J. (2001). *Proteomoics.* The Cromwell Press, Trowbridge, UK.

Petricoin, E.F., Zoon, K.C., Kohn, E.C., Barrett, J.C., Liotta, L.A. (2002). Clinical proteomics, translating benchside promise into bedside reality. *Nat. Rev. Drug Discov.,* 1, 683–695.

Pollack, J.R., Perou, C.M., Alizadeh, A.A., et al. (1999). Genome-wide analysis of DNA copy-number changes using cDNA microarrays. *Nat Genet.,* 23(1), 41–46.

Popkov, M., Lussier, I., Medvedkine, V., Esteve, P.O., Alakhov, V., Mandeville, R. (1998). Multidrug-resistance drug-binding peptides generated by using a phage display library. *Eur J Biochem.,* 251, 155–163.

Proudnikov, D., Timofeev, E., Mirzabekov, A. (1998). Immobilization of DNA in poly-acrylamide gel for the manufacture of DNA and DNA-oligonucleotide microchips. *Anal. Biochem.,* 259 (1), 34–41.

Queen, C., Schneider, W.P., Selick, H.E., et al. (1989). A humanized antibody that binds to the interleukin 2 receptor. *Proc. Natl. Acad. Sci. USA,* 86, 10029–10033.

Raaphorst, F.M., Timmers, E., Kenter, M.J.H., et al. (1992). Restricted utilization of germ-line VH3 genes and short diverse third complementarity–determining regions (CDR3) in human fetal B-lymphocyte immunoglobulin heavy-chain rearrangements. *Eur. J. Immunol.,* 22, 247–251.

Rain, J.C., Selig, L., De Reuse, H. (2001). The protein–protein interaction map of Heli-cobacter pylori. *Nature,* 409, 211–215.

Ramsay, G. (1998). DNA chips—States-of-the-art. *Nat. Biotechnol.,* 16(1), 40–44.

Rechavi, G., Bienz, B., Ram, D., et al. (1982). Organization and evolution of immuno-globulin V-H gene subgroups. *Proc. Natl. Acad. Sci. USA,* 79, 4405–4409.

Reichert, J. (2000). Chip-based optical detection of dna hybridization by means of nanobead labeling, *Anal. Chem.,* 72 (24), 6025–6029.

Reinke, V., Smith, H.E., Nance, J., et al. (2000). A global profile of germline gene expression in C. elegans. *Mol. Cell.,* 6(3), 605–616.

Riechmann, L., Clark, M., Waldmann, H., et al. (1988). Reshaping human antibodies for therapy. *Nature,* 332, 323–327.

Ross, D.T., Scherf, U., Eisen, M.B., et al. (2000). Systematic variation in gene expression patterns in human cancer cell lines. *Nat. Genet.,* 24(3), 227–235.

Routledge, E.G., Lloyd, I., Gorman, S.D., et al. (1991). A humanized monovalent CD3 antibody which can activate homologous complement. *Eur. J. Immunol.,* 21, 2717–2725.

Sahin, N.O., Yalin, S. (2002). Proteomics, a new emerging area of biotechnology. *Acta Pharmaceutica Turcica,* 44, 103.

Sanz, I., Kelly, P., Williams, C., et al. (1989a). The smaller human VH gene families display remarkably little polymorphism. *EMBO J.,* 8, 3741–3748.

Sanz, I., Casali, P., Thomas, J.W., et al. (1989b). Nucleotide sequences of eight human natural autoantibody VH regions reveals apparent restricted use of the VH families. *J. Immunol.,* 142, 4054–4061.

Sasso, E.H., Van Dijk, K.W., Milner, E.C.B. (1990). Prevalence and polymorphism of human VH3 genes. *J. Immunol.,* 145, 2751–2757.

Saul, F.A., Amzel, L.M., Poljak, R.J. (1978). Preliminary refinement and structural analysis of the Fab fragment from human immunoglobulin NEW at 2.0 Å resolution. *J. Biol. Chem.,* 253, 585–597.

Schena, M., Heller, R.A., Theriault, T.P., Konrad, K., Lachenmeier, E., Davis, R.W. (1998). Microarrays, biotechnology's discovery platform for functional genomics. *Trends Biotechnol.,* 16, 301–306.

Schena, M., Shalon, D., Davis, R.W., Brown, P.O. (1995). Quantitative monitoring of gene expression patterns with a complementary DNA microarray. *Science,* 270(5235), 467–470.

Scherf, U., Ross, D.T., Waltham, M., et al. (2000). A gene expression database for the molecular pharmacology of cancer. *Nat. Genet.,* 24(3), 236–244.

Schmidt, W.E., Jung, H.D., Palm, W., et al. (1983). Die primarstruktur des kristallisierbaren monokionalen immunoglobulins IgG1 KOL, I. *Hoppe-Seyler's Z. Physiol. Chem.,* 364, 713–747.

Schroeder, H.W., Hillson, J.L., Perlmutter, R.M. (1987). Early restriction of the human antibody repertoire. *Science,* 238, 791–793.

Schroeder, H.W., Wang, J.Y. (1990). Preferential utilization of conserved immunoglobulin heavy chain variable gene segments during human fetal life. *Proc. Natl. Acad. Sci. USA,* 87, 6146–6150.

Schroeder, H.W., Walter, M.A., Hofker, M.F., et al. (1988). Physical linkage of a human immunoglobulin heavy chain variable region gene segment to diversity and joining region elements. *Proc. Natl. Acad. Sci. USA,* 85, 8196–8200.

Seal, S.N., Hoet, R.M.A., Raats, J.M.H., Radic, M.Z. (2000). Analysis of autoimmune bone marrow by antibody phage display, Somatic mutations and CDR3 Arginines in anti–DNA γ and κ V genes. *Arthritis. Rheum.,* 43, 2132–2138.

Service, R.F. (1998a). Microchip arrays put DNA on the spot. *Science,* 282(5388), 396–399.

Service, R.F. (1998b). Coming soon, the pocket DNA sequencer. *Science,* 282(5388), 399–401.

Shoemaker, D.D., Schadt, E.E., Armour, C.D., et al. (2001). Experimental annotation of the human genome using microarray technology. *Nature,* 409(6822), 922–927.

Sinclair, B. (1999). Everything's great when it sits on a chip—A bright future for DNA arrays. *Scientist,* 13(11), 18–20.

Sowa, A., Kordai M.P., Cavanagh, D.R., et al. (2001). Isolation of a monoclonal antibody from a malaria patient-derived phage display library recognising the Block 2 region of *Plasmodium falciparum* merozoite surface protein-1. *Mol. Biochem. Parasitol.,* 112, 143–147.

Shalaby, M.R., Shepard, H.M., Presta, L., et al. (1992). Development of humanized bispecific antibodies reactive with cytotoxic lymphocytes and tumor cells over expressing the HER2 proto-oncogene. *J. Exp. Med.,* 175, 217–225.

Shearman, C.W., Pollock, D., White, G., et al. (1991). Construction, expression and characterization of humanized antibodies directed against the human alpha,beta-T-cell receptor. *J. Immunol.,* 147, 4366–4373.

Sorlie, T. (2001). Gene expression patterns of breast carcinomas distinguish tumor subclasses with clinical implications. *Proc. Natl Acad. Sci. USA,* 98, 10869–10874.

Souroujoun, M.C., Rubinstein, D.B., Schwartz, R.S., et al. (1989). Polymorphisms in human H chain V region genes from the VHIII gene family. *J. Immunol.,* 143, 706–711.

Söderlind, E., Ohlin, M., Carlsson, R. (1999). CDR implantation, a theme of recombination. *Immunotechnology,* 4, 279–285.

Söderlind, E., Strandberg, L., Jirholt, P., et al. (2000). Recombining germline-derived CDR sequences for creating diverse single-framework antibody libraries. *Nat. Biotechnol.,* 18, 852–856.

Taton, T.A., Mirkin, C.A., Letsinger, R.L. (2000). Scanometric DNA array detection with nanoparticle probes. *Science,* 289(5485), 1757–1760.

Tempest, P.R., Bremner, P., Lambert, M., et al. (1991). Reshaping a human monoclonal–antibody to inhibit human respiratory syncytial virus-infection in vivo. *Bio-Technol.,* 9, 266–271.

Teunissen, S.W.M., Stassen, M.H.W., Pruijn, G.J.M., van Venrooij, W.J., Hoet, R.M.A. (1998). Characterization of an anti–RNA recombinant autoantibody fragment (scFv) isolated from a phage display library and detailed analysis of its binding site on U1 snRNA. *RNA,* 4, 1124–1133.

Thistlethwaite, J.R., Cosimi, A.B., Delmonico, F.L., et al. (1984). Evolving use of OKT3 monoclonal antibody for treatment of renal allograft rejection. *Transplantation,* 38, 695–701.

Tomlinson, I.M., Walter, G., Marks, J.D., et al. (1992). The repertoire of human germline VH sequences reveals about fifty groups of VH segments with different hypervariable loops. *J. Mol. Biol.,* 227, 776–798.

Tramontano, A., Chothia, C., Lesk, A.M. (1990). Framework residue-71 is a major determinant of the position and conformation of the 2nd hypervariable region in the VH domains of immunoglobulins. *J. Mol. Biol.,* 215, 175–182.

Uetz, P., Giot, L., Cagney, G. (2000). A comprehensive analysis of protein–protein interactions in Saccharomyces cerevisiae. *Nature,* 403, 623–627.

Van Dijk, K.W., Schroeder, H.W., Perlmutter, R.M., et al. (1989). Heterogeneity in the human IgVH locus. *J. Immunol.,* 142, 2547–2554.

Van Kuppevelt, T.H., Dennissen, M.A.B.A., Veerkamp, J.H., Van Venrooij, W.J., Hoet, R.M.A. (1996). Anti-heparan sulfate antibody fragments derived from a phage display library. XVth FECTS Meeting, Munich, Germany.

Van Kuppevelt, T.H., Dennissen, M.A.B.A., van Venrooij, W.J., Hoet, R.M.A., Veerkamp J.H. (1998). Generation and application of type-specific anti-heparan sulfate antibodies using phage display technology. Further evidence for heparan sulfate heterogeneity in the kidney. *J. Biol. Chem.,* 273, 12960–12966.

Verhoeyen, M., Milstein, C., Winter, G. (1988). Reshaping human antibodies, Grafting an antilysozyme activity. *Science,* 239, 1534–1536.

Verhoeyen, M.E., Saunders, J.A., Broderick, E.L., et al. (1991). Reshaping human monoclonal-antibodies for imaging and therapy. *Dis. Markers,* 9, 197–203.

Wallace, R.W. (1997). DNA on a chip—Serving up the genome for diagnostics and research. *Mol. Med. Today,* 3, 384–389.

Walter, M.A., Surti, U., Hofker, M.H., et al. (1990). The physical organization of the human immunoglobulin heavy chain gene complex. *EMBO J,* 9, 3303–3313.

Walt, D.R. (2000). Bead-based fiber-optic arrays. *Science,* 287, 451–452.

Weng, N.P., Snyder, J.G., Yu-Lee, L.Y., et al. (1992). Polymorphism of human immunoglobulin VH4 germline genes. *Eur. J. Immunol.,* 22, 1075–1082.

Willems, P.M., Hoet, R.M., Huys, E.L., Raats, J.M., Mensink, E.J., Raymakers, R.A. (1998). Specific detection of myeloma plasma cells using anti-idiotypic single chain antibody fragments selected from a phage display library. *Leukemia,* 12, 1295–1302.

Wilkins, M.R., Williams, K.L., Appel, R.D., Hochstrasser, D.F. (1997). *Proteome Research, New Frontiers in Functional Genomics.* Springer, Heidelberg, Germany.

Woodle, E.S., Thistlethwaite, J.R., Jolliffe, L.K., et al. (1992). Humanized OKT3 antibodies—Successful transfer of immune modulating properties and idiotype expression. *J. Immunol.,* 148, 2756–2763.

Yalin, S., Yalin, A.E., Aksoy, K. (2002). Phage display. *Teknolojisi. Arsiv.,* 11, 15–29.

Yasui, H., Akahori, Y., Hirano, M., et al. (1989). Class switch from m to d is mediated by homologous recombination between sm and Sm sequences in human immunoglobulin gene loci. *Eur. J. Immunol.,* 19, 1399–1403.

# 6 Commonly Used Analytical Techniques for Biotechnology Products

*Wei Tang, Ph.D. and Kadriye Ciftci, Ph.D.*

## CONTENTS

# ANALYTICAL TOOLS FOR PROTEINS

Characterization of proteins and peptides has long been interesting to pharmaceutical scientists. As the progress of the Human Genome Project reveals the human genome sequence, the functional research of discovered genes has moved forward to the systematic analysis of the components that constitute biologic systems. And the most important components relevant to the biological activity of the cells are proteins. Since many biological reactions are greatly affected by proteins such as enzymes, receptors, and intrinsic or extrinsic proteins, the structural analysis of these proteins and the relationship of the structure and functionality are very important in peptide protein research. Several analytical techniques such as amino acid sequencing, x-ray diffraction, x-ray crystallography, circular dichroism (CD), Fourier transform infrared spectroscopy (FTIR), electrophoresis, and mass spectrometry (MS) have been used for this purpose. This chapter describes the methods of detection mentioned above and delves into only a fraction of the applications and variations of these techniques.

# SPECTROSCOPIC AND PHOTOMETRIC METHODS

## Ultraviolet-Visible (UV-VIS)

UV-VIS is one of the spectrophotometric methods that has been extensively used for the identification and quantitative analysis of proteins (Wetlaufer 1962). Spectrophotometry, in general, takes advantage of the quantum mechanical nature of chemical compounds to absorb, scatter, or modify the properties of a beam of electromagnetic radiation (photons) or induce changes in the molecular properties. The UV-VIS technique uses photons in the ultraviolet to visible regions, with wavelengths between 190 nm and 900 nm. Research has shown that proteins contain several chromophores that absorb light in the ultraviolet and infrared regions. Many of them also display fluorescence. The most important chromophores are the aromatic rings of Phe, Tyr, and Trp (Wang 1992). UV absorbance and fluorescence are useful probes of structure and structural changes of proteins. This is due to the fact that chromophores display shifted spectra upon increasing or decreasing polarity of their environment, with changes in wavelength of maximum absorbance (lambda max) and molar extinction coefficient possible. A relation between the level of absorption (absorbance) and beam wavelength can be determined. These spectra are unique for each molecule, since each molecule is unique in quantum structure.

Proteins in solution absorb ultraviolet light with absorbance max at 280 and 200 nm (Wang 1992). Amino acids with aromatic rings are the primary reason for the absorbance peak at 280 nm. Peptide bonds are primarily responsible for the peak at

200 nm. Secondary, tertiary, and quaternary structure all affect absorbance, therefore factors such as pH, ionic strength, and so forth can alter the absorbance spectrum.

Several types of spectrophotometers are available for the UV-VIS range of wavelengths, all of which are single beam- or double beam type-instruments. A single beam spectrophotometer, as the name implies, uses a single beam to adjust the transmittance of a blank solution to 100%. This calibration is exercised prior to measurement at each wavelength. A transmittance is a measure of photonic transmission. At 100% transmittance, the molecule absorbs and then de-excites to release the photon. A more commonly used spectrophotometer is the double beam type. This type of detector uses two beams from a single photon source. One beam passes through a reference cell, and another the sample cell. Measurements of absorbance are taken simultaneously for both cells, eliminating the readjustment of the transmittance at each wavelength. The recording of spectral scans is also simplified (Wetlaufer 1962; Wang 1992).

Some colored compounds have also been linked to peptides or proteins for the detection and measurement within visible spectra ranges. Colorimetric methods are relatively nonspecific but they can be made specific by linking them to enzymatic or immunologic reactions (see Immunoassay section). In these cases, enzymes produce colored products from colorless substrates, rather than radionuclides, are used to label particular antigens and antibodies (Winder and Gent 1971).

## Infrared Spectroscopy

Infrared spectroscopy measures the frequency or wavelength of light absorbed by the molecules caused by transitions in vibrational energy levels (Vollhardt and Schore 1995). The wavelengths of IR absorption bands are characteristic of specific types of chemical bonds. In the past infrared had little application in protein analysis due to instrumentation and interpretation limitations. The development of Fourier transform infrared spectroscopy (FTIR) makes it possible to characterize proteins using IR techniques (Surewicz et al. 1993). Several IR absorption regions are important for protein analysis. The amide I groups in proteins have a vibration absorption frequency of $1630-1670$ cm$^{-1}$. Secondary structures of proteins such as alpha($\alpha$)-helix and beta($\beta$)-sheet have amide absorptions of $1645-1660$ cm$^{-1}$ and $1665-1680$ cm$^{-1}$, respectively. Random coil has absorptions in the range of $1660-1670$ cm$^{-1}$. These characterization criteria come from studies of model polypeptides with known secondary structures. Thus, FTIR is useful in conformational analysis of peptides and proteins (Arrondo et al. 1993).

Through the advanced development of IR instruments, measurements can also be performed in aqueous solutions, organic solvents, detergent micelles as well as in phospholipid membranes. In a study carried out by Vigano et al. (2000), proteins in live cells were characterized by FTIR, and, since it did not introduce perturbing probes, the structure of lipids and proteins were simultaneously studied in intact biological membranes. Information on the secondary structure of peptides can be derived from the analysis of the strong amide I band. Orientation of secondary structural elements within a lipid bilayer matrix can be determined by means of polarized attenuated total reflectance-FTIR spectroscopy (ATR-FTIR). This new technique is widely used in analysis of the structures of proteins from biological materials that cannot be studied by x-ray crystallography and NMR. The principle of ATR-FTIR is based on the measurement of change of the orientation of dipole from the change of electric fields applied (Vigano et al. 2000). FTIR spectroscopy is a sensitive tool to identify $\alpha$-helical and $\beta$-sheet structures and changes in the structure on interaction with charged lipids. Polarized IR spectroscopy reveals that the antiparallel -sheet structures oriented parallel to the membrane surface.

Besides, with the synthesis of peptides becoming increasingly popular nowadays, FTIR spectroscopy has been used to analyze the structure of synthetic peptides corresponding to functionally/structurally important regions of large proteins. The change in secondary structures caused by aggregation can also be detected by FTIR (Haris and Chapman 1995).

## Raman Spectroscopy

Raman spectroscopy is a related vibrational spectroscopic method. It has a different mechanism and therefore can provide complementary information to infrared absorption for the peptide protein conformational structure determination and some multicomponent qualitative and/or quantitative analysis (Alix et al. 1985).

Raman spectroscopy measures the wavelength and intensity of inelastically scattered monochromatic radiation light from molecules. The Raman scattered light occurs at wavelengths that are shifted from the incident light by the energies of molecular vibrations. The spectra result from change in polarizability during the molecular vibrations. The most common light source in Raman spectroscopy is an Ar-ion laser. Raman spectroscopy requires tunable radiation and sources are Ar-ion-laser-pumped dye lasers, or high-repetition-rate excimer-laser-pumped pulsed dye lasers. Raman scattering is a weak process, therefore, the spectrometer must provide a high rejection of scattered laser light in order to get the spectra. New methods such as very narrow rejection filters and Fourier transform techniques are becoming more widespread.

In the literature Raman spectroscopy has been used to characterize protein secondary structure using reference intensity profile method (Alix et al. 1985). A set of 17 proteins was studied with this method and results of characterization of secondary structures were compared to the results obtained by x-ray crystallography methods. Deconvolution of the Raman Amide I band, 1630–1700 $cm^{-1}$, was made to quantitatively analyze structures of proteins. This method was used on a reference set of 17 proteins, and the results show fairly good correlations between the two methods (Alix et al. 1985).

## Atomic Absorption (AA) Spectrophotometry

Atomic absorption spectrophotometry was developed in the 1950s by Dr. Alan Walsh.[1] The instrumentation of this method is shown in Figure 6.1. In general, chemical compounds are converted into their atomic constituents, and then the light absorption at a wavelength characteristic of a particular atomic species is determined

---

[1] http://www.chem.uwa.edu.au/enrolled_students/MAST_sect2/sect2.5.2.html

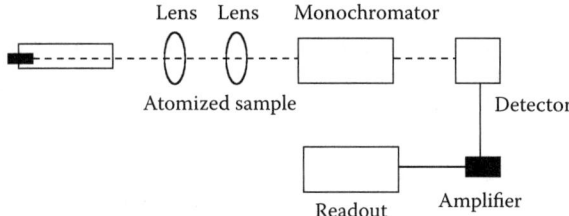

**FIGURE 6.1** The instrumentation of atomic absorption spectrophotometry.

and related to the concentration of the element in the sample from which it originated. The method to convert the analyte into its atomic form and a sensitive method for detecting light absorption over a very narrow wavelength range require sophisticated instrument design and experimental procedures. It exploits the narrowness of atomic absorption lines to avoid the necessity to separate a complex mixture prior to the analysis of its components.

It is a method of elemental analysis for various practical reasons and it is essentially suitable for analysis only of metals. A large number of elements can be analyzed for at trace levels. Therefore, its biotechnologic applications mainly involve in measurement of inorganic elements such as alkali, trace, and heavy metals in biological investigations. It is also used in industry to monitor contaminating inorganic elements in bioreactors fermentation process and preparation of culture media.

## Mass Spectrometry (MS)

The scope of the use of mass spectrometry in the protein analysis has grown enormously in the past few decades. MS has become an important analytical tool in biological and biochemical research. Its speed, accuracy and sensitivity are unmatched by conventional analytical techniques. The variety of ionization methods permits the analysis of peptide or protein molecules from below 500 Da to as big as 300 Da (Biemann 1990; Lahm and Langen 2000). Basically, a mass spectrometer is an instrument that produces ions and separates them in the gas phase according to their mass-to-charge ratio (m/z). The basic principle of operation is to introduce sample to volatilization and ionization source, and then the molecular fragments from the ionization of the sample are detected by various kinds of detector and the data are analyzed with computer software.

There are many different ionization techniques available to produce charged molecules in the gas phase, ranging from simple electron (impact) ionization (EI) and chemical ionization (CI) to a variety of desorption ionization techniques with acronyms such as fast atom bombardment (FAB), plasma desorption (PD), electrospray (ES), and matrix-assisted laser desorption ionization (MALDI) (Mano and Goto 2003).

In the early 1980s, ionization techniques such as FAB, PD, and thermospray (TSP) made it possible to use MS in analysis of high-mass macromolecules since the production of gas phase ions from charged and polar compounds can be done without prior chemical derivatization. FAB is a soft ionization technique that performs well for polar and thermally labile compounds. FABMS can analyze polypeptides up to 3–3.5 kDa and can extend the capability to 10–15 kDa with improvement of the instrument (Beranova-Giorgianni and Desiderio 1997). FABMS and liquid secondary ion mass spectrometry (LSIMS) techniques are useful in characterization of disulfide bonds and the glycosylation sites in proteins, which are very difficult to do by other techniques (Monegier et al. 1991).

Over the last decade or so, new ionization techniques such as ES and MALDI have been introduced and have increased still further the use of mass spectrometry in biology. Identification of proteins and characterization of their primary structure is a rapidly growing field in the postgenomic era. ES ionization was the first method to extend the useful mass range of instruments to well over 50 kDa (Mano and Goto 2003). The sample is usually dissolved in a mixture of water and organic solvent, commonly methanol, isopropanol, or acetonitrile. It can be directly infused, or injected into a continuous flow of this mixture, or be contained in the effluent of an HPLC column or CE capillary. First introduced in late 1980s, MALDI is a soft ionization technique that allows the analysis of intact molecules of high masses. It allows determination of the molecular mass of macromolecules such as peptides and proteins more than 300 kDa in size.

Matrix-assisted laser desorption ionization-time of flight (MALDI-TOFF) mass spectrometry, which was first invented about a decade ago, has become, in recent years, a tool of choice for large molecule analyses (Leushner 2001). In the past, the mass of larger biomolecules (>1–2k Da) could only be approximately determined using gel electrophoresis or gel permeation chromatography. The MALDI soft ionization technique coupled with a TOFF detector measuring flight time of the charged molecules in applied electric field make it possible to determine the exact mass of large protein molecules (Kevin 2000). This platform is ideal for analysis of protein and nucleic acid sequence, structure and purity. MALDI-TOFF is the method of choice for quality assurance in oligonucleotides and peptide synthesis (Leushner 2001). MALDI-TOFF MS is a very fast and sensitive technique, implemented on small, relatively inexpensive instruments that do not require extensive expertise in mass spectrometry. Such instruments are ideally suited for biological scientists who need molecular mass information more quickly and more accurately than can be obtained by gel electrophoresis.

MS is the most important tool to study post-translational modifications including partial proteolytic hydrolysis, glycosylation, acylation, phosphorylation, cross-linking through disulfide bridges, etc of the proteins (Jonsson 2001). These modifications usually result in the functional complexity of proteins.

Characterization of noncovalent bonding of the proteins can also be done using MS. For example MALDI MS has been used in measurement of the molecular mass of the noncovalently linked tetramer of glucose isomerase, a complex consisting of identical monomers of 43.1 kDa each. MALDI-TOFF peptide mass fingerprinting combined with electrospray tandem mass spectrometry can efficiently solve many complicated peptide protein analysis problems.

Affinity capture-release electrospray ionization mass spectrometry (ACESIMS) and isotope-coded affinity tags (ICAT) are two recently introduced techniques for the quantitation of protein activity and content with applications to clinical enzymology and functional proteomics, respectively. Another common feature of the ACESIMS and ICAT approaches is that both use conjugates labeled with stable heavy isotopes as internal standards for quantitation (Turecek 2002).

Phosphorylation on serine, threonine, and tyrosine residues is an extremely important modulator of protein function. Phosphorylation can be analyzed by mass spectrometry with enrichment of compounds of interest using immobilized metal affinity chromatography and chemical tagging techniques, detection of phosphopeptides using mass mapping and precursor ion scans, localization of phosphorylation sites by peptide sequencing, and quantitation of phosphorylation by the introduction of mass tags (McLachlin and Chait 2001).

Recent developments in instrument design have led to lower limits of detection, while new ion activation techniques and improved understanding of gas-phase ion chemistry have enhanced the capabilities of tandem mass spectrometry for peptide and protein structure elucidation. Future developments must address the understanding of protein–protein interactions and the characterization of the dynamic proteome (Chalmers and Gaskell 2000).

New instrumentation for the analysis of the proteome has been developed including a MALDI hybrid quadrupole time of flight instrument which combines advantages of the mass finger printing and peptide sequencing methods for protein identification (Andersen and Mann 2000).

Electrospray in the mid 1980s revolutionized biological mass spectrometry, in particular in the field of protein and peptide sequence analysis. Electrospray is a concentration-dependent, rather than a mass-dependent process, and maximum sensitivity is achieved at low flow rates with high-concentration, low-volume samples (Griffiths 2000). Joint NMR, x-ray diffraction, electrophoresis, and chromatography techniques with mass spectrometry (MS) techniques would be a trend in the future.

## NUCLEAR MAGNETIC RESONANCE SPECTROSCOPY (NMR)

NMR analysis allows characterization of proteins to an atomic level. The most frequently used nuclei on protein NMR are $^1H$, $^2H$, $^{13}C$, $^{15}N$, and $^{17}O$ with proton NMR (Jefson, 1988). The use of NMR methods for protein sequence and conformational studies was limited to the small proteins or peptides because high magnetic fields were required but not widely available to study larger molecules and it was very time consuming with the capability of instruments in the past.

As the recent development of the high-field superconducting magnet systems and improvement in computer hardware and software, the use of 2-dimensional or 2D-NMR techniques becomes feasible and has made it possible to analyze larger protein molecules (Nagayama 1981). Furthermore, the line broadening effect from large proteins and the nonisotropic motion of proteins in the lipid matrix will affect the resulting chromatogram, therefore, the identification of structures. Solid-state NMR spectroscopy can be used to analyze structures of proteins or peptides interact with lipid membranes (Bechinger 1999). Detailed structure analysis of these oligosaccharides conjugated to the proteins can be accomplished by high field NMR, providing anomeric configurations and linkage information.

## ELECTRON SPIN RESONANCE

Electron spin resonance (ESR) spectroscopy has been used for more than 50 years to study a variety of paramagnetic species. ESR or electron paramagnetic resonance (EPR) spectroscopy measures the presence of free radicals in solution and also gains information about the environment surrounding the free radical species (Sealy 1985). The electron spin resonance spectrum of a free radical or coordination complex with one unpaired electron is the simplest of all forms of spectroscopy. The degeneracy of the electron spin states characterized by the quantum number, $m_S = \pm 1/2$, is lifted by the application of a magnetic field and transitions between the spin levels are induced by radiation of the appropriate frequency. The integrated intensity in the spectrum is proportional to the radical concentration from both unpaired and free electrons. An unpaired electron interacting with its environment also affects the details of ESR spectra. Several studies investigated the role of certain metal ions in enzyme mechanisms (Pellegrini and Mierke 1999; Hoffman, 1991). These studies are further enhanced by the technique known as electron nuclear double resonance (ENDOR) of metalloenzymes. An ENDOR experiment provides an NMR spectrum of those nuclei that interact with the electron spin of a paramagnetic center. ESR has its limitation in characterization of proteins that are of interest as biotechnology products. It can only analyze proteins that contain a paramagnetic metal ion or other source of unshared electrons. Therefore, ESR spectroscopy is useful to characterize enzymes that require a paramagnetic metal for its catalytic activity. The information about the presence of the metal and its oxidation state can be obtained through ESR.

## OPTICAL ROTARY DISPERSION AND CIRCULAR DICHROISM SPECTROSCOPY

Circular dichroism (CD) is a spectroscopic analytical technique used for conformational analysis of peptides and proteins (Johnson 1988). It uses the principles of chirality and absorption; specifically the different absorption profiles demonstrated by a system for left as opposed to right circularly polarized light. For a system to exhibit CD activity, it must contain a chiral (asymmetric) center that is linked in some way to the chromophore responsible for the absorption.

Rotation of the polarization plane (or the axes of the dichroic ellipse) by a small angle $a$ occurs when the phases for the two circular components become different, which requires a difference in the refractive index $n$ (Pearlman and Nguyen 1991). This effect is called circular birefringence. The change of optical rotation with wavelength is called optical rotary dispersion (ORD).

Biologically active molecules often have both requirements. Disulfide groups, amide groups in the peptide backbone, and aromatic amino acids provide the UV absorption required to acquire a CD spectrum (Ettinger and Timasheff 1971). The units are measured in $cm^2/dmol$. Two major benefits to using this technique are that it only requires a small amount of protein and it takes less that 30 minutes to process the sample. The application of circular dichroism with regard to the analysis of peptides and proteins focuses on determining secondary structure: $\alpha$-helices and $\beta$-sheets, and gives some information regarding changes in tertiary structure (Pearlman and Nguyen 1991; Ettinger and Timasheff 1971a,b). The far UV region (190–250 nm) usually corresponds to the secondary structure and changes in the near UV region are considered to correspond to changes in the tertiary structure. α-Helices have a distinct positive shoulder ~175 nm, with a crossover ~170 nm and a negative peak below 170 nm. β-Sheets have a crossover at ~185 nm and a strong negative peak between 170 and 180 nm (Pearlman and Nguyen 1991; Ettinger and Timasheff 1971a).

Further information is available from the vacuum ultraviolet region (VUV, below 190 nm), although measurement in this region is difficult with most conventional lab CD equipment due to the low intensity of the light source and the high absorption of the sample, buffer, and solvent. Synchrotron radiation CD has been used to determine protein spectra in aqueous solutions to wavelengths as low as 160 nm; it is hoped that this will find a useful application in structural genomics. CD spectra are sensitive to changes in temperature, pH, solvent, and ionic strength among others. With optimized conditions, CD spectra can serve as unique identifiers for biological molecules (Kenney and Arakawa 1997).

## Fluorescence Spectroscopy

### Chemiluminescence and Bioluminescence

Colorimetric assays are commonly used in molecular biology and biotechnology laboratories for determining protein concentrations because the procedures and their instrumentation requirements are simple. Two forms of assays are used. The first involves reactions between the protein and a suitable chemical to yield a colored, fluorescent, or chemiluminescence product. Second, a colored dye is bound to the protein and the absorbance shift is observed. Disadvantages of both these methods include limited sensitivity at below 1 μg/mL, interferences from buffers, and unstable chromophores (Jain et al. 1992).

Luminescence spectroscopy involves three related optical methods: fluorescence, phosphorescence, and chemiluminescene. These methods utilize excited molecules of an analyte to give a species whose emission spectrum can provide information about the molecule. In fluorescence, atoms can be excited to a higher energy level by the absorption of photons of radiation. Some features of luminescence methods are increased sensitivity (in the order of three magnitudes smaller than absorption spectroscopy), larger linear range of concentration, and method selectivity (Parsons 1982).

Fluorescence determinations are important to analyze cysteine, guanidine, proteins, (LSD), steroids, a number of enzymes and coenzymes, and some vitamins, as well as several hundred more substances. A fluorometer can be used to verify conformational changes in multipartite operator recognition by λ-repressor as explained in a journal article by Deb et al. (2000). Upon titration with single operators site, the tryptophan fluorescence quenches to different degrees, suggesting different conformations of the DNA-protein complexes.

Phosphorimetric methods have been used to determine such substances as nucleic acids, amino acids, and enzymes. However, this is not a widely used method since it cannot be run at room temperature. Measurements are usually performed with liquid nitrogen to prevent degradation due to collision deactivation. Fluorometric methods are used to determine both inorganic and organic species. Instruments used for measuring fluorescence and phosphorescence are fluorometers and spectrofluorometers, respectively. These instruments are similar to ultraviolet and visible spectrometers, except for some differences in both the source and detector. The source must be more intense than the deuterium or tungsten lamps and therefore usually use either mercury or xenon arc lamps or lasers to excite the source. The detector signal usually requires amplification due to the low intensity of the fluorescent signal.

Chemiluminescence chemical reactions are found in several biological systems in which light is emitted from the excited species as it returns to the ground state.

Often, treatment of samples with fluorescence labeling agent reacts with primary and secondary amines to give a fluorescent compound. This is especially important for detecting amino acids in protein hydrolyzates. Fluorescence detectors may also be integrated into a high performance liquid chromatographic (HPLC) system.

## X-Ray Crystallography

X-ray crystallographic analysis is a spectroscopic experimental method utilized for the three-dimensional analysis of proteins. This technique involves an x-ray beam bombarding a crystalline lattice in a given orientation and the measurement of the resulting defraction pattern and intensity which is attributable to the atomic structure of the lattice (Palmer and Niwa 2003). It is important to note that in order to resolve the structure of a macromolecule using x-rays, it must be crystallized and the crystals must be singular. Molecules with highly hydrophobic portions such as the transmembrane portions of proteins are very difficult to crystallize, although crystals may be obtained in some cases using detergents.

An x-ray analysis will measure the diffraction pattern (positions and intensities) and the phases of the waves that formed each spot in the pattern. These parameters combined result in a three-dimensional image of the electron clouds of the molecule, known as an electron density map. A molecular model of the sequence of amino acids, which must be previously identified, is fitted to the electron density map and a series of refinements are performed. A complication arises if disorder or thermal motion exist in areas of the protein crystal; this makes it difficult or impossible to discern the three-dimensional structure (Perczel et al. 2003).

Time-resolved x-ray crystallography (TC) is a more recent advanced application of x-ray crystallography. It uses an intense synchrotron x-ray source and data collection methods to reduce crystallographic exposure times. This allows multiple exposures to be taken over time at near-physiological, crystalline conditions to determine the structures of intermediates. A typical problem with this method is that the existence of the intermediates is brief, resulting in difficulty in interpreting the resulting electron density maps.

## Dynamic Light Scattering

Dynamic light scattering (DLS), also called photon correlation spectroscopy (PCS) or quasi-elastic light scattering (QELS), can be used for comparing the stability of different formulations, including monitoring of changes at increasing temperatures (Martindale et al. 1982). Upon degradation many proteins start to form aggregates. The appearance of these aggregates may take several months. DLS is used to detect the precursory aggregates too small to be detected by the naked eye. Researchers must know the rate at which proteins start to degrade and experiment with various solutions in order to find the formulation that has the longest shelf life. However, if the particulates can be detected and identified early then formation may be prevented or minimized (Tsai et al. 1993).

DLS uses the scattered light from the protein to measure the rate of diffusion of the protein particles. The instrument, DynaPro MS/X measures the time dependence of the light scattered from a small region of solution over a short time range. The instrument is based on Brownian motion. Brownian motion is the observed movement of small particles that are randomly bombarded by the molecules of the surrounding medium. The changes in the intensity of the light are related to the rate of diffusion of the molecules in and out of that small region of solution. The information is plotted in a graph of the size of the particle radius (nm) vs. intensity (%) (Kadima et al. 1993). One of the largest shortcomings of this technique is that it is difficult to quantitate the amount of aggregates. Therefore, it is used primarily for the comparison of different formulations.

## OTHER METHODS

### Differential Scanning Calorimetry (DSC)

Differential scanning calorimetry is primarily used to determine changes in proteins as a function of temperature. The instrument used is a thermal analysis system, for example a Mettler DSC model 821e. The instrument coupled with a computer can quickly provide a thermal analysis of the protein solution and a control solution (no protein). The instrument contains two pans with separate heaters underneath each pan, one for the protein solution and one for the control solution that contains no protein. Each pan is heated at a predetermined equal rate. The pan with the protein will take more heat to keep the temperature of this pan increasing at the same rate of the control pan. The DSC instrument determines the amount of heat (energy) the sample pan heater has to put out to keep the rates equal. The computer graphs the temperature as a function of the difference in heat output from both pans. Through a series of equations, the heat capacity ($C_p$) can be determined (Freire 1995).

As the temperature of a protein solution is increased the protein will begin to denature. Various protein formulations can be compared to see which has the greatest resistance to temperature. The formulation may vary by testing different agents to stabilize the protein and inhibit denaturation. Creveld et al. (2001) used DSC to determine the stability of *Fusarium solani pisi* cutinase (an enzyme used in laundry detergent) when used in conjunction with surfactants. They concluded that sodium taurodeoxycholate (TDOC) stabilized the unfolding state of the enzyme and in return lowered the unfolding temperature and made the unfolding reversible.

## CHROMATOGRAPHIC METHODS

Chromatography is a broad range of techniques that study the separation of molecules based on differences in their structure and/or composition. In general, chromatography involves a sample being dissolved in a mobile phase (which may be a gas, a liquid, or a supercritical fluid). The mobile phase is then forced through an immobile, immiscible stationary phase. The molecules in the sample will have different interactions with the stationary support leading to separation of similar molecules. Test molecules, which display tighter interactions with the support will tend to move more slowly through the support than those molecules with weaker interactions. In this way, different types of molecules can be separated from each other as they move over the support material.

Based on the mechanism of the separation, several different types of chromatography (partition, adsorption, ion exchange [IEC], and size exclusion [SEC]) are available to biotechnologists for verifying protein stability. These techniques are summarized below.

## Adsorption Chromatography

Adsorption chromatography relies on the different affinity of components of a mixture for a liquid moving phase and a solid stationary phase. The separation mechanism depends upon differences in polarity between the different feed components. The more polar a molecule, the more strongly it will be adsorbed by a polar stationary phase (Varki et al. 1999). Similarly, the more nonpolar a molecule, the more strongly it will be adsorbed by nonpolar stationary phase. It is often employed for relatively nonpolar, hydrophobic materials so that the solvent tends to be nonpolar while the stationary phase is polar. Proteins have a high affinity to polar chromatographic media and the recovery of the sample is usually difficult. Therefore, this method is not commonly used to purify and characterize proteins.

Recently, a new developed adsorption chromatography named expanded bed adsorption chromatography has been proved a very powerful technique for capture of proteins directly from unclarified crude sample. Ollivier and Wallet (19XX) described the successful high-yield purification of a recombinant therapeutic protein in compliance with current Good Manufacturing Practice (cGMP) using expanded bed adsorption chromatography.

## Gas-Liquid Chromatography (GLC)

Gas-liquid chromatography is a chromatographic technique that can be used to separate volatile organic compounds (48). As shown in Figure 6.2, a gas chromatograph consists of a flowing mobile phase, an injection port, a separation column containing the stationary phase, and a detector. The organic compounds are separated due to differences in their partitioning behavior between the mobile gas phase and the stationary phase in the column. Gas-liquid chromatography makes use of a pressurized gas cylinder and a carrier gas, such as helium, to carry the solute through the column. In this type of column, an inert porous solid is coated with a viscous liquid, which acts as the stationary phase. Diatomaceous earth is the most common solid used. Solutes in the feed stream dissolve into the liquid phase and eventually vaporize. The separation is thus based on relative volatilities. The most common detectors used in this type of chromatography are thermal conductivity and flame ionization detectors. In the latter, the compounds are ionized in a flame, the ions are collected on electrodes, and the quantity of ions collected is determined by the

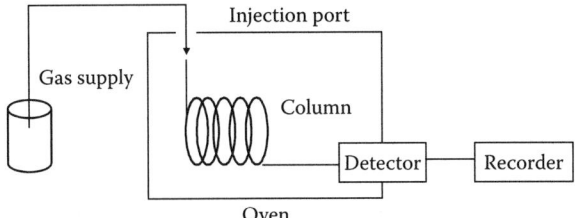

**FIGURE 6.2** Schematic of a gas chromatograph (GC). Mobile phases are generally inert gases such as helium, argon, or nitrogen. The injection port consists of a rubber septum through which a syringe needle is inserted to inject the sample. The injection port is maintained at a higher temperature than the boiling point of the least volatile component in the sample mixture. Since the partitioning behavior is dependant on temperature, the separation column is usually contained in a thermostat-controlled oven. Separating components with a wide range of boiling points is accomplished by starting at a low oven temperature and increasing the temperature over time to elute the high-boiling point components. Most columns contain a liquid stationary phase on a solid support. Separation of low-molecular weight gases is accomplished with solid adsorbents. Separate documents describe some specific GC columns and GC detectors.

perturbation of the voltage in the electrode circuit. The chromatogram consists of peaks relative to time after injection of the sample onto the column. Peak areas (in arbitrary units) are usually determined with an integrator. Personal computer programs are available for online identification and quantitation of compounds.

Gas-liquid chromatography used for the determination of C-terminal amino acids and C-terminal amino acid sequences in nanomolar amounts of proteins was described in 1976 by Davy and Morris. Based on carboxypeptidase A digestion of the protein, the partially digested protein was removed and the amino acids released after known time intervals were analyzed by quantitative gas-liquid chromatography. Sequences deduced from the kinetics of release of specific amino acids are compared with the known C-terminal sequences of well-characterized proteins. Thus the amino acid sequences were determined.

## PARTITION CHROMATOGRAPHY

The basis of this chromatography is the partition of solutes between two immiscible liquid phases, one stationary and the other mobile (Figure 6.3). In general, if two phases are in contact with one another and one phase contains a solute, the solute will distribute itself between the two phases according to its physical and chemical properties, which is called the partition coefficient (the ratio of the concentrations of the solute in the two phases). With mobile liquid phases, there is a tendency for the stationary liquid phase to be stripped or dissolved (Muller 1990). Therefore, the stationary liquid phase has to be chemically bonded to the solid bonding support. In partition chromatography, the stationary liquid phase is coated onto a solid support such as silica gel, cellulose powder, or hydrated silica. Assuming that there is no adsorption by the solid support, the feed components move through the system at

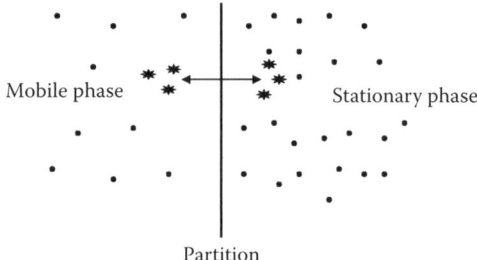

FIGURE 6.3 Schematical diagram of the principle of partition chromatography.

rates determined by their relative solubility in the stationary and mobile phases. In general, it is not necessary for the stationary and mobile phases to be totally immiscible, but a low degree of mutual solubility is desirable. Hydrophilic stationary phase liquids are generally used in conjunction with hydrophobic mobile phases (referred to as "normal-phase chromatography") (Abbott 1980), or vice versa (referred to as a "reverse-phase chromatography") (Smith 1967). Suitable hydrophilic mobile phases include water, aqueous buffers, and alcohols. Hydrophobic mobile phases include hydrocarbons in combination with ethers, esters, and chlorinated solvents. Partition chromatography is used primarily for proteins of small molecular weight.

Partition chromatography is of great importance for ascertaining the sequence of amino-acid residues in the peptide chains of proteins. If a peptide chain is partially degraded to dipeptide and tripeptide fragments, and so forth it should be possible, by identifying these, to recognize the original compound from which they are derived. Sanger and colleagues (1959) elucidated by partition-chromatographic methods what may be the entire peptide sequences in the structure of ox insulin, the minimum molecule of which embodies 51 amino-acid residues.

## ION EXCHANGE CHROMATOGRAPHY

Ion exchange chromatography is applicable to charged solutes which can be separated on the basis of their strength of binding of oppositely charged groups presented on the stationary phase (Figure 6.4). It is a special type of adsorption chromatography in which the adsorption is very strong (Gold 1997). Charged molecules adsorb to ion exchangers readily and can be eluted by changing the ionic environment.

Ion exchange chromatography is commonly used in the purification of biological materials. There are two types of exchange: cation exchange in which the stationary phase carries a negative charge, and anion exchange in which the stationary phase carries a positive charge. Charged molecules in the liquid phase pass through the column until a binding site in the stationary phase appears. The molecule will not elute from the column until a solution of varying pH or ionic strength is passed through it. The degree of affinity between the stationary phase and feed ions dictates the rate of migration and hence degree of separation between the different solute species. The most widely used type of stationary phase is a synthetic copolymer of styrene and divinyl benzene (DVB), produced as very small beads in the micrometer range. Careful

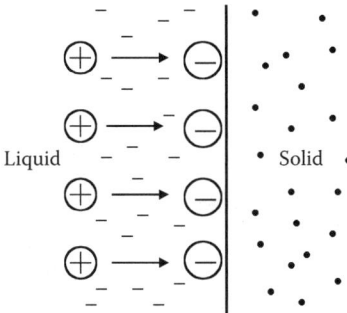

**FIGURE 6.4** Schematic illustration of the principle of ion exchange chromatography.

control over the amount of DVB added dictates the degree of cross-linking and hence the porosity of the resinous structure (Yang and Regnier 1991). Resins with a low degree of cross-linking have large pores that allow the diffusion of large ions into the resin beads and facilitate rapid ion exchange. Highly cross-linked resins have pores of sizes similar to those of small ions. The choice of a particular resin depends on a given application. Cation or anion exchange properties can be introduced by chemical modification of the resin.

Ion exchange chromatography is probably the most widely used large-scale chromatographic process. Separation of proteins by this method is highly selective since the resins are fairly inexpensive and high capacities can be used. It has widespread uses in industrial processes. Ion-exchange chromatography is an extremely common protein purification technique. The majority of proteins are separated using anion, rather than cation IEC, probably because most proteins have an acidic pI. De et al. (2002) used anion IEC to separate a lethal neurotoxic protein from cobra venom. In their study, proteins were loaded onto the CM-Sephadex column at around pH 7 and eluted by increasing the amount of salt. Changing the pH could elute the targeted proteins. Thus the toxin was isolated and purified.

## SIZE-EXCLUSION CHROMATOGRAPHY (SEC)

Size-exclusion chromatography (SEC) also known as gel permeation chromatography, uses porous particles to separate molecules of different sizes or molecular weight. The use of SEC and IEC is well suited for use with biologically active proteins since each protein has its own unique structure and the techniques may be performed in physiological conditions (Liu et al. 2002). The retention of components is based on the size in solution. The largest molecules are excluded from the stationary phase pores and elute earlier in the chromatogram. Molecules that are smaller than the pore size can enter the particles and therefore have a longer path and longer time than larger molecules that cannot enter the particles (Barth et al. 1994) (Figure 6.5). The stationary phase of SEC consists of a porous cross-linked polymeric gel. The pores of the gel vary in size and shape such that large molecules tend to be excluded by the smaller pores and move preferentially with the mobile

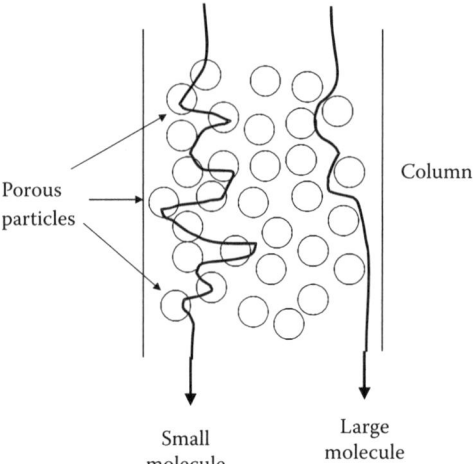

**FIGURE 6.5** Schematic of a size-exclusion chromatography column. Exclusion chromatography separates molecules on the basis of size. A column is filled with semisolid beads of a polymeric gel that will admit ions and small molecules into their interior but not large ones. When a mixture of molecules and ions dissolved in a solvent is applied to the top of the column, the smaller molecules (and ions) are distributed through a larger volume of solvent than is available to the large molecules. Consequently, the large molecules move more rapidly through the column, and in this way the mixture can be separated (fractionated) into its components.

phase (Beattie 1998). The smaller molecules are able to diffuse into and out of the smaller pores and will thus be slowed down by the system. The components of a mixture therefore elute in order of decreasing size or molecular weight. The stationary phase gels can either be hydrophilic for separations in aqueous or polar solvents, or hydrophobic for use with nonpolar or weakly polar solvents. Sephadex® (Sigma Aldrich Co, St. Loius, MO) a cross-linked polysaccharide material available in bead form, is widely used with polar/hydrophilic mobile phases. The degree of cross-linking can be varied to produce beads with a range of pore sizes to fractionate samples over different molecular weight ranges. Hydrophobic gels are made by cross-linking polystyrene with divinyl benzene (DVB) and are therefore similar to ion exchange resins but without the ionic groups.

The principal feature of SEC is its gentle noninteraction with the sample, enabling high retention of biomolecular enzymatic activity while separating multimers that are not easily distinguished by other chromatographic methods. SEC is used extensively in the biochemical industry to remove small molecules and inorganic salts from valuable higher molecular weight products such as peptides, proteins and, enzymes. SEC is also used to determine the molecular weight and molecular weight distribution of a number of polymers such as polycarbonate, polyurethane and organopolysiloxanes.

Size-exclusion chromatography can be used to analyze protein–protein interactions. Bloustine et al. (2003) presented a method to determine second virial coefficients (B2) of protein solutions from retention time measurements in size-exclusion chromatography. B2 was determined by analyzing the concentration dependence of the chromatographic partition coefficient. This method was able to track the evolution of B2 from positive to negative values in lysozyme and bovine serum albumin solutions. The size-exclusion chromatography results agree quantitatively with data obtained by light scattering.

## AFFINITY CHROMATOGRAPHY

Affinity chromatography involves the use of packing that has been chemically modified by attaching a compound with a specific affinity for the desired molecules, primarily biological compounds (Wilchek and Chaiken 2000). The packing material used, called the affinity matrix, must be inert and easily modified. Agarose is the most common substance used, despite its cost. The ligands that are inserted into the matrix can be genetically engineered to possess a specific affinity. Successful separation by affinity chromatography requires that a biospecific ligand is available and that it can be covalently attached to a matrix. It is important that the biospecific ligand (antibody, enzyme, or receptor protein) retains its specific binding affinity for the substance of interest (antigen, substrate, or hormone) (Kent 1999) (Figure 6.6).

Affinity chromatography is widely used as a means of separation and purification with specific properties. It represents one of the most effective methods for the purification of proteins as well as many other molecules. For example, Loog et al. (2000) used affinity ligands, which consist of ATP-resembling part coupled with specificity determining peptide fragment, to purify protein kinases. Affinity sorbents, based on two closely similar ligands AdoC-Aoc-Arg4-Lys and AdoC-Aoc-Arg4-NH(CH2)6NH2, were synthesized and tested for purification of recombinant protein kinase A catalytic subunit directly from crude cell extract. Elution of the enzyme

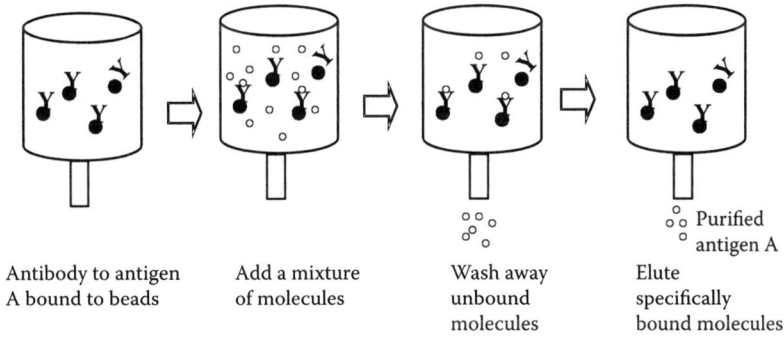

| Antibody to antigen A bound to beads | Add a mixture of molecules | Wash away unbound molecules | Elute specifically bound molecules |

**FIGURE 6.6** Affinity chromatography uses antigen–antibody binding to purify antigens or antibodies. To purify a specific antigen from a complex mixture of molecules, a monoclonal antibody is attached to an insoluble matrix, such as chromatography beads, and the mixture of molecules is passed over the matrix. The specific antibody binds the antigen of interest; other molecules are washed away. Altering the pH, which can usually disrupt antibody–antigen bonds, then elutes specific antigen. Antibodies can be purified in the same way on beads coupled to antigen.

with MgATP as well as L-arginine yielded homogeneous protein kinase A preparation in a single purification step. The affinity ligand was highly selective. Protein kinase with acidic specificity determinant (CK2) as well as other proteins possessing nucleotide binding site (L-type pyruvate kinase) or sites for wide variety of different ligands (bovine serum albumin) did not bind to the column.

Purification of antibodies from animal sera is another common use of affinity chromatography. Sun et al. (2003) produced a peptide affinity column by employing intein-mediated protein ligation (IPL) in conjunction with chitin affinity chromatography. Peptide epitopes possessing an N-terminal cysteine were ligated to the chitin bound CBD tag. The resulting peptide columns permit the highly specific and efficient affinity purification of antibodies from animal sera.

## HIGH-PERFORMANCE LIQUID CHROMATOGRAPHY

The application of HPLC to biomolecules began in the mid-1970s and is preferred because of its speed and resolution. Since the native protein (bioactive) molecule has a well-defined tertiary or quaternary structure, during method validation the HPLC conditions are dependent on the protein conformation (Figure 6.7). Varying HPLC conditions such as the HPLC method, mobile phase composition, the nature of the packing material, flow rate, and separation temperature are all critical to the analysis of the proteins stability. Changes in conformation that occur during or after separation usually result in an unsatisfactory separation. The ionic charge of the peptide molecule has an important role in the separation and can be controlled by varying mobile phase pH. pKa values of side chains and terminal amino acids may be altered in the native protein. The perseverance of the native conformation (and bioactivity) is the primary

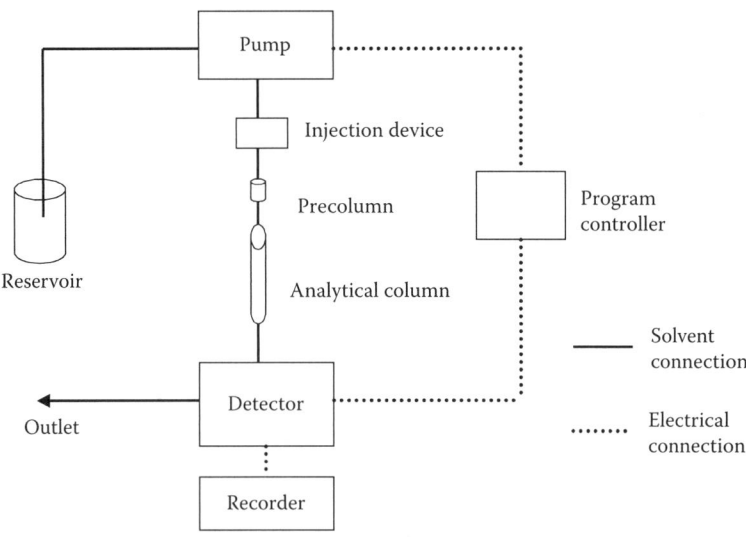

**FIGURE 6.7** A block diagram of the arrangement of the components in HPLC.

consideration in method development. The following are examples of how bioactivity can be preserved: restricted pH range of the mobile phase, limited concentration of organic solvent, low temperatures, ionic strengths, presence or absence of metal ions, substrates, or products, detergents, and lastly, range of the bulk protein. Reasons for poor recovery can be attributed to partial denaturation and irreversible attachment to particle matrix. The detectors utilized by chromagraphic systems are ultraviolet absorption, infrared absorption, fluorometry and LC coupled with mass spectrometry as well as others. Fluorescence detection can be more sensitive and selective but is used less universally (Fukushima et al. 2003).

HPLC was used by Lu and Chang (2001) to identify conformational isomers of mPrP(23-231). The data produced from the study indicated that the reduced form of mouse prion protein was able to exist as at least four diverse isoforms and were able to be separated by HPLC. More importantly, this technique faciliated the isolation of the isomers and confirmation of the protein molecule.

## ELECTROPHORESIS

Electrophoresis is a commonly used technique in the analysis of peptides and proteins. This technique refers to the migration of charged particles when dissolved or suspended in an electrolyte through which an electric current is passed. Cations migrate toward the negatively charged electrode (cathode), while anions are attracted toward the positively charged electrode (anode). Neutral particles are not attracted toward either electrode. The migration observed for proteins is dependent on its size, shape, electrical charge, and molecular weight, as well as characteristics and operating parameters of the system. Parameters of the system include the pH, ionic strength, viscosity and temperature of the electrolyte, density or cross-linking of any stabilizing matrix such as a gel, and the voltage gradient employed. The rate of migration is directly related to the magnitude of the net charge on the particle and is inversely related to the size of the particle, which in turn is directly related to its molecular weight (Goldenberg and Creighton 1984; West et al. 1984).

Common types of electrophoresis include gel electrophoresis, isoelectric focusing and capillary methods (Strege and Lagu 1993). In gel electrophoresis, analytical processes employ a gel such as agar, starch, or polyacrylamide as a stabilizing medium. The method is particularly advantageous for protein separations because of its high resolving power. The separation obtained depends upon the electrical charge to size ratio coupled with a molecular sieving effect dependent primarily on the molecular weight.

One of the most widely used forms of gel electrophoresis is known as sodium dodecyl sulfate-polyacrylamide gel electrophoresis (SDS-PAGE). Polyacrylamide gel has several advantages that account for its extensive use. It has minimal adsorptive properties and produces a negligible electroosmotic effect (Hjelmeland and Chrambach 1981). In identity tests, for the determination of molecular weight, SDS-PAGE has been shown to be an appropriate, fast, and easy method that is often used in quality control laboratories. The use of SDS-PAGE followed by a densitometric analysis, such as MS, is a helpful technique for the determination of peptide or protein mass (Westbrook et al. 2001). Blots (Southern, Northern, and Western) are an extension of SDS-PAGE and are especially useful in biotechnology. Among them Western blots are used for proteins rather than nucleic acids. In this method, following electrophoresis the resolved proteins are transferred onto a nitrocellulose or polyvinylidene difloride membrane and reacted with an antibody. The complex then can be visualized using an enzymatically labeled or radioactively labeled antibody as explained in immonassay section.

As the focus of the pharmaceutical industry turns toward the functional use of proteins, the use of two-dimensional (2D) gel electrophoresis followed by mass spectrometry has become the norm for the quantitative measurement of gene expression at the protein level (Davy and Morris 1976). Recent developments in this field aim to enhance the staining technology and the recovery of peptides from gel digests for mass spectrometric identification. Advances in this area include the use of a novel, ruthenium-based fluorescent dye, SYPRO Ruby Protein Gel Stain, for the detection of proteins in SDS-PAGE. Since the linear dynamic range of SYPRO Ruby Protein is orders of magnitude greater than conventional stains, quantitative differences, especially in low-abundance proteins are easier to detect (Nishihara and Champion 2002). This improved capability is of significant importance in differential protein expression studies where a small increase or decrease in the abundance of a protein isoform might allow the early detection of disease.

Isoelectric focusing separates proteins according to their respective isoelectric points (pI, the pH at which proteins have no charge). It can also be used to test the stability of a protein since the deamidation leads to the production of a new carboxylic acid group, resulting in a shift of pI toward the acidic side (Welinder 1971; Kosecki 1988).

Another electrophoretic method that has become quite popular recently is capillary electrophoresis (CE). The use of capillaries as a migration channel in electrophoresis has enabled analysts to perform electrophoretic separations on an instrumental level comparable to that of high-performance liquid chromatography (Strege and Lagu 1993; Rabel and Stobaugh 1993). CE-based analytical procedures have advantages such as high separation efficiency, short run-time, instrumentation simplicity, minimum operation costs, and compatibility with small sample volumes. And as with 2D gel electrophoresis, the combination of a CE separation system with densitometric analysis has proved to be a useful combination. Researchers observed that the coupling of CE to matrix-assisted laser desorption/ionization TOFF (time of flight flow) mass spectrometry (MALDI-TOFF) greatly increases the total number of identifiable peptides when compared to MALDI-TOFF techniques alone (Rubakhin et al. 1993). Even in the dairy industry, where the composition and integrity of proteins or peptide-derived proteins is important in determining a products value, CE proves to a central analytic technique as it provides rapid, high resolution separation and quantification of proteins and peptides without a need for prior separation (Righetti 2001).

During typical CE operation with an uncoated capillary filled with an operating buffer, silanol groups on the inner wall of the glass capillary release hydrogen ions to the buffer and the wall surface becomes negatively charged, even at a fairly low pH. Solutes having partial positive charges in the medium are attracted to the negatively charged wall, forming an electrical double layer. Applying voltage across the length of the capillary causes the solution portion of the electrical double layer to move toward the cathode end of the capillary, drawing the bulk solution. This movement is known as the electro-osmotic flow (EOF). The degree of ionization of the inner-wall capillary silanol groups depends mainly on the pH of the operating buffer and on the modifiers that may have been added to the electrolyte.

Currently, there are five major modes of operation of CE: capillary zone electrophoresis (CZE), also referred to as free solution or free flow capillary electrophoresis; micellar electrokinetic chromatography (MEKC); capillary gel electrophoresis (CGE); capillary isoelectric focusing (CIEF); and capillary isotachophoresis (CITP). Of these, the most commonly utilized capillary techniques are CZE and MEKC (Rabel and Stobaugh 1993; Issaq 1999; Smyth and McClean 1998).

In CZE, separations are controlled by differences in the relative electrophoretic mobilities of the individual components in the sample or test solution. The mobility differences are functions of analyte charge and size under specific method conditions. They are optimized by appropriate control of the composition of the buffer, its pH, and its ionic strength.

In MEKC, the supporting electrolyte medium contains a surfactant at a concentration above its critical micelle concentration (CMC). The surfactant self-aggregates in the aqueous medium and forms micelles whose hydrophilic head groups and hydrophobic tail groups form a nonpolar core into which the solutes can partition. The micelles are anionic on their surface, and they migrate in the opposite direction to the electroosmotic flow under the applied current. The differential partitioning of neutral molecules between the buffered aqueous mobile phase and the micellar pseudostationary phase is the sole basis for separation as the buffer and micelles form a two-phase system, and the analyte partitions between them (Smyth and McClean 1998).

In CIEF, proteins are separated on the basis of their relative differences in isoelectric points. This is accomplished by achieving steady-state sample zones within a buffer pH gradient, where the pH is low at the anode and high at the cathode. With the application of an electric field, the proteins then migrate according to their overall charge. This technique has proven to be useful in the characterization and quality control of pharmaceutical peptides for separating and estimating the classes of differently charged isoforms, as well as the detection of post-translational modifications like glycosylation or phosphorylation (Shimura 2002).

CITP employs two buffers in which the analyte zone is enclosed between. Either anions or cations can be analyzed in sharply separated zones. In addition, the analyte concentrations are the same in each zone; thus, the length of each zone is proportional to the amount of the particular analyte. And finally, CGE, which is analogous to gel filtration, uses gel-filled capillaries to separate molecules on the basis of relative differences in their respective molecular weight or molecular size. It was first used for the separation of proteins, peptides, and oligomers. Advantages of the gel include decreasing the electro-osmotic flow and also reducing the adsorption of protein onto the inner wall of the capillary (von Brocke et al. 2001).

Capillary electrophoresis has become very successful in the bioanalysis of proteins but researchers are often looking for ways to improve the various techniques, as seen with the advances being researched in SDS-PAGE staining techniques. One complaint about CE is that, due to its low loading capacity, there is an intrinsically poor concentration sensitivity. Several solutions to increase sample loadability without resolution reduction, such as on-capillary preconcentration, have been made. The need for high-throughput screening systems (HTS) is becoming more significant and advances in CE such as the development of high-efficiency separation carriers for CGE methods. As the human genomic project moves into the postgenomic sequencing era, and the need for the accumulation of information on genes and proteins will be ever more increasing, advances such as the development of HTS systems for protein analysis could possibly one day allow for the bedside analysis of disease.

# BIOASSAYS

## IMMUNOPRECIPITATION

Immunoprecipitation (IP) is one of the most widely used immunochemical techniques. It involves the interaction between an antigen and its specific antibody. Antigen–antibody interactions may produce a network of many antigen molecules cross-linked by antibody molecules, which result in insolubilization and precipitation of the complex (Williams 2000).

Immunoprecipitation followed by SDS-PAGE and immunoblotting, is routinely used in a variety of applications: to determine the molecular weights of protein antigens, to study protein–protein interactions, to determine specific enzymatic activity, to monitor protein post-translational modifications and to determine the presence and quantity of proteins. The IP technique also enables the detection of rare proteins which otherwise would be difficult to detect since they can be concentrated up to 10,000-fold by immunoprecipitation (MacMillan-Crow and Thompson 1999). As usually practiced, this technique provides a rapid and simple means to analysis a protein of most interest. However the name of the procedure is a misnomer since removal of the protein antigen from solution does not depend upon the formation of an insoluble antibody–antigen complex. Rather, antibody–antigen complexes are removed from solution by addition of an insoluble form of an antibody binding protein such as protein A, protein G, or second antibody (Figure 6.8). Typically, the antigen is made radioactive before the immunoprecipitation procedure. Having a radioactive antigen is not required but interpretation of data is simplified since the antigen, and not the antibody, is radiolabeled.

The success of immunoprecipitation depends on the affinity of the antibody for its antigen as well as for protein G or protein A. In general, while polyclonal antibodies are best, purified monoclonal antibodies (MAb), ascites fluid, or hybridoma supernatant can also be used.

Immunoprecipitation can be used to study protein–DNA interactions (Kuo and Allis 1999). For instance, the basic chromatin immunoprecipitation technique is remarkably versatile and has now been used in a wide range of cell types, including budding yeast, fly, and human cells. This technique has been successfully employed to map the boundaries of specifically modified (e.g., acetylated) histones along target

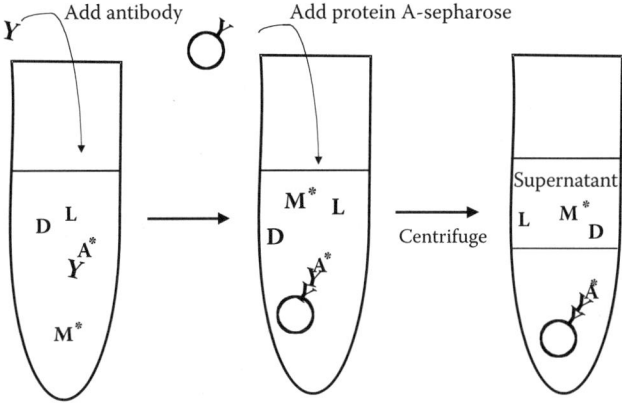

**FIGURE 6.8** Schematic representation of the principle of immunoprecipitation. An antibody added to a mixture of radiolabeled (*) and unlabeled proteins binds specifically to its antigen (A) (left tube). Antibody–antigen complex is absorbed from solution through the addition of an immobilized antibody binding protein such as protein A—sepharose beads (middle panel). Upon centrifugation, the antibody–antigen complex is brought down in the pellet (right panel). Subsequent liberation of the antigen can be achieved by boiling the sample in the presence of sulfate-polyacrylamide gel electrophoresis (SDS).

genes, to define the cell cycle-regulated assembly of origin-dependent replication and centromere-specific complexes with remarkable precision, and to map the *in vivo* position of reasonably rare transcription factors on cognate DNA sites.

## COMPLEMENT ASSAY

Complement is one of the triggered enzyme systems of serum proteins, its action usually initiated by the combination of antigen–antibody complex. The central reaction of the complement system can be regarded as the cleavage of C3 to C3a and C3b (Andersen and Mann 2000). C3 is the bulk component of the system and all the reactions of the system can be conveniently related to the cleavage, inactivation and subsequent part in the causation of the lytic lesion and other biological phenomena. The classical pathway of complement is activated by the binding of the C1q subunit of the C1 macromolecule to IgG or IgM that is aggregated with the antigen. This results in a cascade of proteins (Figure 6.9). Complement activation is customarily determined by measuring fixation of complement or complement-mediated haemolysis. Complement fixation is a test measured by the capacity of complement, after incubation with antigen–antibody complexes to hemolyze sensitized sheep red blood cells. The procedure involves some, if not all, of the components that comprise the complement system.

Complement fixation assays can be used to look for the presence of specific antibody or antigen in a patient's serum. The test uses sheep red blood cells (SRBC), anti-SRBC antibody and complement, along with specific antigen (if looking for antibody in serum) or specific antibody (if looking for antigen in serum). If antibody

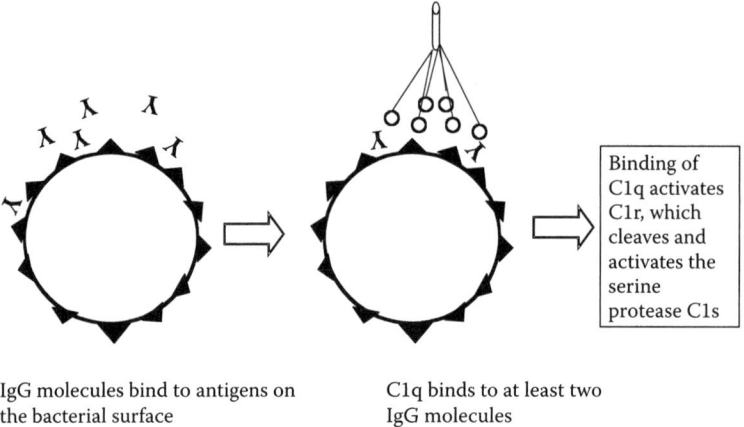

IgG molecules bind to antigens on
the bacterial surface

C1q binds to at least two
IgG molecules

Binding of
C1q activates
C1r, which
cleaves and
activates the
serine
protease C1s

**FIGURE 6.9** The classical pathway of complement activation is initiated by binding of C1q to antibody on a surface such as a bacterial surface. Multiple molecules of IgG bound on the surface of a pathogen allow the binding of a single molecule of C1q to two or more Fc pieces. The binding of C1q activates the associated C1r, which becomes an active enzyme that cleaves the proenzyme C1s, generating a serine protease that initiates the classical complement cascade.

(or antigen) is present in the patient's serum, then the complement is completely utilized and SRBC lysis is minimal. However, if the antibody (or antigen) is not present in the patient's serum, then the complement binds anti-SRBC antibody and lysis of the SRBCs ensues. Color changes in the solution induced by RBC lysis can be measured spectrophotometrically (Figure 6.10).

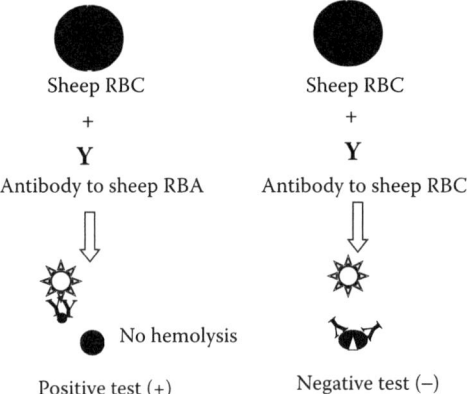

Sheep RBC

+

Y
Antibody to sheep RBA

No hemolysis

Positive test (+)

Sheep RBC

+

Y
Antibody to sheep RBC

Negative test (−)

**FIGURE 6.10** Complement fixation assay. Left: Antigen is present in the patient's serum, then the complement is completely utilized and SRBC lysis is minimal. Right: Antigen is not present in the patient's serum, then the complement binds anti-SRBC antibody and lysis of the SRBCs ensues. SRBC, sheep red blood cells.

## AGGLUTINATION

Agglutination tests are based on the presence of agglutinating antibodies that can react with specific antigens to form visible clumps. When antibodies are mixed with their corresponding antigens on the surface of large, easily sedimented particles such as animal cells, erythrocytes, or bacteria, the antibodies cross-link the particles, forming visible clumps (Janeway et al. 2001). This reaction is termed agglutination. Agglutination is a serological reaction and is very similar to the precipitation reaction. Both reactions are highly specific because they depend on the specific antibody and antigen pair. The main difference between these two reactions is the size of antigens. For precipitation, antigens are soluble molecules, and for agglutination, antigens are large, easily sedimented particles. Agglutination is more sensitive than a precipitation reaction because it takes a lot of more soluble antigens and antibody molecules to form a visible precipitation. To make the detection of soluble antigen and antibody reaction more sensitive, a precipitation reaction can be transformed into an agglutination reaction by attaching soluble antigens to large, inert carriers, such as erythrocytes or latex beads (Van Oss 2000).

Agglutination reactions have many applications in clinical medicine that can be used to type blood cells for transfusion, to identify bacterial cultures, and to detect the presence and relative amount of specific antibody in a patient's serum. For example agglutination of antibody-coated latex beads has become a popular commercial method for the rapid diagnosis of various conditions such as pregnancy and streptococcal infections.

## ENZYME IMMUNOASSAY (EI)

Methods of using the specific reactivity of antibody with antigen to reveal the presence of antibodies in sera or to identify antigens in tissues and cells differ according to the mechanisms by which the antigen–antibody interaction is revealed. Agglutination and precipitation are classical examples of gross effects produced by antigen–antibody interaction that are more or less immediately visible to the naked eye. Other methods exploit secondary devices as a means of converting antigen–antibody interaction to observable forms. One such device is to label one of the reactants, antigen or antibody, and to use it as a topographical tracer.

Enzyme immunoassay (EIA) is one of such methods that label antigen or antibody with enzyme. The most representative form of EIA is the enzyme-linked immunosorbent assay (ELISA) in which bound antigen or antibody is detected by a linked enzyme that converts a colorless substrate into a colored product (Figure 6.11).

For ELISA, an enzyme is linked chemically to the antibody. The labeled antibody is allowed to bind to the unlabeled antigen, under conditions where nonspecific adsorption is blocked, and any unbound antibody and other proteins are washed away. Binding is detected by a reaction that converts a colorless substrate into a colored reaction product. The color change can be read directly in the reaction tray, making data collection very easy, and ELISA also avoids the hazards of radioactivity. This makes ELISA the preferred method for most direct-binding assays (Plested et al. 2003).

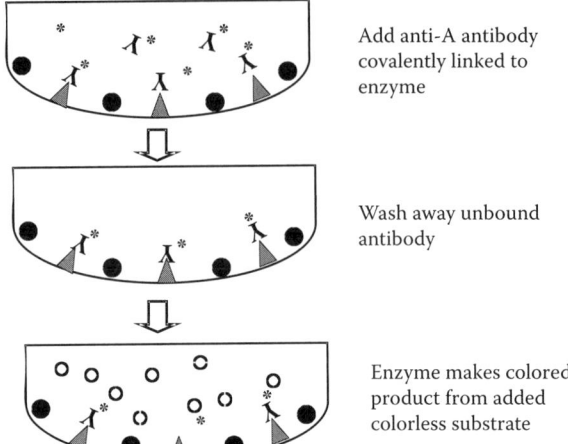

Add anti-A antibody covalently linked to enzyme

Wash away unbound antibody

Enzyme makes colored product from added colorless substrate

| Reagents: | POSITIVE SAMPLE High level of hormone | NEGATIVE SAMPLE Low level of hormone |
|---|---|---|
| **Y**: Ab specific for hormone (coating the filter)<br>•: Unkown sample with  hormone<br>Allow time to react and wash away unbound substances | | |
| Reagents:<br><br>**\***: $^{125}$I-labeled hormone<br>Allow time to react and wash away unbound radiolabeled hormone | | |
| **Procedure:**<br>Measure radioactivity in a gamma counter<br><br>**Result:**<br>Amount of radioactivity is inversely proportional to the concentration of hormone in the sample. | | |

**FIGURE 6.11** The principle of the enzyme-linked immunosorbent assay (ELISA). To detect antigen A, purified antibody specific for antigen A is linked chemically to an enzyme. The samples to be tested are coated onto the surface of plastic wells. The labeled antibody is then added to the wells under conditions where nonspecific binding is prevented, so that only the labeled antibody binding to antigen A was retained on the surface. Unbound labeled antibody is removed from all wells by washing, and bound antibody is detected by an enzyme-dependent color-change reaction.

ELISA can also be carried out with unlabeled antibody stuck to the plates and a second labeled antibody added. Rather than the antigen being directly attached to a plastic plate, antigen-specific antibodies are bound to the plate. These are able to bind antigen with high affinity, and thus concentrate it on the surface of the plate, even with antigens that are present in very low concentrations in the initial mixture. A separate labeled antibody that recognizes a different epitope to the immobilized first antibody is then used to detect the bound antigen. This is a modification of ELISA known as "sandwich ELISA" which can be used to detect secreted products such as cytokines (Itoh and Suzuki 2002).

Whittier et al. (19XX) used sandwich ELISA to measure the amount of protein S in patients' plasma. Monoclonal antibody to human free protein S is coated as a capture antibody to the bottom and sides of microplates of a polystyrene multiple-well plate (96-wells/plate). The wells are stabilized, and noncoated areas of the plastic are blocked to decrease nonspecific binding. Diluted patient plasma is incubated in the wells, allowing patient free protein S to bind specifically to the monoclonal capture antibody. The detection antibody, a solution of polyclonal antihuman protein S antibody conjugated to an enzyme is added to quantitate patient free protein S that is bound to the wells. After incubation, any unbound enzyme-conjugated antibody is washed from the wells. The degree of conjugate binding is measured by adding a chromogenic substrate system (tetramethylbenzidine and hydrogen peroxide) resulting in a soluble colored product. Color development (absorbance) is measured with a spectrophotometer, in optical density units (ODs). Patient ODs are used to determine free protein S antigen levels in relative percent concentrations from a reference curve produced by testing multiple dilutions of assayed reference plasma.

ELISA does not allow one to measure directly the amount of antigen or antibody in a sample of unknown composition, as it depends on the binding of a pure labeled antigen or antibody (Reen 1994). One way to solve this problem is to use a competitive inhibition assay. In this type of assay, the presence and amount of a particular antigen in an unknown sample is determined by its ability to compete with a labeled reference antigen for binding to an antibody attached to a plastic well. A standard curve is first constructed by adding varying amounts of a known, unlabeled standard preparation; the assay can then measure the amount of antigen in unknown samples by comparison with the standard. The competitive binding assay can also be used for measuring antibody in a sample of unknown composition by attaching the appropriate antigen to the plate and measuring the ability of the test sample to inhibit the binding of a labeled specific antibody.

## Radio Immunoassay

Another method studying antigenñantibody interaction by labeled reactant is radioimmunoassay (RIA) in which antigen or antibody is labeled with radioactivity. The basic principle of a radioimmunoassay (RIA) is the use of radiolabeled Abs or Ags to detect Ag:Ab reactions (Goldberg and Djavadi-Ohaniance 1993). The Abs or Ags are labeled with the $^{125}$I (iodine-125) isotope, and the presence of Ag:Ab reactions is detected using a gamma counter. The unlabeled component, which in this case would be antigen, is attached to a solid support, such as the wells of a plastic multiwell plate, which will adsorb a certain amount of any protein. The labeled antibody is allowed to bind to the unlabeled antigen, under conditions where non-specific adsorption is blocked, and any unbound antibody and other proteins are washed away. Antibody binding in RIA is measured directly in terms of the amount of radioactivity retained by the coated wells. Labeled anti-immunoglobulin antibodies can also be used in RIA to detect binding of unlabeled antibody to unlabeled antigen-coated plates. In this case, the labeled anti-immunoglobulin antibody is used in what is termed a "second layer." The use of such a second layer also amplifies the signal, as at least two molecules of the labeled anti-immunoglobulin antibody are able to bind to each unlabeled antibody.

RIAs are highly sensitive and quantitative, capable of detecting small amounts of Ag or Ab. As a result, they are often used to measure the quantities of hormones or drugs present in a patient's serum. In this case, RIAs are performed in a manner similar to the competitive ELISA. The presence of the hormone in the serum sample inhibits binding of the radiolabeled hormone. Thus, the amount of radioactivity present in the test is inversely proportional to the amount of hormone in the serum sample. A standard curve using increasing amounts of known concentrations of the hormone is used to determine the quantity in the sample.

Another use of the RIA is to measure the quantities of serum IgE antibodies specific for various allergens in a patient's serum, in which case it is called a Radio Allergo Sorbent Test (RAST) (Nalebuff 1985). In this case, the test is performed similar to an ELISA for Ab, using radiolabeled antiglobulins specific for IgE, rather enzyme-labeled antiglobulins. However, this test has been almost completely replaced by ELISA.

Because of the requirement to use radioactive substances, RIAs are frequently being replaced by other immunologic assays, such as ELISA and fluorescence polarization immunoassays (FPIA) (Niemann et al. 1985). These have similar degrees of sensitivity. FPIAs are highly quantitative, as are RIAs, and ELISAs can be designed to be quantitative.

## REFERENCES

Abbott, S.R. (1980). Practical aspects of normal-phase chromatography. *J. Chromatogr. Sci.,* 18 (10), 540–550.

Alix, A.J.P., Berjot, M., Marx, J. (1985). Determination of the secondary structure of proteins. In: Alix, A.J.P., Bernard, L., Manfait, M., eds., *Spectroscopy of Biological Molecules,* Wiley, Chichester, UK, 149–154.

Andersen, J.S., Mann, M. (2000). Functional genomics by mass spectrometry. *FEBS Lett.,* 480(1), 25–31.

Arrondo, J.L., Muga, A., Castresana, J., Goni, F.M. (1993). Quantitative studies of the structure of proteins in solution by Fourier-transform infrared spectroscopy. *Prog. Biophys. Molec. Biol.,* 59, 23–56.

Barth, H.G., Boyes, B.E., Jackson, C. (1994). Size exclusion chromatography. *Anal. Chem.,* 66(12), 595R–620R.

Beattie, J. (1998). Size-exclusion chromatography. Identification of interacting proteins. *Methods Mol. Biol.,* 88, 65–69.

Bechinger, B. (1999). The structure, dynamics and orientation of antimicrobial peptides in membranes by multidimensional solid-state NMR spectroscopy. *Biochim. Biophys. Acta,* 1462(1–2), 157–183.

Beranova-Giorgianni, S., Desiderio, D.M. (1997). Fast atom bombardment mass spectrometry of synthetic peptides. *Methods Enzymol.,* 289, 478–499.

Biemann, K. (1990). Applications of tandem mass spectrometry to peptide and protein structure. In: Burlingame, A.L., McCloskey, J.A., editors. *Biological Mass Spectrometry.* Elsevier, Amsterdam, 176.

Bloustine, J., Berejnov, V., Fraden, S. (2003). measurements of protein–protein interactions by size exclusion chromatography. *Biophysic. J.,* 85, 2619–2623.

Chalmers, M.J., Gaskell, S.J. (2000). Advances in mass spectrometry for proteome analysis. *Curr. Opin. Biotechnol.,* 11(4), 384–390.

Creveld, L.D., Meijberg, W., Berendsen, H.J., Pepermans, H.A. (2001). DSC studies of Fusarium solani pisi cutinase: consequences for stability in the presence of surfactants. *Biophys. Chem.,* 92(1–2), 65–75.

Davy, K.W., Morris, C.J. (1976). Applications of gas-liquid chromatography in protein chemistry. Determination of C-terminal sequences on nanomolar amounts of proteins. *J. Chromatogr.,* 116(2), 305–314.

De, P., Dasgupta, S.C., Gomes, A. (2002). A lethal neurotoxic protein from Indian king cobra (*Ophiophagus hannah*) venom. *Indian J. Exp. Biol.,* 40(12), 1359–1364.

Deb, S., Bandyopadhyay, S., Roy, S. (2000). DNA sequence dependent and independent conformational changes in multipartite operator recognition by lambda-repressor. *Biochemistry,* 39(12), 3377–3383.

Ettinger, M.J., Timasheff, S.N. (1971a). Optical activity of insulin. I. On the nature of the circular dichroism bands. *Biochemistry,* 10, 824–830.

Ettinger, M.J., Timasheff, S.N. (1971b). Optical activity of insulin. II. Effect of nonaqueous solvents. *Biochemistry,* 10, 831–840.

Freire, E. (1995). Differential scanning calorimetry. *Methods Mol. Biol.* 40, 191–218.

Fukushima T., Usui N., Santa T., Imai K. (2003). Recent progress in derivatization methods for LC and CE analysis. *J. Pharm. Biomed. Anal.,* 30(6), 1655–1687.

Glod, B.K. (1997). Ion exclusion chromatography: parameters influencing retention. *Neurochem. Res.,* 22(10), 1237–1248.

Goldberg, M.E., Djavadi-Ohaniance, L. (1993). Methods for measurement of antibody/antigen affinity based on ELISA and RIA. *Curr. Opin. Immunol.,* 5(2), 278–281.

Goldenberg, D.P., Creighton, T. E. (1984). Gel electrophoresis in studies of protein conformation and folding. *Anal. Biochem.,* 138, 1–18.

Griffin, T.J., Goodlett, D.R., Aebersold, R. (2001). Advances in proteome analysis by mass spectrometry. *Curr. Opin. Biotechnol.,* 12(6), 607–612.

Griffiths, W.J. (2000). Nanospray mass spectrometry in protein and peptide chemistry. *EXS,* 88, 69–79.

Haris, P.I., Chapman, D. (1995).The conformational analysis of peptides using Fourier transform IR spectroscopy. *Biopolymers,* 37(4), 251–263.

Hjelmeland, L.M., Chrambach, A. (1981). Electrophoresis and electrofocusing in detergent containing media: A discussion of basic concepts. *Electrophoresis,* 2, 1–11.

Hoffman, B.M., (1991) Electron nuclear double resonance (ENDOR) of metalloenzymes. *Acc. Chem. Res.,* 24, 164–170.

Huybrechts, T., Dewulf, J., Van Langenhove, H. (2003). State-of-the-art of gas chromatography-based methods for analysis of anthropogenic volatile organic compounds in estuarine waters, illustrated with the river Scheldt as an example. *J. Chromatogr. A,* 1000(1–2), 283–297.

Issaq, H.J. (1999). Capillary electrophoresis of natural products–II. *Electrophoresis,* 20(15–16), 3190–3202.

Itoh, K., Suzuki, T. (2002). Antibody-guided selection using capture-sandwich ELISA. *Methods Mol. Biol.,* 178, 195–199.

Jain, S., Kumar, C.V., Kalonia, D.S. (1992). Protein–peptide interactions as probed by tryptrophan fluorescence: Serum albumins and enkephalin metabolites. *Pharm. Res.,* 9, 990–992.

Janeway, C.A., Travers, P., Walport, M., Shlomchik, M. (2001). *Immunobiology.* Garland Publishing, New York and London.

Jefson, M. (1988). Applications of NMR spectroscopy to protein structure determination. *Ann. Rep. Med. Chem.,* 23, 275–283.

Johson, W.C. (1988). Secondary structure of proteins through circular dichroism spectroscopy. *Ann. Rev. Biophys. Biophys. Chem.,* 17: 145–166.

Jonsson, A.P. (2001). Mass spectrometry for protein and peptide characterisation. *Cell. Mol. Life Sci.,* 58(7), 868–884.

Kadima, W.L. Ogendal, R., Bauer, N., Kaarsholm, K., Brodersen, Hansen, J.F., Porting, P. (1993). The Influence of ionic strength and ph on the aggregation properties of zinc-free insulin studied by static and dynamic laser light scattering. *Biopolymers,* 33, 1643–1657.

Kenney. W.C., Arakawa. T. (1997). Biophysical and biochemical analyses of recombinant proteins: Structure and analyses of proteins. In: Crommelin D.J.A., Sindelar R.D., eds., *Pharmaceutical Biotechnology.* Harwood Academic Publishers, Amsterdam, 27–51.

Kent, U.M. (1999). Purification of antibodies using affinity chromatography. *Methods Mol. Biol.,* 115, 23–28.

Kevin, M.D. (2000). Contributions of mass spectrometry to structural immunology. *J. Mass. Spectrom.,* 35, 493–503.

Kosecki, R. (1988). Recirculating isoelectric focusing: A system for protein seperations. *BioPHARM,* 1(6), 28–31.

Kuo, M.H., Allis, C.D. (1999). In vivo cross-linking and immunoprecipitation for studying dynamic protein: DNA associations in a chromatin environment. *Methods,* 19(3), 425–433.

Lahm, H.W., Langen, H. (2000). Mass spectrometry: A tool for the identification of proteins separated by gels. *Electrophoresis,* 21, 2105–2114.

Leushner, J. (2001). MALDI TOF mass spectrometry: An emerging platform for genomics and diagnostics. *Expert Rev. Mol. Diagn.,* 1(1), 11–18.

Liu, H., Lin, D., Yates, J.R., 3rd. (2002). Multidimensional separations for protein/peptide analysis in the post-genomic era. *Biotechniques,* 32(4), 898, 900, 902 passim.

Loog, M., Uri, A., Jarv, J., Ek, P. (2000). Bi-substrate analogue ligands for affinity chromatography of protein kinases. *FEBS Lett.,* 480(2–3), 244–248.

Lu, B.Y., Chang, J.Y. (2001). Isolation of isoforms of mouse prion protein with PrP(SC)-like structural properties. *Biochemistry,* 40(44), 13390–13396.

MacMillan-Crow, L.A., Thompson, J.A. (1999). Immunoprecipitation of nitrotyrosine-containing proteins. *Methods Enzymol.,* 301, 135–145.

Mano, N., Goto, J. (2003). Biomedical and biological mass spectrometry. *Anal Sci.,* 19(1), 3–14.

Martindale, H., Marsh, J., Hallett, F.R., Albisser, A.M. (1982). Examination of insulin formulations using quasi-elastic light scattering. *Diabetes,* 31, 364–366.

McLachlin, D.T., Chait, B.T. (2001). Analysis of phosphorylated proteins and peptides by mass spectrometry. *Curr. Opin. Chem. Bio.l,* 5(5), 591–602.

Monegier, B., Clerc, F.F., Van Dorsselaer, A., Vuihorgne, M., Green, B., Cartwright, T. (1991). Using mass spectrometry to characterize recombinant proteins. Part II. *Pharm. Technol.,* 15(4), 28–40.

Muller, W. (1990). Liquid-liquid partition chromatography of biopolymers in aqueous two-phase systems. *Bioseparation,* 1(3–4), 265–282.

Nagayama, K. (1981). Two-dimensional NMR spectroscopy: an application to the study of flexibility of protein molecules. *Adv. Biophys.,* 14, 139–204.

Nalebuff, D.J. (1985). PRIST, RAST, and beyond. Diagnosis and therapy. *Otolaryngol. Clin. North Am.,* 18(4), 725–744.

Niemann, A., Oellerich, M., Schumann, G., Sybrecht, G.W. (1985). Determination of theophylline in saliva, using fluorescence polarization immunoassay (FPIA). *J. Clin. Chem. Clin. Biochem.,* 23(11), 725–732.

Nishihara, J.C., Champion, K.M. (2002). Quantitative evaluation of proteins in one- and two-dimensional polyacrylamide gels using a fluorescent stain. *Electrophoresis,* 23(14), 2203–2215.

Ollivier, P.B., Wallet, J.C. (19XX). Rhône Poulenc Rorer GENCELL*, 13 quai Jules Guesde, 94403 Vitry sur Seine, France.

Palmer, R.A., Niwa, H. (2003). X-ray crystallographic studies of protein-ligand interactions. *Biochem. Soc. Trans.,* 31(Pt 5), 973–979.

Parsons, D.L. (1982). Fluorescence stability of human albumin solutions. *J. Pharm. Sci.,* 71, 349–351.

Pearlman, R., Nguyen T.H., (1991). Analysis of protein drugs. In: Lee, V.H.L., ed., *Peptide and Protein Drug Delivery.* Marcel Dekker, Inc., New York, 247–301.

Pellegrini, M., Mierke, D.F. (1999). Structural characterization of peptide hormone/receptor interactions by NMR spectroscopy. *Biopolymers (Peptide Sci.),* 51, 208–220.

Perczel, A., Jakli, I., Csizmadia, I.G. (2003). Intrinsically stable secondary structure elements of proteins: A comprehensive study of folding units of proteins by computation and by analysis of data determined by x-ray crystallography. *Chemistry,* 9(21), 5332–5342.

Plested, J.S., Coull, P.A., Gidney, M.A. (2003). ELISA. *Methods Mol. Med.,* 71, 243–261.

Rabel, S.R., Stobaugh, J.F. (1993). Application of capillary elecrophoresis in pharmaceutical analysis. *Pharm. Res.,* 10, 171–186.

Reen, D.J. (1994). Enzyme-linked immunosorbent assay (ELISA). *Methods Mol. Biol.,* 32:461–466.

Righetti, P.G. (2001). Capillary electrophoretic analysis of proteins and peptides of biomedical and pharmacological interest. *Biopharm. Drug Dispos.,* 22(7–8), 337–351.

Rubakhin, S.S., Page, J.S., Monroe, B.R., Sweedler, J.V. (2001). Analysis of cellular release using capillary electrophoresis and matrix assisted laser desorption/ionization-time of flight-mass spectrometry. *Electrophoresis,* 22(17), 3752–3758.

Sanger, F. (1959). Chemistry of insulin; determination of the structure of insulin opens the way to greater understanding of life processes. *Science,* 129(3359), 1340–1344.

Sealy, R.C., Hyde, J.S., Antholine, W.E. (1985). Electron spin resonance. In: Neuberger, A. and Van Deenen, L.L.M., eds., *Modern Physical Methods in Biochemistry,* Part A. Elsevier, New York, 11A, 69–148.

Shimura, K. (2002). Recent advances in capillary isoelectric focusing: 1997–2001. *Electrophoresis,* 23(22–23), 3847–3857.

Smith, E. (1967). Application of reverse phase partition chromatography to the analysis of testosterone propionate in oil injectables. *J. Pharm. Sci.,* 56(5), 630–634.

Smyth, W.F., McClean, S. (1998). A critical evaluation of the application of capillary electrophoresis to the detection and determination of 1,4-benzodiazepine tranquilizers in formulations and body materials. *Electrophoresis,* 19(16–17), 2870–2882.

Strege, M.A., Lagu, A.L. (1993). Capillary electrophoresis as a tool for the analysis of protein folding. *J. Chromatogr. A,* 652, 179–188.

Sun, L., Ghosh, I., Xu, M.Q. (2003). Generation of an affinity column for antibody purification by intein-mediated protein ligation. *J. Immunol. Methods,* 282(1–2), 45–52.

Surewicz, W.K., Mantsch, H.H., Chapman, D. (1993). Determination of protein secondary structure by Fourier transform infrared spectroscopy: A critical assessment. *Biochemistry,* 32, 389–394.

Tsai, P.K., Volkin, D.B., Dabora, J.M., et al. (1993). Formulation design of acidic fibrolast growth factor. *Pharm. Res.,* 10, 649–659.

Turecek, F. (2002) Mass spectrometry in coupling with affinity capture-release and isotope-coded affinity tags for quantitative protein analysis. *J. Mass. Spectrom.,* 37(1), 1–14.

Van Oss, C.J. Precipitation and agglutination. (2000). *J. Immunoassay,* 21(2–3), 143–164.

Varki, A., Cummings, R., Esko, J., Freeze, H., Hart, G., Marth, J. (1999). *TITLE?* Cold Spring Harbor Laboratory Press, Cold Spring Harbor, NY.

Vigano, C., Manciu, L., Buyse, F., Goormaghtigh, E., Ruysschaert, J.M. (2000). Attenuated total reflection IR spectroscopy as a tool to investigate the structure, orientation and tertiary structure changes in peptides and membrane proteins. *Biopolymers,* 55(5), 373–380.

Vollhardt, K.P.C., Schore, N.E. (1995). *Organic Chemistry,* 2nd ed. W. H. Freeman & Company, New York.

von Brocke, A., Nicholson, G., Bayer, E. (2001). Recent advances in capillary electrophoresis/electrospray-mass spectrometry. *Electrophoresis,* 22(7), 1251–1266.

Wang, Y.J. (1992). Parenteral products of peptides and proteins. In: Avis, K.E., Lieberman, H.A., Lachman, L., eds., *Pharmaceutical Dosage Forms: Parenteral Medications.* Marcel Dekker, Inc., New York, 283–319.

Welinder, B.S. (1971). Isoelectric focusing of insulin and insulin derivatives. *Acta Chem. Scand.,* 25, 3737–3742.

West, M.H.P., Wu, R.S., Bonner, W.M. (1984). Polyacrylamide gel electrophoresis of small peptides. *Electrophoresis,* 5, 133–138.

Westbrook, J.A., Yan, J.X., Wait, R., Welson, S.Y., Dunn, M.J. (2001). Zooming-in on the proteome: very narrow-range immobilised pH gradients reveal more protein species and isoforms. *Electrophoresis,* 22(14), 2865–2871.

Wetlaufer, D.B. (1962). Ultraviolet spectra of proteins and amino acids. *Adv. Protein Chem.,* 17, 303–390.

Whittier, A.M., Taylor D.O., Fink C.A., Lopez L.R. (19XX). The measurement of free Protein S by a monoclonal ELISA. American Clinical Laboratory Association, Washington, DC.

Wilchek, M., Chaiken, I. (2000). An overview of affinity chromatography. *Methods Mol. Biol.,* 147, 1–6.

Williams, N.E. (2000). Immunoprecipitation procedures. *Methods Cell. Biol.,* 62, 449–453.

Winder, A.F., Gent, W.L.G. (1971). Correction of light scattering errors in spectrophotometric protein determinations. *Biopolymers,* 10, 1243–1252.

Yang, Y.B., Regnier, F.E. (1991). Coated hydrophilic polystyrene-based packing materials. *J. Chromatogr.,* 544(1–2), 233–247.

# 7 Formulation of Proteins and Peptides

*Michael J. Groves, Ph.D., D.Sc.*

## CONTENTS

## INTRODUCTION

In nature proteins and some simpler polypeptides are agents for controlling most, if not all, body functions. It will be evident that having these materials present in the correct quantities at the appropriate body site and at the correct time are essential requirements for good health. Nevertheless, from a formulators' perspective, it is necessary to remember that these materials are synthesized in nature in minuscule quantities, usually at only one site and diffuse or are actively transported to their site of action, which may only be a few microns away. There they may react with one specific molecule or a group of molecules at a receptor site in order to produce the appropriate physiological response. Unfortunately, modern drug delivery methods involve delivering relatively massive amounts of drug or biological response modifier throughout the body once or twice a day. Sites all over the body may be affected in addition to the one that is being targeted and it is hardly surprising that toxicity is often manifested to modern drugs or their delivery systems.

What is urgently needed at present is a pharmaceutical delivery system that mimics the natural process as closely as possible and delivers the necessarily small quantity of active material close to its target tissue. The need for targeting in many cases is recognized but the difficult part is achieving this ideal. Any delivery system has to move through the physiological medium and survive challenges on its way to the target site. That we will eventually achieve this aim is not doubted. That it will take a few more years to achieve cannot be doubted either. An example to illustrate this point may be seen in recent vigorous efforts to make gene engineering function in this area. A great deal of effort—and money—has been spent on taking advantage of the fact that genes producing certain proteins can be made to enter some target cells with the aid of appropriate delivery systems and produce benefits from subsequent excretion of that protein or modified protein. This approach has produced some claimed benefits for a limited number of patients but, at the time of writing, little of significant value has emerged to be of benefit to the general population. Part of this issue is the difficulty of getting the gene or its delivery system to penetrate the appropriate cells, sometimes targeting is an issue and, more recently, the observation that some genes are capable of synthesizing more than one protein or a totally different one from that observed *in vitro* has been an occasional surprise. Thus, while the specific proteins synthesized by many genes have been identified, the net effect from a practical standpoint has been relatively small.

Perhaps successful application of the previous work will start from proteomics. Here specific proteins associated with various diseases are being identified and this may lead to the development of specific drug delivery systems being designed to overcome a deficiency. If these active proteins can be identified with greater certainty then we might be more confident in how we can target the appropriate tissues. We already have some successful drug delivery systems; what we will need to do is to make them more specific and sensitive.

# MAKING SMALL PROTEIN PARTICLES: PRECIPITATION OF PROTEINS FROM SUPERCRITICAL FLUIDS

Technically there is a need for small uniform protein particles that are easier to handle and formulate while, at the same time, retaining their biological activity. Various granulation techniques have been tried but, in general, these are not effective and the proteins concerned tend to loose their biological actvity on reconstitution.

A way around this issue may have been found with the use of supercritical fluids. These materials, such as liquid carbon dioxide, have many interesting properties from the point of view of pharmacutical processing since they combine liquid-like solvent properties with gas-like transportation properties. Small changes in the applied pressure or temperature can result in large changes of the fluid density and, correspondingly, the solvent capacity and properties of the resultant particles.

As noted, the commonest supercritical fluid in practice is liquid carbon dioxide since it is readily available, inexpensive and nontoxic. York and Hanna (1996) demonstrated how changes in the resultant particles produced by dissolving materials in liquid carbon dioxide and allowing them to come to ambient conditions could be achieved by making small changes in the environmental conditions. Mechanical precipitation was usually simply achieved by allowing the solution to expand to atmospheric conditions, the carbon dioxide evaporating into the atmosphere to leave the solid particles behind. The process was illustrated by Rehman et al. (2003) when considering the preparation of dry powders for inhalation therapy.

When considering proteins, these same authors demonstrated that these materials were generally not soluble in liquid carbon dioxide alone but solubility could be improved by adding small quantities of dimethylsulfoxide or even water. The former solvent was shown to produce some denaturation of a few proteins but a later investigation showed that, for example, lysosyme could be successfully dissolved in a mixture of deimethylsulfoxide and water before being mixed with the liquid carbon dioxide. This final stage was kept brief before recovering the dried protein from an expansion chamber. After further drying, particles of between 0.8 and 12 μm diameter were obtained and these retained their original biological activity on being reconstituted in water.

Solvents such as dimethylsulfoxide were shown to reduce the biological activity, probably due to subtle effects on the conformation of the protein. Larger proteins remain to be evaluated but the process obviously has considerable potential for further development.

# PARENTERAL DRUG DELIVERY SYSTEMS

## GENERAL INTRODUCTION

Parenteral products, without exception, are designed to be administered to the patient by circumventing the systems that the body has in place in order to survive the hostile environment it is in. For example, the air surrounding us is filled with dust particles which in many cases contain bacteria. Not all of these bacteria are pathogenic but most will produce some form of reaction or disease when introduced inside the body. It therefore makes sense that a solution or suspension of a drug or even a device introduced under the skin or directly into the blood stream should not contain any of these foreign organisms. In other words these systems must be sterile to prevent bacterial invasion.

All parenteral products have basic product requirements in addition to those required for all pharmaceutical products, such as accuracy and uniformity of contents and dosage. These can be summarized as

1. Sterility
2. Apyrogenicity
3. Safety
4. Efficacy
5. Reproducibility in performance

Sterility is not always considered by reviewers and formulators but, in fact, it is the prime requirement of any material intended for parenteral administration, although there are situations where it may not be easy to achieve. For example, if a drug is thermolabile, such as most proteins, and is unable to resist the application of the heat stress usually required to achieve sterility of a solution, then an alternative process must be designed to achieve the same objective. On occasion this will, of necessity, affect the formulation itself and influence the choice of manufacturing process used to prepare the formulation on a large scale. The administration of a contaminated product to a patient who is already debilitated must be avoided since they have a much lower resistance to any insult. However, the same consideration may apply to a healthy individual who is, for example, receiving a vaccine or diagnostic agent.

Apyrogenicity is another requirement which is associated with the need for sterility. Pyrogens are materials shed by living or dead organisms that can produce a febrile reaction in humans (and many animals) on administration. This pyrogenic reaction is rarely fatal but can be extremely unpleasant for the patient. Today, pyrogens are readily measured with exquisite sensitivity and precision with limulus lysate (Novitsky 1996). If a product has a low level of pyrogens it is safe to assume that it also contains a very low level of bacterial cells, dead or alive, and this provides another degree of quality assurance for the product.

*Safety* and *efficacy* are two more general pharmaceutical product requirements, as is *reproducibility of performance*. This last-named requires that the product produces the same response on multiple administrations of the same batch or from batch to batch and to some extent implies there is uniformity of content and reproducibility. The same implication applies to products that have some form of controlled release properties; the inference is that the rates of release of the incorporated drug are similar under all circumstances.

## Heat Sterilization

Application of heat stress to sterilize foods is an old and well-established process but is rather less well developed for pharmaceutical products. Pharmacopoeias of just 50 years ago tended to regard the official sterilization processes as if they were in cookery books and there were some obvious dangers associated with this approach.

Any material introduced into the body, human or animal, parenterally (that is *para-enteron* or "beyond the gut") must be sterile because the body defenses have been bypassed. This concept is absolute in the sense that one cannot have "partially sterile"—a product is either sterile or not. Sterility means "rendered free of living microorganisms" (*The Shorter Oxford Dictionary,* 1993) or, quoting the United States Pharmacopeia (USP) 28 (2005), "completely free of viable microbial contamination."

The induction of disease or even death by administration of a contaminated, nonsterile, parenteral product to a patient is not unknown and obviously this is not acceptable in practice.

The absolute need for sterility in a parenteral product is not difficult to understand but, unfortunately, microorganisms are ubiquitous in the surrounding environment and are difficult to remove or, more to the point, kill. In a closed system such as a solution or suspension of a product in a sealed container, organisms can be killed by the application of heat stress. The same process, designed to kill any contaminating microorganisms, can also degrade the pharmaceutical product in the container. Until very recently the cookery book approach to sterilization required that the product be heated, for example, at 121°C for 15 minutes. It is generally recognized that these conditions will be sufficient to produce a sterile product, although it is also recognized that this is an "overkill" situation in that damage to the product can occur at the same time. However, questions now arise. For example, how much will the heating up and cooling down from ambient room temperature to a steady state temperature of 121°C contribute to the overall kinetics of the sterilization process? Is a product sterile if it was only heated for 14 minutes? Or 13 minutes, and so on.

## BACTERIAL DEATH

Answers to the above questions can only be obtained by close study of the kinetics of bacterial death. One main issue here is that the product inside the sealed container is never contaminated with a single species of organism; there is a spectrum of different organisms present, each with its own unique susceptibility to the applied heat stress. Today the overall hazard due to viable organisms being present in the product prior to the application of heat stress is diminished considerably by repeated filtration through porous membranes with mean pore diameters of 100 nm or less. This reduces the overall numbers of viable organisms present in the system, the so-called "bioburden," considerably and enables a lower degree of heat stress to be supplied to obtain the necessary "absolute" sterility.

Taking this filtration analogy further, it should be technically feasible to sterile filter a liquid product and this is done when a product is said to be prepared *aseptically.* Here the solution is repeatedly filtered through filters considered to be "absolute" (that is, they remove all microorganisms) directly into presterilized containers and sealed in an atmosphere claimed to be clean since it too has been repeatedly filtered. Today the whole filling operation is often separated from the human operators by using so-called "isolators." This type of operation is considered to be state of the art and is claimed to result in a "sterile" product. However, absolute sterility cannot be guaranteed because, paradoxically perhaps, test methods are insufficiently sensitive to detect small numbers of viable microorganisms.

However, this statement deserves to be examined much more closely. The absolute certainty of an overkill heating process is diminished considerably when the product is not heated or not given sufficient heat to sterilize the container. In addition, there is inevitable manipulation involved before the container is sealed. This uncertainty has resulted in a discussion of the acceptable element of risk in an aseptic process. Unfortunately, this risk cannot be measured directly and various guess factors have been vectored into the discussion, all without scientific foundation. Going back to fundamentals, it is necessary to remind oneself that there are no degrees of sterility—a product is either sterile or it is not.

This issue has been discussed by Gilbert and Allison (1996) who pointed out that a filtration process is actually anything but absolute and any resulting filtrate may not be totally devoid of microorganisms. The next question is the likely pathogenicity of the few survivors since not all bacteria are *pathogens* and in most cases the average human can deal with this insult provided that they are not immunosuppressed. Beyond that, how are such small numbers of bacteria detected, preferably rapidly because a production process is involved? Answers to some of these questions are beginning to appear. For example, with the possible exception of anthrax where disease can be produced by a single organism, a finite number of microorganisms are usually required to produce any manifestation of disease. Rapid methods for the detection of very small concentrations of viable microorganisms are becoming available and are being applied industrially. Scientifically based *measurements* are being made on which sound judgments can be made rather than the empirical statements made earlier. This development is being made in parallel with the food industry where fatalities do occur; fatalities due to contaminated product within the pharmaceutical industry seem to be increasingly rare although the risk is greater.

Because many of the newer biologicals such as DNA, genes or proteins are very heat labile, it has become necessary to explore the sterilization procedures from the point of view of the product. Overkill heating sterilization methods would generally destroy these products so a graded approach is now necessary, using the minimum amount of energy required to destroy any undesirable microorganisms present in the product without, at the same time, significantly damaging the product. The issue has now become one involved with defining bacterial "death." In essence, it is accepted that some bacteria may remain after an antibacterial filtration process. The question is not necessarily how to kill them but rather how can their proliferation be prevented so that, effectively, the product is "sterile" or appears to be "sterile" when tested.

Bacterial death, in fact, is not easy to define or to measure. Heat, for example, may damage critical bacterial DNA sequences which stop the organism from functioning. Many organisms are capable of repairing damaged DNA and over time will begin to function again. When exposed to stress such as heat or desiccation some organisms will form endospores which enable them to survive under stressful conditions. Endospores can be killed but require more extreme conditions than those required to kill the original vegetative form. On the other hand, conceptually at least, it is possible to visualize that a key metabolic protein in the organism may become conformationally irreversibly changed so that the entire organism becomes dysfunctional. Unfortunately such examples appear to be rare.

Another factor that requires to be taken into account is the intrinsic variability of organisms within what is in effect a mixed culture. There are different species of organism present and even the same species will have cells at different stages of cellular division. Some cells in a mixed system will be very susceptible to the applied stress and others will be more resistant. Obviously it is the resistant cells that will require sufficient stress to "kill" them, no matter how that term is applied. Filtration to remove as many microorganisms as possible before any sterilizing stress is applied is clearly a sound practice.

These considerations have been extensively explored by producers of canned foods and some simplified kinetics have been derived to allow better control of sterilization procedures. For example, the overall death process in a mixed culture can be described by an exponential decay curve. The equation will follow the form

$$kt = \ln(Nt/No)$$

where k is the proportionality constant or rate constant, i.e., the thermal death rate, and No and Nt are the numbers of organisms at the beginning of the process and after time t, respectively. This confirms the above supposition that the fewer the number of organisms at the beginning of the process the less likely are there going to be survivors at the end.

A detailed discussion of these kinetics is out of place in this present text but the above equation does allow some simplified considerations to be made. For example, a useful parameter easily derived is the decimal reduction time, D, for a mixed or single culture is the time it takes for a set of conditions such as heat to reduce the number of organisms to one-tenth of the initial value, i.e.,

$$kt = tD = \ln(1/10) = 2.303 \log 0.1$$

or

$$D = 2.303/k$$

The D value is the determining factor if the death rate is, indeed, exponential, an assumption which is not necessarily always valid. Although only conditions which supply a single D value would be sufficient to completely sterilize a solution containing, say, one organism per unit volume, most heat sterilization processes are designed to administer 12D. This is an example of overkill and becomes more evident the fewer organisms there are in the untreated product in the first place.

There is another, nonkinetic parameter widely used when evaluating a heat sterilization process. This is the $F_o$ value, defined as the number of minutes required to kill all *endospores* present in a system held at a temperature of 250°F or 121°C. It should be noted that, in principle, the kinetic approach implies that sterilization cannot be achieved (because there is no zero on a logarithmic plot) so the nonkinetic offers more hope. The $F_o$ can be measured and there is an interconnection between $F_o$ and D if $F_o$ is slightly redefined as the number, n, of D values required to sterilize a system, i.e., $F_o = nD$.

It has become an accepted practice to attempt to achieve a $10^{12}$ reduction in viable organisms, including endospores, i.e., $F_o = 12D$. This implies to all intents and purposes that there will be only one surviving organism in a single container in a batch of *1,000,000,000,000* ($10^{12}$) containers. Since the batch size is usually significantly smaller than this number, the batch can be regarded as sterile.

Hence, the value of extensive prefiltration in order to reduce the overall initial bioburden of the product being treated cannot be over emphasized.

## STERILIZATION METHODS

Bearing in mind the previous discussion, there are basic official or unofficial sterilization procedures, all of which are overkills, designed to kill or get rid of the very last and most resistant organism in the system being treated. Filtration is, of course, designed to physically remove all bacteria present. It does not usually remove viruses or mycoplasms and, as noted above, some of the assumptions made during a filtration process need to be very carefully evaluated by the operator.

The basic sterilization procedures involve the application of dry or moist heat, the use of sterilizing radiation or, more recently, the use of intense bright light. A valuable source of information is the section on Sterilization and Sterility Assurance of Compendial Articles in the General Information Chapter (<1211>) of the United States Pharmacopeia, now in its 28th revision. Since it is now revised on an annual basis the interested reader is referred to the latest edition for up-to-date information.

### Moist Heat

The cookery book approach to heat sterilization simply dictated that the product be heated (in an autoclave) at 121°C for 15 minutes. Looking at the process more carefully, one can see that, as the product starts to warm up, enzymic processes in the contaminating bacteria will also speed up their activity until, ultimately, the hydrogen bonding holding the molecule together is affected. The protein is unable to stay in solution and becomes denatured. As the temperature continues to rise the proteins in general become dysfunctional, together with the DNA and RNA in the bacterial cells. However, these processes proceed at different rates and the one that is more readily affected than the others can be said to be rate limiting. Since the cell is now unable to function for all intents and purposes it can be said to be dead although the process is designed to be an overkill, thereby preventing recovery at any later stage. Marker organisms are sometimes used to ensure that the process is functioning as designed. One example is the use of *Geobacillus stearothermophilus* endospores, which are known to be more resistant to moist heat than most average microorganisms. Unfortunately, there is now evidence that other organisms in nature are much more resistant to heat, such as those found in natural geysers and deep sea underwater hot vents. However, for all intents and purposes these are unlikely to be encountered in practice.

### Dry Heat

Dry heat sterilization, carried out in an oven, is defined exposing the product to at least 150°C for 1 hour and is required for the sterilization of surfaces, such as metal components, glassware, and some stoppers. Here the marker organisms are endospores of *Bacillus atropheaus* (formerly known as *Bacillus subtilis*, var. *niger*) and the process is probably more efficient than moist heat since it is more rapidly applied. Obviously the surface is also desiccated thoroughly which destroys functionality of most of the internal and external cellular proteins, but there is likely to be some considerable degree of oxidation involved and this will also interfere with functionality. This method is applied for the most part to the sterilization of surfaces of glass or metal such as components which are to be subsequentally used in the aseptic assembly of a product. The process is also applied to the depyrogenation of glass and metal surfaces. However, it should be noted that oils and hydrophobic products which are not readily destroyed by oxidation under these conditions are also treated in this manner (Groves and Groves 1991).

## Sterilizing Gases

The use of dry heat for the most part is too destructive to be used on anything other than metal or glass surfaces that are not affected by the heat process. Gases such as ozone, formaldehyde, chlorine, ethylene oxide, or hydrogen peroxide can be used to sterilize materials such as surgical dressings and other labile substances. The process can be carried out at ambient room temperature but is affected by temperature and humidity as well as the length of time the gas is in contact with the surface. Normally gas sterilization is carried out in a sealed vessel such as a modified autoclave and the gas is allowed to be in contact with the product for up to 6 hours. After this the gas is substantially removed by exposure to vacuum, preferably overnight. The important stage in the process is the subsequent removal of unreacted gas by the application of repeated cycles of reduced pressure and, often, exposure to ambient conditions over time (for example, storage in a warehouse for several weeks following the sterilization in order to allow the gas to dissipate). In some cases the gas is also flushed out with sterile-filtered carbon dioxide or nitrogen.

Because these sterilizing gases are very reactive chemically, their uncontrolled use could be hazardous to the operators of the process and any residual gas could irritate the patient or affect any product that it comes into contact with. Other gases have been introduced such as chlorine dioxide but the main issue remains that of the residual gas and, in some cases, their degradation products. Hydrogen peroxide has some advantages since the degradation product is water. Moreover, ethylene oxide and high concentrations of hydrogen peroxide represent a considerable explosion hazard and precautions must be taken to prevent this. Ethylene oxide, for example, is usually supplied diluted with carbon dioxide. The use of sterilizing gases is therefore usually best avoided unless there is no other alternative.

## Ionizing Radiation

Heat is only one form of radiant energy and other forms of energy can also be used to kill microorganisms. A wide range of electromagnetic radiation can be applied to the sterilization of pharmaceutical products, ranging from low-frequency ultraviolet radiation through x-rays to the hard, high-frequency gamma rays. Gamma irradiation is becoming more widely used because, although expensive from the installed capital point of view, it is readily applied on a continuous basis that reduces per item costs.

The main use for ionizing radiation is for the sterilization of dressings and plastic syringes. In practice a major advantage is that sealed paper or plastic packages can be sterilized *in situ* unopened. Radiation usually produces free radicals in solutions which may increase the degradation rate of the product and often causes discoloration of glasses. For this reason radiation is only rarely applied to ampouled products. Glass itself is also usually discolored although this color often fades with time. Doses of absorbed radiation are generally at the level of 25 kGrays (kGy), 2.5 mega rads, but lower doses have been used successfully in practice.

Recently an alternative form of radiation has been offered, namely the use of an intense source of bright white light. The claims made for this process are impressive although whether or not it proves to be practical in the real world has yet to be demonstrated.

## Sterile Filtration

Many of the drugs produced by the new biotechnology have proved to be labile and can only be sterilized by cold filtration. These methods depend critically on physically removing microorganisms and general debris from the product by passage through an appropriate filtration medium, often a porous membrane. The membrane itself requires to be supported in a suitable holder made of plastic, glass or stainless steel and the assembly needs to be considered as a whole rather than just the porous membrane by itself. As noted by Groves (1998), the criteria for an ideal filtration medium depend to some extent on the nature of the product being filtered but it is probably reasonable to consider the following attributes as being generally desirable.

1. The filtration matrix should be absolute in the sense that there should be a predetermined and known limit to the diameter of particles in the filtrate.
2. An independent method for checking or validating the performance of the filter.
3. The efficiency of the filtration process should not be significantly affected by the pressure differential across the surface of the membrane or pressure fluctuations produced by the pumping of fluids through it.
4. The filtration medium should not produce any extractibles or otherwise affect the product being passed through it.
5. The filtration system as a whole (i.e., the filtration membrane and its supporting system) should be capable of being sterilized prior to being exposed to the product. This usually involves the application of dry heat, moist heat, or, in some cases, sterilizing gases.
6. In use, the system should be easy to use and economical.

Unfortunately, in practice there is no ideal filtration medium and as described by Groves (1998) and Olson (1998), the whole arena has been confused by over enthusiastic salesman and underqualified consultants. The neophyte coming into this subject should be careful when reviewing sales literature.

Physically one can envisage the movement of particles suspended in a liquid passing through a porous matrix until the pores become too small to allow any further passage. This has become known as the sieving mechanism and, intuitively, makes the process easy to envisage. Another adsorptive mechanism also exists in which charged particles adhere to sites of opposite charge on the walls of the filtration matrix. Depending on the pH or constitution of the medium, this mechanism may become dominant so that a filter will collect particles much smaller than the nominal pore size of the filter itself. Membrane filters are now available in a wide variety of materials, including cellulosic derivatives and polymers and come with a variety of compatibilities, pore diameters, and available surface for filtration. Usually there is a spread of pore diameters but it is usually prudent to select a filter in which the largest pores present are unlikely to allow the passage of any undesirable particulate materials. This would suggest that selection of a matrix with a narrow size distribution would be desirable although methods for the evaluation of this parameter are not necessarily well developed. The subject is well reviewed by Jornitz and Meltzer (2004).

Validation of filter performance has remained an issue for some years but is now approaching some sort of resolution. The consensus is that a filter should prevent the passage of the microform *Berkholderia* or *Brevundimonas diminuta,* an organism which grows to a diameter of around 250–300 nm under selected environmental conditions. This is obviously a worst-case scenario but, as discussed by Olson (1998) and others, if a filter is challenged with more than 60 organisms per cm$^2$, the chances of microorganisms passing through the filter are enhanced. Passage of particles can be considered to be a probabilistic process, the higher the number of particles per unit volume challenging the surface, the more likely it is that some will pass completely through.

This situation is analogous to the heat sterilization situation in that, the cleaner the solution presented to the sterilizing filter, the less likely are particles (microorganisms) to pass through. The initial or final bioburden of the product become major issues. Industrially it has become a routine practice to pass the solution through at least two filters. This covers the possibility of a filter breaking or leaking but, more to the point, the first filter acts as a prefilter so that the final filtrate has a greater chance of producing sterile filtrate.

In recent years the collective wisdom of producers and regulators has been that sterility will be achieved using filters with mean pore diameters from 200–250 nm. This was a step up from the situation 20 years ago or so when it was believed that filters with nominal diameters of 450 nm would achieve sterility. As noted by Jornitz and Meltzer (2004), there may be situations that require 100 nm porous filters but this causes problems with the flow rates which may become very low on a large scale.

Using *B. diminuta* was sufficient to demonstrate that small microorganisms could pass through any particular filter. The use of 450 nm filters therefore fell into disuse. However, recently Trotter et al. (2002) have raised the possibility of using 450 nm filters as prefilters. They demonstrated how different 450 nm filters could significantly reduce the initial bioburden which would therefore provide a better prefilter to a final 200 nm filter. These authors demonstrated the importance of controlling the pressure differential across the filter, removing >10$^8$ cfu/cm$^2$ at a pressure differential of 30 psig. This is a major difference from the older standard of 60 organism/cm$^2$ and may indicate the importance of having a higher bioburden to close off some of the pores in the filter. Nevertheless, this approach is more attractive than having a 200 nm prefilter and a 100 nm filter in terms of cost, effectiveness, and flow rate of product. Sight should not be lost of the fact that nonbacterial organisms may, or may not, be removed from a product. Few claims to remove viruses, for example, have been substantiated and there is little doubt that sterile filtration cannot be considered to be an absolute process.

### The Concept of "Size" as Related to a Filtration Process

Most bacteria have "sizes" larger than 1 μm, but spherical forms of microorganisms are rare in nature and spherical equivalents are not always easy to measure. A general exception may be found in some bacterial endospores approximate in shape to spheres, with diameters varying from 400 nm to 1 μm. Viruses are considerably smaller, with diameters from 50 to 100 nm and fungi are much larger than 20 μm. Thus, over three orders of magnitude need to be considered when designing a filtration sterilization process although it must be admitted that removing smaller viruses offers a considerable challenge. This consideration immediately suggests that the solid phase separation requires to be carried out in stages, removing the larger particles, first with one screen and the smaller ones with another, smaller, screen. The work of Trotter et al. (2002) noted above would certainly be consistent with this philosophy. The size quoted for a sterile filtration porous medium should be that of the largest pores present although some producers talk about a mean size which may be misleading.

However, large holes or tears can be introduced into an otherwise effective sterilizing filter and performance tests have to be designed to ensure that the complete assembly of the filter membrane and its supporting equipment are devoid of imperfections. These include bubble testing in which gas is forced under pressure through the wetted filter and the pressure required for bubbles to first appear measured. In principle this identifies the largest hole present in the complete system.

Some microorganisms are capable of growing through a filter if allowed enough contact time. Protoplasts lack the rigid walls associated with most bacterial cells and are also capable of passing through a membrane filter. This is one reason why 100 nm filters are being evaluated, especially in the biopharmaceutical industry, and it is considered that at least large viral particles can be substantially removed from a product. Smaller pore size filters are available but the flow rates through them are miniscule and not helpful for large scale production methods.

Overall it will be evident that filtration methods are not necessarily absolute and the best that can be claimed for the process is that it is aseptic.

### Aseptic Assembly

There are situations where the application of energy in any form is impractical because the formulation or the active ingredient is extremely labile and is not be able to withstand an otherwise effective sterilization process. Under these constraints an alternative would be to sterilize all the components of the product separately and assemble them in an aseptic environment. For example, all glass or metal components could be sterilized by dry heat, rubber or plastic by hydrogen peroxide or ozone, and the solution packaged by filtration repeatedly through appropriate presterilized filters directly into the cooled containers.

The assembly itself would be carried out in an isolator designed to provide a clear separation of the human operators and the product, at least until such time as the container is sealed. Ideally, robotic assembly, using machines to carry out all the assembly stages under thoroughly, sterile, filtered air cover is used, again in an isolator. However, for most practical purposes this is rarely used because the batch sizes are too small and the mix of product for the assembly line is too large to justify the often considerable capital expenditure. The object of the exercise is to avoid or at best minimize exposure of the product to the human operative because they are usually the major sources of airborne particulate matter and dust. Operators are suitably dressed to reduce the contamination but the risk associated with human handling of the product is always finite and must be minimized. Newer designs of isolator are now allowing operators to wear street clothing because they work outside the isolator, manipulating product through porthole gloves. These appear to be more successful at protecting the product and are more popular with the workers who can work comfortably without the constraints of special clothing and procedures. It is generally accepted that under properly controlled operations, a modern aseptic process will result in a product that will have a very low probability of containing contaminating microorganisms. As noted earlier, this does not mean that the product is sterile in the absolute sense.

In recent years the tendency has been to move toward the use of various plastic containers. They are lighter and cheaper than glass containers and are not as easily broken. Factors such as water transpiration rates and surface adsorption of formulation components need to be monitored when investigating the practicality of using any particular plastic as an alternative container. If the product volume throughput justifies again the considerable capital cost of the machine a useful packaging system would be the use of a "form-fill-seal" device, sometimes called a "blow-fill-seal" machine. Available for the past 40 years, the pharmaceutical industry has been slow to accept this technology, although there are signs that this cautious approach is changing.

Available from a number of sources (e.g., Rommelag, Edison, NJ; APL Engineering, Woodstock, IL; or Weiler Engineering, Elgin, IL), this machine takes plastic granules, melts them and forms a bottle inside a mold by blowing in sterile-filtered air. The filtered product is then filled into the resulting cavity which has the effect of cooling the body of the bottle. The neck is then sealed and the filled bottle ejected. The whole operation is carried out in an atmosphere of sterile-filtered air. Validation studies have demonstrated that these systems, operated correctly, are capable of producing sterile product with a very high throughput. Filling volumes down to 0.2 mL and up to 50 or 100 mL, these machines require considerable capital investment and are difficult to adjust for optimal efficiency. However, once functional, at no time during the filling and sealing process is it necessary for an operator to become involved. Indeed, manpower requirements are low since one operator can monitor several machines and, ultimately, this saves costs.

It should be noted that aseptic assembly of a sterile product becomes difficult with large volume containers and the Food and Drug Administration, for example, would be unlikely to approve an aseptic process for container volumes in excess of 100 mL, i.e., by definition only small volume injectables are acceptable for aseptic assembly.

## QUALITY CONTROL ISSUES

Like all pharmaceutical products, parenterals require quality tests designed to ensure they contain the correct ingredients in the appropriate doses. Stability of the product is measured and suitability for its stated purpose determined. In some quarters this is a definition of "quality." Perhaps it should be pointed out that this definition does not go far enough. There are nonquantifiable issues associated with this definition. An example would be automobiles since, intuitively, one can perceive the difference between, say, an Aston Martin and a Ford Focus, both, as it happens, made by the Ford organization. They certainly are suitable for their stated purpose with engine, wheels, and other factors associated with an automobile but, intuitively and given the choice, the average person would be likely to choose the Aston Martin because of a perceived *higher quality*.

However, parenterals have other requirements in that they must be sterile and free from pyrogens. Sterility testing, especially when the initial bioburden has been lowered by repeated filtration, becomes very problematic. The statistics of sampling as well as the methodologies for detecting small numbers of bacteria ensure that the chances of detecting a contaminated container are very low. The only real value of the so-called sterility tests is to satisfy regulatory authorities although, in fairness, it has to be admitted that the tests might detect a situation when a batch has not been heated at all or where the heat treatment is inadequate. Parametric release, where a product is exempted from the lengthy and rather dubious sterility testing, is allowed if it can be shown that the process will successfully result in a sterilized product. Generally applied to high throughput large volume parenterals such as saline or dextrose solutions, it can be applied to any product if the batch size is sufficiently large as to justify the validation processes required.

Sterility testing is designed to detect bacterial contamination. Although viruses and other small microorganisms will pass through filters they are not usually tested for unless the product has a special requirement that demands this additional testing.

Combined with testing for bacteria, the pyrogen test has become important, especially now that exquisitely sensitive methods for *measuring* pyrogens are available. Although not all bacteria produce pyrogenic materials, low levels of pyrogen in the final product provides additional assurance that the initial bioburden in the product, prior to any applied sterilization process, was correspondingly low.

Finally, additional quality test being applied to many but not all pharmacopoeial parenteral solutions are the instrumental tests for particulate matter. Although almost impossible to remove in its entirety, particulate matter, however it is defined, are another direct measure of "quality," the lower the number per unit volume the higher the quality. Fortunately instruments are now available that will accurately count the numbers of suspended particles in a solution down to 2.0 μm diameter. The interested reader is referred to <788> in the current USP for a complete account of the methodology and standards.

## Lyophilization (Freeze-Drying)

Lyophilization is a process in which a solution of a drug is frozen to a solid and the solvent, usually water, removed by sublimation on exposure to a vacuum. The process has been studied intensively because it can be applied to the preservation of labile drugs or materials such as proteins which would otherwise be adversely affected by the solvent over a period of time.

In practice the entire process is required to be carried out under aseptic conditions, including operating in an atmosphere of filtered sterile air. This provides for difficulties in the design and application of suitable equipment although the situation is slowly improving in practice, mainly at the insistence of the FDA.

In the operation, a sterile-filtered solution of the drug together with appropriate formulation aids and bulking agents such as mannitol or lactose is pumped into a presterilized container and the sterile closure placed loosely on the neck of the vial. The vial is then placed on a shelf that is hollow in cross-section or has cooling coils underneath to allow the circulation of a refrigerant or heating fluid. The contents of the vial are then cooled to at least $-40°C$, enough to freeze the liquid contents to a solid cake. This is then exposed to a vacuum of around 0.1 Torr (mm Hg) and the shelf temperature allowed to warm up above the melting point of the solid in order to start the sublimation process. The water vapor is allowed to condense on a cold surface at some distance from the vial. This is often designed to be the walls of the chamber which are maintained at $-60°C$. All environmental conditions inside the chamber are monitored to ensure that the contents of the vial do not go above the so-called triple point, the point at which there is an equilibrium between solid, liquid, and gas phases of the solvent. Below this point the solid sublimes directly to the gaseous phases without melting. A challenge here is to prevent the solid cake inside the vial from melting or approaching the melting point since the ideal characteristics of a dried powder cake are affected, including appearance and ease of solution when being reconstituted.

If the process is carried out correctly, a dry, porous solid cake will be produced that essentially retains the shape of the original solution. Often a secondary heating stage is applied to the solid cake in order to remove the final traces of water. Since the cake is effectively formed there is no need to worry about the solid structure changing and conditions can be allowed to rise above the triple point. This stage is limited by the need to avoid thermal damage to the product and the shelf temperature under these conditions will rise to as much as $40°C$, before the final traces of water are driven off. The overall process is slow, taking as much as 3 days for the drying to go to completion.

The need for two stages in the lyophilization process is that the free water, frozen to crystals, is readily removed under low pressure by sublimation. However, water bound by hydrogen-bonding to the crystalline lattice or molecular matrix of the product, especially if it is proteinaceous, is more difficult to remove and requires a much longer time and increased heat exposure.

After the drying is judged to be complete the closures are pressed into place by hand or by hydraulically raising the shelf to the roof or adjacent shelf. The retaining metal bands or rims are then fixed into place. However, until this final stage, the whole process has to be maintained under a laminar flow of sterile filtered air at a carefully controlled temperature and humidity.

As noted, lyophilization has been subjected to a number of academic studies, which have helped to understand the basic physics of the process. Many of the observed effects are subtle and sometimes difficult to explain. Loss of biological activity of proteins has been observed and this may be due to conformational changes that occur during the drying process or subsequent storage and handling (Chi et al. 2003). Some proteins dimerize or aggregate to higher molecular weight forms, even forming denatured or insoluble compounds. These reactions usually take place during the initial drying stage. Selection of a different bulking agent or different environmental conditions during the freezing or drying process will often overcome these issues. However, when dealing with a new product or even a reformulated product it will be evident that a considerable amount of investigational work is required to ensure a successful outcome.

## DRUG DELIVERY THROUGH THE SKIN

### SKIN

The skin is the largest organ of the body and offers a lot of opportunities for passive delivery of drugs under appropriate conditions into the underlying systemic circulation. The surface area of the skin for a typical adult weighing, say, 90 kg and 180 cm tall is approximately 2 m$^2$ (data derived from the Geigy Scientific Tables, 3, 1981). Not all of the skin is of equal thickness as will be evident when comparing, for example, the sole of the foot with the scrotum. There are other advantages associated with delivery of drugs through the skin. For example, first pass metabolic effects seen on passage through the liver are generally avoided and constant levels of drug in the bloodstream are often maintained for prolonged periods of time. If these levels are below levels that manifest toxicity, side effects associated with the drug are often avoided or at least minimized. In turn this allows the administration of lower doses and bioavailability is enhanced. This often increases patient compliance. On the other hand, once delivery has commenced it is not easy to terminate delivery of the drug if unwanted side effects are seen following initial administration and, as with all drugs, there is always the possibility of skin irritation or local sensitivity reactions.

Structurally, the skin consists of the outermost epidermis which is essentially composed of dead squamous cells sloughed off from the underlying dermis (Figure 7.1). The dermis lies on top of subdermal layers, which further down gives access to nerve endings and capillaries of the circulatory system.

The *stratum corneum*, located as a discrete layer just underneath the outermost part of the epidermis, offers the main resistance to water loss from the body and to the ingress of toxic materials such as solvents or other insults from the external environment. The skin as a whole offers a natural resistance to the application of

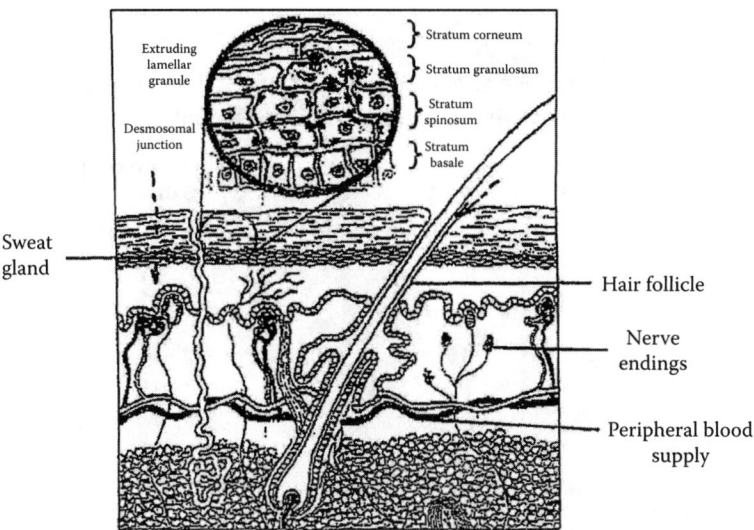

**FIGURE 7.1** Cross-section of human skin showing drug pathways.

externally applied drugs or toxins but there are alternative pathways for the movement of materials applied to the skin, for example, down the sweat glands or hair
follicles. Some drugs applied in oleaginous vehicles such as ointments and creams
are known to penetrate through the skin layers, reaching the peripheral blood vessels
and then the systemic circulation.

Scientifically, however, one would like to have greater control over the absorption
process. This means applying external materials or drugs to a defined area of the
skin surface where the diffusion process would be more likely to be uniform and
therefore more controllable. Individuals vary considerably in their body hair, as one
example, and obviously there are additional differences between men and women
from this perspective. Some materials readily penetrate the skin but a more systematic exploration of the factors influencing skin permeability over the past two decades
has determined many of the properties of materials that influence the movement of
drugs through the skin. Intuitively, for example, properties such as charge, molecular
weight, solubility in oil or water (or the o/w distribution coefficient), area of application, concentration of the applied drug, and specific biological activity on a mg/kg
basis would all have some influence and, for the most part, this has proved to be
correct experimentally.

This area of exploration and development has resulted in the successful clinical
and commercial development of a number of transdermally applied drugs. Some of
these are shown in Table 7.1, but it will be noted that there appears to be a limiting
molecular weigh below approximately 500 Da.

Most of the drugs shown in Table 7.1 are administered passively in the now
familiar skin "patches." These devices are carefully engineered in order to ensure
that the drug is released at a predetermined rate, often with the use of rate-controlling

## TABLE 7.1
### The Molecular Weights of Drugs Delivered Transdermally, Either Commercially or Under Clinical Evaluation

| Drug | Molecular Weight (Daltons) |
|------|---------------------------|
| Nicotine | 162 |
| Ibuprofen | 206 |
| Nitroglycerine | 227 |
| Clonidine | 230 |
| Methylphenidate | 233 |
| Lidocaine | 234 |
| Flurbiprofen | 244 |
| Estradiol | 272 |
| Testosterone | 288 |
| Diclofenac | 296 |
| Ethinylestradiol | 296 |
| Scopolamine | 303 |
| Dihydrotestosterone | 322 |
| Norelgestromin | 327 |
| Fentanyl | 337 |
| Alprostadil | 354 |
| Oxbutynin | 358 |
| Dextroamphetamine | 369 |
| Dexamethazone | 392 |

membranes. There are basically three major types of patch or transdermal drug delivery system (TDDS) as shown in Figure 7.2, adapted from Hopp (2002). These are all based on the technology associated with spreading materials on a base layer of plastic sheet or paper and consists of the drug dispersed or dissolved in a suitable solvent, the rate-limiting membrane, and an adhesive to ensure that the drug delivery device is maintained in intimate contact with the skin surface to which it is applied. The actual identity of the materials composing the patch are usually determined by the nature and properties of the drug involved.

The familiar use of creams or ointments as alternative means of delivering drugs to the skin has been systematically explored in the case of, for example, nitroglycerin where the controlling factor in delivering a known, controlled, and reproducible dose is the area of application. Generally, if the drug is effectively an insoluble solid dispersed in a matrix such as petrolatum or oil-in-water cream, we find that the particle diameter of the dispersed insoluble drug begins to exert an influence as well as that of the vehicle. Oil droplets in water-miscible creams are between 10 and 100 µm diameter. Droplets in excess of 100 µm diameter tend to remain on the surface of the skin but those below 3 µm can penetrate down hair follicles at random. There have therefore been reports in the literature that submicron emulsion systems improve

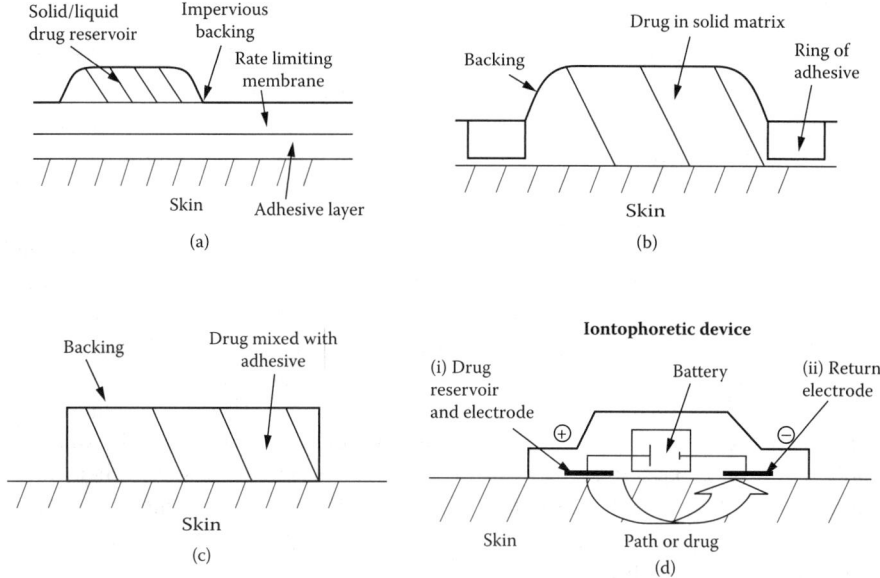

**FIGURE 7.2** Typical transdermal drug delivery system.

drug penetration through the skin. The *stratum corneum* is hydrated, allowing the formation of gaps in this protective layer through which the small emulsion particles can penetrate to form a drug reservoir within the skin. These observations do not appear to have been exploited commercially but clearly the approach could be promising given a suitable drug candidate.

As recently reviewed by Gupta and Garge (2002), there are some materials known to penetrate the skin readily and appear to be capable of acting as penetration enhancers for certain selected drugs. These enhancers sometimes work more effectively in the presence of solvents such as ethanol or propylene glycol. A well-known example is the use of the insect repelent DEET (N,N diethyl-m-toluamide) as an enhancer for corticosteroids or the use of isopropyl myristate and propylene glycol for diclofenac sodium. Indeed, cyclodextrins have also been employed as penetration enhancers for hydrocortisone although how this system functions is not easy to visualize (see later section on cyclodextrins).

When considering transdermal delivery of proteins it is likely that the strongest and controlling influence would be the size of the molecule. The use of enhancers such as DEET has been claimed to considerably broaden the range of molecular weights of penetrating compounds (Walters 1989). Experimentally it has proved possible to drive some complex proteins and other immunologically active materials such as vaccines into and through the skin but, as yet, none have been utilized on a large scale in the clinic. However, it must be said that some methods look extremely promising.

## Iontophoresis

Iontophoresis is a method for assisting the transport of materials through the intact skin by applying a mild electrical field of opposite sign to that of the diffusate (Guy, 1998). The diffusing species must therefore carry a charge under the optimal conditions and many low molecular weight polypeptides such as insulin have been successfully administered to the body in this fashion. Controlling the exact dose is not easy but the prospect for the design of electronically controlled feedback drug delivery systems is promising. In the case of insulin, for example, glucose excreted through the skin can be measured before applying enough insulin back through the skin to correct the situation. Prototypes for this type of device have been constructed and demonstrated clinically so obviously these systems are, indeed, successfully approaching the marketplace and should provide a valuable function in the future. Claims have been made for the successful application of electrical currents to increase the natural closures between cells, to open up gaps through which externally applied drugs can pass. Reading some of these claims raise doubts as to their likely application in the future. For example, high although transitory currents are required and there is always the prospect for causing pain to the patient which will affect the compliance to a given therapy.

# MULTIPHASE DRUG DELIVERY SYSTEMS

## MICROEMULSIONS

Microemulsions have been controversial since they were first defined by Schulman in 1959, although Schulman himself had first described such a system in 1943 (Rosano and Clausse 1987). Various authors have offered different explanations for the spontaneous formation of thermodynamically stable, isotropic, transparent compositions when mixtures of surfactant, oil, and water are titrated with a cosurfactant, usually a medium chain aliphatic alcohol. These systems, not always formed with pharmaceutically acceptable surfactants, have proved to be of exceptional interest to the oil-drilling industry but recently have started to be explored pharmaceutically. At least one microemulsion system is being used commercially to deliver the exceptionally hydrophobic cyclic oligopeptide drug cyclosporine and similar drug delivery systems were being evaluated at the time of writing.

The term "microemulsion" has been disputed by a number of authors. A consensus is slowly appearing although some authors prefer the terms "transparent emulsion," "micellar emulsion," or "swollen micellar emulsion." As Rosano and Clausse (1987) pointed out, "microemulsion" is most often used to describe any multicomponent fluid made of water (or a saline solution), a hydrophobic liquid (oil), and one or several surfactants. The resulting systems are stable, isotropic, and transparent and have low viscosities. In addition, microemulsions can themselves be compounded to be hydrophilic or hydrophobic in which the continuous phase can be either water or oil, respectively. Indeed, some authors have described waterless systems as microemulsions or media where formamide or glycerol have been used instead of water.

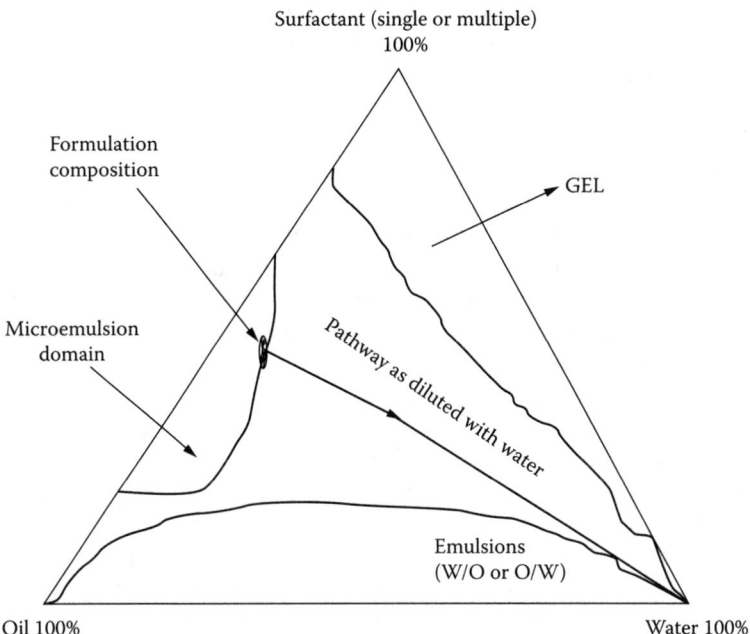

**FIGURE 7.3** Hypothetical three-phase diagram of surfactant/oil/water composition illustrating the rapid change in the constitution as a microemulsion phase is diluted in water.

From the pharmaceutical perspective it should be noted that although appropriate systems may be very stable, they are strongly dependent upon the composition of the system, usually characterized in the form of a tertiary or quaternary phase diagram. In addition, most systems contain high concentrations of at least one of the surfactants present. When the composition is changed, in some cases even slightly, the microemulsion structure and properties are changed, no longer being isotropic or transparent and increasing in viscosity. Although they have been claimed to be valuable drug delivery systems, when injected or administered orally into an aqueous environment the systems rapidly destabilize as their composition moves to the point represented as 100% water on the phase diagram (Figure 7.3). An important formulation issue is that the order of mixing is an essential requirement during manufacture and this should be determined at a very early stage of development. This consideration brings to mind the comment by Schulman that the cosurfactant requires to be added by titration.

Microemulsions form spontaneously in much the same way as structural elements, such as surfactant micelles, rearrange themselves following the addition of the cosurfactant. Because the water may be incorporated into the hydrophilic structures of reverse micelles, when examined by x-ray analysis spherical droplets with diameters of 6–80 nm have been reported.

In an attempt to explain the spontaneous movement Schulman introduced the concept of *negative interfacial tension*. This was used to explain why the positive interfacial tension suddenly dropped to zero or, at least, very low values when the order of mixing to the same composition point was changed. The same concept has been used to explain the spontaneous formation of emulsion systems and is discussed under that heading. Basically, it was suggested that there was a point in the composition of the interface between the oil and water at which the cosurfactant, usually a short chain acyl alcohol, readily penetrated and redistributed itself between the aqueous and oily phases. Negative interfacial *tensions* have been reported in some situations, although this can only be understood as an extension of positive surface *pressures.*

All things being equal, microemulsions only form under very specific conditions and compositions. Temperature is a critical condition as is the initial composition of surfactant, oil, and water. The cosurfactant and manner of mixing are both critical to the formation of a microemulsion but, once it is formed, the system will be remarkably stable. However, microemulsions are adversely affected by any attempt to dilute the composition with one of its components such as oil or water. Again, from a pharmaceutical perspective, the fact that the composition almost invariably contains a high concentration of the main surfactant is generally not favorable to either the cost of the basic admixture or to the probable toxicity because concentrated surfactants can be irritating when applied to the skin or irritating on or around an injection site. Once applied to the skin, injected or ingested the composition changes in its identity and is no longer structurally a microemulsion. Considerable care is therefore required when selecting a suitable surfactant for a particular purpose, especially when considered from the aspect of toxicity.

Indeed, the selection of raw materials from the pharmaceutical perspective is severely restricted by toxicity concerns, and Attwood and Florence (1998) suggested that only a few nonionic surfactants such as polysorbates 80 and 20 may be suitable for oral administration, with the possibility of some phospholipids serving the same function. Since that time a small number of other nonionic surfactants (e.g., cremophores) have been evaluated.

At issue here is the question of toxicity and the willingness of manufacturers to undertake the extensive and costly testing required by regulatory agencies. There may be other suitable materials but the perceived need and economic return on investments, of necessity, control whether new materials will appear on the market place.

The review by Attwood and Florence (1998) is valuable because it provides a comprehensive survey of the earlier pharmaceutical literature. From this we can learn that microemulsions are formed spontaneously from fixed compositions of surfactant(s), oil, and water but these compositions change with temperature. As a stable system is diluted progressively with, for example, water, the microemulsion phase will spontaneously revert to other phases on the diagram, including unstable emulsions or a solution. This becomes relevant when considering what happens to the system when it is administered. No matter how it is administered the microemulsion phase will be diluted, most likely with water in some form or another, and revert to some other composition in the body.

Some authors have claimed that their system survived as microemulsions almost down to the 100% water point on the triple-phase diagram. Unfortunately this claim does not appear to be substantiated by the published phase diagrams and the claim is counterintuitive. Experimentally there are considerable practical difficulties in establishing an accurate phase boundary with a viscous component and this question must remain open for the present.

## Microemulsions as Particulate Systems

There has been a tendency to defend the definition of a microemulsion as a system—composition of matter—that consists of a stable dispersion of small (or water) droplets with diameters of 50–140 μm in a matrix usually consisting of of a surfactant/cosurfactant admixture. This definition ignores the importance of the high surfactant concentration in the subsequent performance of the system, and some authors have attempted to measure the droplet size by diluting the system down with water before analysis, thereby totally ignoring the significance of the phase composition. Interestingly, these studies suggested that the resultant emulsion particles were indeed small (~150 nm), but whether these particles were in the initial matrix is questionable.

These types of studies illustrate the confusion amongst scientists that has existed for generations and it is essential that these issues are resolved as we study the *in vivo* and *in vitro* behavior of drug delivery formulations based on microemulsions.

## Spontaneous Microemulsion Formation

There are two main characteristics of microemulsions that are interesting to the pharmaceutical formulator, assuming all other attributes are acceptable such as the toxicity of the component materials. The first is the spontaneous formation and dispersion of small (<1 μm) oil droplets in the continuum as soon as the water concentration has reached an optimum value during a mixing process. This composition is remarkably stable isothermally but does revert to other structures if the temperature is significantly changed one way or another.

The other significant factor concerns the viscosity of the transparent system which is low although, as one group claimed, it is unlikely to be non-Newtonian. Whether these considerations are relevant to the formation of spontaneous emulsions (later) remains to be seen but this whole area is one of considerable scientific interest, quite apart from its pharmaceutical application as drug delivery systems.

# PROTEIN COMPACTION

Tablets or compacted powders and granules are widely utilized for the oral delivery of drugs. In addition, tablets may also be used for delivery to other body sites such as the vagina or as sterile implants.

The subject of protein compaction started to become of interest when some proteinaceous enzymes were offered as oral treatments for indigestion and resulted in published studies of the effects of compaction on biological activity.

More recently the use of compacted bovine somatotropin pellets has been introduced. These have been administered to cattle by using a compressed air gun to shoot the pellets into intramuscular tissues. This was a major advantage since the drug could be administered at a distance, cows being notoriously shy of individuals in white coats waving syringes. In addition, compressed hormonal pellets are used for pigs and fowl and this has become a new venture for both human and veterinary medicine.

Before discussing the effects of pressure on biological activity it will be necessary to briefly review the behavior of powders or granules under applied pressure. Powder and granular particles tend to compact in three main stages.

1. Particle rearrangement and closer packing as the available volume becomes reduced
2. As they move closer together the particles undergo plastic and elastic deformation, and, finally
3. There is cold working in which the particles begin to adhere to each other, with or without fragmentation

At the onset of the compression cycle, particles undergo some rearrangements to reduce particle–particle distances without undergoing excessive deformation. This initial movement results in the decrease of spaces or voids between particles. As the applied compactional stress is increased, the available volume becomes less and the individual particles begin to deform, thereby filling inter- and intraparticular spaces. At this stage there is a progressive increase in the compact density and bonds can form between adjacent particles so that, should the compaction be removed, the shape of the final compacted space will be retained. Further increase in the compaction pressure will result in more particle fracture, the formation of more interparticulate bonds, and reduction of the void spaces between particles. Ultimately there is irreversible work hardening as the voids disappear altogether.

However, it should be noted that these processes may not be sequential and will often occur simultaneously within the same compact bed. In some formulations the plasticity is such that deformation and adhesion stages will be influenced and the compact—the shape of the compaction volume that the composition forms—is retained more readily.

Several processes for the formation of compacts have been proposed as it will be evident that links or bonds can only form across adjacent surfaces. Some of these mechanisms for bond formation are

1. Mechanical interlocking
2. Formation of solid bridges by the fusion of adjacent crystals ("cold welding")
3. Adhesive and cohesive forces at binder bridges which are not freely movable
4. Dispersion forces being available between particles
5. Interfacial forces and capillary pressures become available through access to freely movable liquid surfaces

Groves and Teng (1992) investigated the effect of compactional pressure on biologically active proteinaceous enzymes such as $\alpha$-amylase, $\beta$-glucuronidase, lipase, and urease. Assaying the "activity" of these enzymes before and after the compaction process was relatively straightforward. These materials were amorphous and were compacted at pressures in excess of those normally used to prepare formulated powders containing diluents, lubricants, and other compaction aids. Mechanically stable compacts were formed and it seems reasonable to conclude that these were due to the bonds formed from molecular dispersion forces.

The pressure applied was sufficient to remove most if not all voids from the compacts. The original protein particles could be thought of as molecular aggregates, not individual molecules. However, the properties of the aggregates themselves needed to be considered since interparticular voids will exist and may also collapse during the compaction process.

Another important factor proved to be the influence of trace quantities of moisture in the compacts, from both the environment and also within the protein itself. "Dry" proteins adsorb water with avidity because of the exposed hydrophilic regions within the protein structure, changing the properties of the hydrogen bonding within the structure and, therefore, the flexibility of the overall structure. Excessive drying, on the other hand, can result in the collapse of an otherwise water-soluble protein to the point where it is effectively denatured and will no longer dissolve.

Evaluation of the data obtained by Groves and Teng (1992) suggested that biological activity was not lost initially during the initial compaction process but in most cases was suddenly lost over a critical compaction pressure range. This indicated that activity did not begin to be lost until most of the void spaces had been lost and there was an increase in the particle interfacial reactions. Earlier work had suggested that thermal inactivation was involved but while energy is undoubtedly released during the compaction process in the form of heat as surfaces slid over each other, this effect was not confirmed as an inactivation factor. Detailed examination suggested that different, nonthermal, interactions were involved.

It should be pointed out that these studies were carried out on pure proteins, without the benefit of formulation aids. Additives such as diluents and lubricants would certainly function to reduce the energy applied during the compaction process. However, although suitable for oral use, many orally acceptable diluents are toxic when used parenterally. Thus, the only variable that could be utilized for the compaction of a pure protein might be the residual water level in the protein itself. This information would probably be needed when formulating a compressed pellet for subdermal administration or some other parenteral route.

## SELF-EMULSIFYING DRUG DELIVERY SYSTEMS

Self-emulsifying drug delivery systems (SEDDS) are anhydrous solutions of the drug in oil containing surfactant and cosurfactant, which spontaneously emulsify when added to an excess of water.

When used as drug delivery systems these formulations have advantages because it is not necessary to transport water from the production facility to the application site, the drug is not exposed to an aqueous environment until immediately before application, or, in the case of a SEDDS–filled gelatin capsule, after swallowing. The final product is generally a finely divided submicron emulsion that forms very rapidly when added to water. Thus, stability to hydrolysis is increased during storage and transportation of the anhydrous oil phase. In addition, the oily vehicle can be designed to dissolve hydrophobic drugs. Because the droplets form so quickly and are so small, mass-transfer from oil to water phase is usually very rapid.

This phenomenon of self-emulsification was first observed by Johannes Gad in 1878 when he gently layered a solution of lauric acid on top of an aqueous alkaline solution, thereby making a soap *in situ* but also forming an emulsion without the aid of external agitation. A laboratory curiosity for the next 50 or so years, the principle became recognized as being valuable for the formulation of herbicides and insecticides such as DDT. The concentrate could be reconstituted with ditch water and sprayed without the need to carry water to the site.

Spontaneous emulsification of oils carrying drugs make SEDDS good candidates for the oral delivery of hydrophobic drugs with adequate oil solubility since the drug will be presented as a fine (submicron) emulsion that has a large surface area across which diffusion can take place rapidly, thereby facilitating absorption into the body.

For drugs subject to dissolution rate limited absorption such as solid dosage forms, SEDDS may offer improvement in both the rate and the extent of absorption (Constantinides 1995). In at least one published example, for the experimental drug WIN 54954 (Charman et al. 1992) both the dissolution rate and the pharmacokinetic parameters on oral administration to nonfasted dogs from a SEDDS were improved. Spontaneous emulsions may be formed with mild agitation when added to water in order to ensure uniformity of mixing but when given orally in a soft gelatin capsule, SEDDS tend to be mixed with the aqueous environment by the normal motility of the stomach and the intestine.

SEDDS have their main advantage when delivering drugs such as peptides which are otherwise hydrolyzed or degraded by moisture or bulk water. They are particularly useful for the formulation and presentation of drugs such as proteins and peptides which would otherwise become degraded by the extreme temperatures and shearing conditions encountered during homogenization. A SEDDS can be made by simple, direct mixing under low shear conditions and packed into soft gelatin capsules without undergoing exposure to extreme conditions. A major difference between SEDDS and microemulsions would appear to be that the order of mixing for the SEDDS concentrate is unimportant since there are no structures formed at this stage.

Indeed, recently SEDDS themselves have been delivered as liquids absorbed onto powders such as colloidal silicon dioxide or microcrystalline cellulose (Nazzal et al., 2002). Selection of the absorbent was obviously critical to the performance of the system but, as an aside, it seems that this approach negates the rapid release properties of a SEDDS. It will be interesting to follow the future of this technology.

## EVALUATION OF SEDDS PERFORMANCE PARAMETERS

Few methods exist in the literature for evaluation of performance parameters of SEDDS although some basic evaluations have been published (Groves and Mustafa, 1973). The ability of a SEDDS formulation to form small droplets on dilution is an essential feature because this determines the rates of both drug release and extent of the absorption process. This parameter is largely controlled by the nature and concentration of the primary emulsifier. Thus, phase diagrams at ambient temperature have become useful as a means of identifying compositions with optimal performance.

The rate and extent of emulsion formation, the *spontaneity*, is important although there are no widely accepted methods suitable for measurement of this parameter. A standardized visual test was proposed by the Collaborative Pesticide Analytical Committee of Europe (CPAC) and the United Nations Food and Agricultural Organization (FAO). Groves and Mustafa (1973) devised a laser-light scattering device attached to a flowing fluid system in which the time of transit was measured volumetrically. This allowed a transit time to be determined with some precision as the dispersing emulsion was being carried through the tube, thereby allowing a dispersion time, or spontaneity, to be measured. Iranloye and Pilpel (1984) used a simple laser nephelometer to measure the change in intensity of light as the system was being gently mixed and a similar system was used by Pouton (1985). These authors considered this method to be simple and suitable for comparing the self-emulsifying ability of various SEDDS.

Emulsion droplets formed by the aqueous dilution process are often weakly charged and the origin of this charge is usually the free lipid acids in the oil or nonionic surfactant employed in the system. Cells in the body are generally negatively charged, and Gershanik and Benita (2000) showed that positively charged emulsion droplets formed from a SEDDS electrostatically interacted with the mucosal lining of an inverted rat intestine. The formulation was then shown to enhance the oral bioavailability of progesterone when compared with a corresponding but negatively charged formulation.

The polarity of the oil phase may also influence biological behavior by affecting the rate of dispersion in to the aqueous phase. Polarity is determined by the length of the acyl or lipid chains, the degree of unsaturation and the molecular weight of the hydrophilic portion of the molecule and the concentration of the emulsifier. In effect the polarity reflects the ability of the drug to interact with the oil initially and then the water/oil combination after emulsification. Constantinides (1995) suggested that optimum drug release parameters would be obtained by a combination of small droplet size and a low oil/water partition coefficient.

Emulsion stability of SEDDS is usually good because the droplets are small and have narrow size distributions. However, stability can be measured by determining the flocculation rates, degree of separation, or changes in the diameter of droplets formed on dilution over time during storage under various conditions.

The final performance characteristic of any formulation, including SEDDS, is the effect produced physiologically in terms of the drug absorption. The extent of adsorption might be determined by the metabolism of the oil once ingested, the droplet size and charge and the resultant partition of the drug between the oil droplets and the excess of the aqueous phase.

## SELECTION OF COMPONENTS FOR A SEDDS

To some degree the selection of components for a SEDDS will be determined by the properties of the drug itself but, ideally, the final system concentrate prior to dilution should be

1. Composed of readily available, nontoxic ingredients
2. Transparent and monophasic at ambient temperature
3. Providing a solvent for the drug, not as a suspension

For these reasons the selection of an appropriate oil is limited to some extent by the need to use metabolizable long-chain triglycerides. These are represented by natural oils such as olive or soy oil or the short-chain triglycerides obtained from coconut oils. The chain length and degree of unsaturation of the oil, as well as the molecular weight, influence the droplet diameter and polarity once formed.

Initial work on self-emulsifying oils suggested that two surfactants may be required, one slightly hydrophilic and the other slightly hydrophobic. In the phosphated systems evaluated by Groves (1978) an interaction between the two surfactants could be detected as anomalies in the respective phase diagrams and the lowering of the surface tension between the two. However, many modern systems use ethanol as the cosurfactant and current pharmaceutical systems tend to use only a single surfactant. Surfactants used in modern pharmaceutical SEDDS mainly consist of ethoxylated nonionic long-chain derivatives. This is claimed to be because these materials are more stable, have low irritation potential, and generally do not need a cosurfactant. Wakerly et al. (1986) screened a number of medium-chain and long-chain surfactants for their ability to form pharmaceutically acceptable SEDDS and found that the most efficient were surfactants with predominantly unsaturated acyl chains. Oleates with HLBs (hydrophil–lypophil balance) of around II were found to be effective, as were sorbitan esters and the ethoxylated triglycerides such as Tagat TO.

## POSSIBLE MECHANISMS FOR THE FORMATION OF SELF-EMULSIFIED EMULSIONS

What is intriguing about spontaneous emulsions is the fact that they form dispersed system with very small droplets over a short time span, usually within seconds, so that a series of complex events, requiring internal energy sources, must take place as soon as the oil concentrate is added to an excess of water. Emphasis here must be on the excess of water since, if only relatively small amounts of water are involved, the dispersing system rearranges itself and forms a part of the concentrated phase diagram without necessarily forming droplets. The question then arises as to what is actually going on at the very rapidly expanding interfacial area between the oil phase and the water.

Based on the 19th century work of Gad, Gopal (1968) in his review suggested that three main mechanisms were taking place simultaneously, as follows:

1. Agitation of the interface, resulting in the formation of lamellae that break up into small droplets
2. Diffusion of one phase into the other so that solute droplets become stranded
3. Areas of negative interfacial tension develop between the two phases, causing spontaneous breakup of the disperse phase

On their own, these descriptions are insufficient and do not provide an explanation of the energy source. In the 1950s Schulman and coworkers had offered the explanation of negative interfacial tension as part of an explanation for the spontaneous formation of microemulsions but, again, this suffers from a lack of reality.

Groves (1978) provided an intuitive explanation based on a mechanical model in which water penetrates into the oil/surfactant system, forming liquid crystals but, more to the point, considerably expanding the interface. This is the reason why it is necessary to postulate that water is inconsiderable excess. The surface expands so that instead of a negative interfacial tension what we have is a positive surface pressure. At this point it is not unreasonable to visualize the surface expanding and stranding as postulated in the Gopal model.

At this stage Groves (1978) suggested that hexagonal liquid crystalline systems were rapidly forming from the surfactant molecules, the lipid esters, and the water. The net effect is the conversion of a disordered, random mixture of the essential components into an organized, ordered interfacial structure which can continue growing until the components are no longer available or are taking too long to diffuse from the interior of the droplet. The explosive instability due to the rapid increase in the surface area comes to a stop when the constitution of the liquid crystalline phase changes from an easily penetrated hexagonal structure to a laminar phase which is less easy to penetrate and does not easily spread. This laminar phase forms the interfacial structure of the final, stabilized, emulsion droplets.

Here, the source of the energy for these interactions is supposed to come from internal molecular forces such as the free energy of mixing that occurs during the violent mass transfer. One of the main characteristics of a spontaneously emulsifying system is that there is no requirement for the application of external energy.

In recent years some SEDDS containing ethanol have been explored and have proved to be useful in the case of cyclosporine. There is insufficient evidence at present to connect these ethanol-containing systems and the more classic self-emulsifying systems but again the possibility exists that the self-emulsification is another manifestation of the free energy of mixing but mainly involving the surfactants since the oil is not very water soluble. An intuitive depiction of this process is illustrated in Figure 7.4.

## INCLUSION COMPOUNDS AND CYCLODEXTRINS AS DELIVERY SYSTEMS

It is probably fair to suggest that drugs with poor water solubilities are poorly absorbed *in vivo* and have low bioavailabilities. One way of improving the bioavailability of such compounds is to reduce the particle diameter, thereby increasing the surface area for solution to take place. A classical example of this approach was the almost insoluble drug griseofulvin, the bioavailability of which is critically affected by the mean crystal diameter. Chiou and Riegelman (1971) took this approach to the next stage by forming a solid solution of griseofulvin and other water insoluble drugs in polyethylene glycol, which was a true solution, unaffected by the inevitable lot-to-lot variations that occur in the unformulated crystalline raw drug. These authors coined the term *solid dispersions* for this type of solid solution. Chiou and his students have explored the use of solid solvents such as succinic acid for this and other drugs.

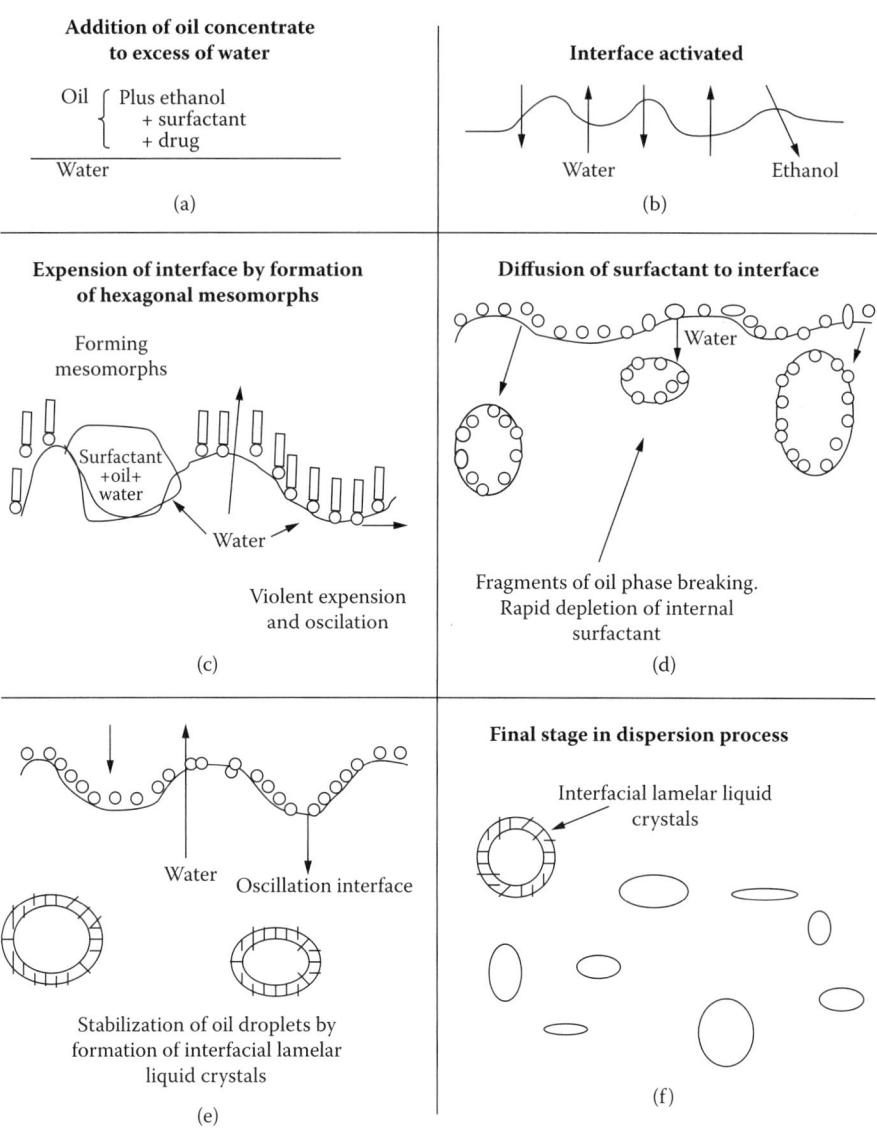

**FIGURE 7.4** Intuitive explanation of the spontaneous formation of emulsion droplets from a SEDDS.

Drugs presented as solid solutions often have advantages in that the drug is absorbed more rapidly and often more completely although the exact mechanism for absorption is not always clearly defined. Some, for example, are complexes. Others are glass-solutions or suspensions but most are metastable, readily forming brittle or unstable materials on storage, especially in the presence of moisture.

Solid dispersions have been explored as drug delivery systems for over 30 years, initially starting with various forms of polyethylene glycols, citric and succinic acids, and sugars. However, more recent success has been achieved using hydroxypropy-lcellulose (HPC), ethylcellulose (EC), and the commercial forms of methylacrylic acids and their copolymers sold under the name Eudragits. In addition, chitosans have been evaluated for this purpose.

In the meantime, attention has turned to the more stable *inclusion compounds*, which are materials capable of binding or holding a second chemical species or guest molecule inside their internal molecular structure. The host molecule usually has cavities, channels, or tunnels within the structure that are capable of accepting guest molecules within the hydrophilic space. There is no covalent bonding between the two, just much weaker van der Waals forces so the guest molecules can escape readily if the complex is placed in a solvent, for example. If the guest molecule is completely trapped the complex is termed a *clathrate* or a *cage compound*. Widely found in nature these materials are also called crown complexes and are rarely of interest to the pharmaceutical formulator because they are toxic.

However, some inclusion compounds are of considerable interest to pharmacists. For the most part these are cyclodextrins (CDs) as shown in Figure 7.5. Chemically they are *cyclic oligosaccharides* containing 6, 7, or 8 glucopyranose units, respectively, α, β, or γ-CDs. These compounds are hydrogen bonded internally to create characteristic three-dimensional truncated cones (Figure 7.6). With different dimensions, these cones are capable of accepting various other molecular entities according to the available space inside the cone. Effectively another form of solid dispersion, CDs have proved to be more durable than previous attempts to stabilize and enhance the solubility of drugs with otherwise low intrinsic aqueous solubility.

Not all CDs are suitable for the required purpose and because they can also absorb biological entities when administered to the body, the toxicity of some CDs may be questionable. For example, β-CD has proved interesting and some products containing this compound have appeared on the market. Unfortunately, parenterally

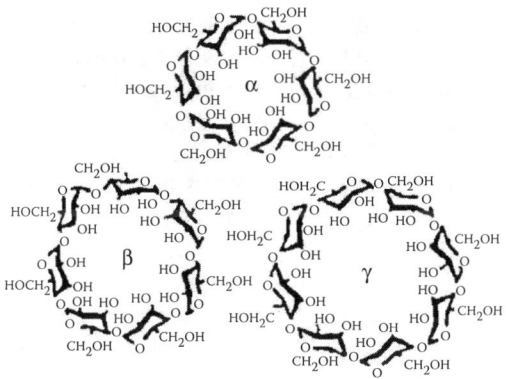

**FIGURE 7.5** Natural cyclodextrins α, β, and γ. (Reprinted with the permission of Aster Publishing Corp. These figures originally appeared in *BioPharm.*, 1991, 4(10), 44–51.)

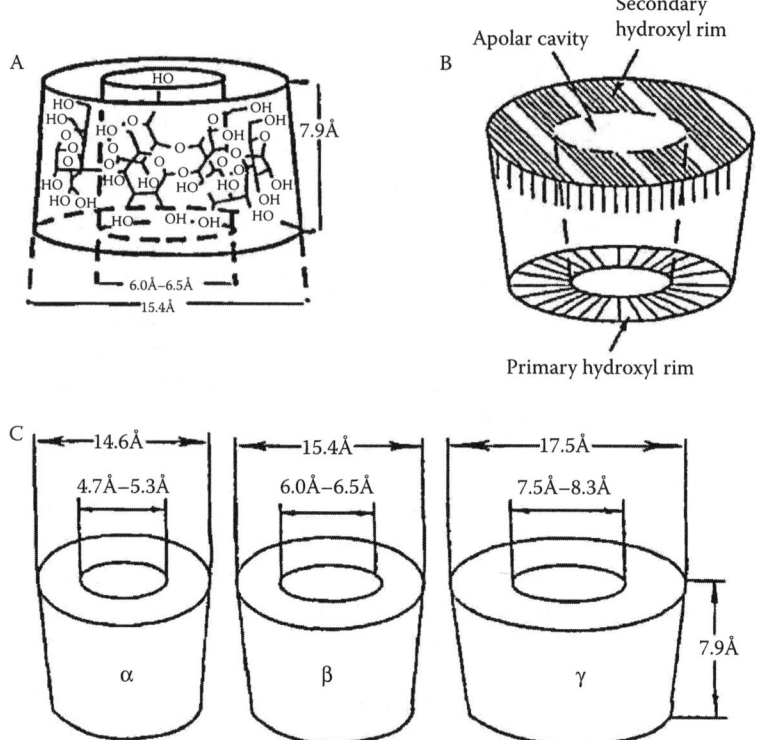

**FIGURE 7.6** Truncated ionical form of cyclodextrins: (A) β-cyclodextrin; (B) basic feature of cone. The hydroxyl rim is made up of hydroxyls attached to the C6 carbons, and the secondary hydroxyl rim, of hydroxyls attached to the C2 and C3 carbons; (C) approximate molecular dimensions of the three major cyclodextrins. (Reprinted with the permission of Aster Publishing Corp. These figures originally appeared in *BioPharm.*, 1991, 4(10), 44–51.)

it may have some direct renal toxicity and it has a relatively low aqueous solubilty, 1.8 g/100 mL. When formulated with garlic oil it produces a completely odorless and tasteless inclusion compound, and other drugs are known to lose their bitterness or unpleasant taste which makes them more acceptable to the consumer.

Recently some modified CDs have been offered on the market that appear to be more useful as drug delivery systems. One of these, the Janssen hydroxypropyl β-CD, appeared very promising but Janssen Pharmaceuticals are no longer in business and although another company is offering this material, there may be unresolved proprietory issues involved. An alternative compound, sulfobutyl ether β-CD, Captisol (CyDex), (SBE–CD), has a much larger capacity to dissolve insoluble compounds (Table 7.2) and a significantly higher water solubility, 90 g/100 mL. Moreover, unlike β-cyclodextrin, this compound does not manifest parenteral toxicity. Recently (October 2002) this cyclodextrin was approved by the FDA for the solubilization of the Pfizer compound voriconazole and it has been approved for use in other countries.

**TABLE 7.2**
**Properties of Some Cyclodextrins (CDs)**

| Cyclodextrin | α-CD | β-CD | γ-CD | HPCD | SBE-CD |
|---|---|---|---|---|---|
| Molecular Weight (Da) | 973 | 1135 | 1297 | ~1402 | ~2160 |
| Molarity (mol/L) | 0.149 | 0.016 | 0.179 | 0.428 | 0.370 |
| CD solubility (g/100 mL) | 14.5 | 1.85 | 23.2 | 60 | 80 |
| Maximum drug concentration, assuming a guest molecular weight of 500 Da and all cavities are filled (mg/g) | 75 | 8 | 90 | 214 | 185 |

Adapted from Thompson, D. and Chaubal, M.V. (2002). Excipient update. *Drug Delivery Technol.*, 2(7), 34–38.

Drugs are incorporated into CDs by using a number of simple procedures such as the direct addition of an aqueous solution or suspension of the drug to a solution of the CD, followed by a period for equilibration and subsequent freeze drying. An alternative process is to spray dry an aqueous suspension of the two components, but direct mixing of the two powders with water and kneading together as a paste or slurry before drying is feasible in most cases and attractive as a method for large scale manufacture. If operated at temperatures above 100°C this method can produce granules suitable for compression into tablets. Some compositions can also be made by grinding and exposing the mixture to temperatures in the range of 60–90°C in a closed vessel. A disadvantage of CDs is that the molecular cavities are too small to incorporate significant quantities of drug in excess of a ratio of 1:1 on a weight or molecular weight basis and this can result in large, bulky, formulations (Table 7.2).

One other possible disadvantage associated with the use of CDs is that they cannot incorporate an entire protein molecule, Stratton (1991). This is relatively unimportant if the hydrophilic sections of the protein molecule could be entrapped in the CD cavity, thereby leaving the hydrophobic portions outside the CD (Figure 7.7). Experimentally some exceptions to this rule have been noted, one being insulin entrapped

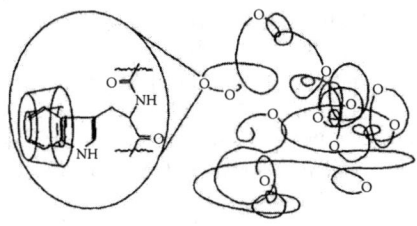

**FIGURE 7.7** Amino acid side chain of a protein interacting with a cyclodextrin. (Reprinted with the permission of Aster Publishing Corp. These figures originally appeared in *BioPharm.*, 1991, 4(10), 44–51.)

in 2-hydroxypropyl–β-cyclodextrin. In this case not only is the protein incorporated but the formulated drug is more stable and the solubility enhanced without affecting the biological activity. This approach may find itself being used in the next generation of implantable insulin pumps, which have been under development for the past decade.

Regulatory agencies in the United States, Japan, and Europe have approved the use of cyclodextrins for the delivery of drugs by the parenteral, oral, and topical routes. There are usually few proprietary protections such as composition of matter claims, with the exception of the SBE–CD derivatives (above) and material can generally be purchased from a number of suppliers in various grades of technical, food, or pharmaceutical qualities. Care needs to be taken when selecting a material for use in parenteral formulations. Since the association of drug and CD breaks up as soon as it is delivered to the blood stream, formulations are approved on the basis of a drug with CD, not as a drug/CD complex.

## REFERENCES AND FURTHER READING

Attwood, D.A. and Florence, A.T. (1983). Aspects of surfactant toxicity. In: *Surfactant Systems: Their Chemistry, Pharmacy and Biology,* D.A. Attwood and A.T. Florence, eds. Chapman & Hall, London, 614–697.

Attwood, D.A. and Florence, A.T. (1998). *Physicochemical Principles of Pharmacy, Third Edition,* Palgrave, New York.

Bourell, M. and Schechter, R.S. (1988). *Microemulsions and Related Systems: Formulation, Solvency and Physical Properties.* Marcel Dekker, New York.

Charman, S.A., Charman, W.N., Rogge, M.C., Wilson. T.D., Dutko, F.J., and Pouton, C.W. (1992). Self-emulsifying drug delivery systems: Formulation and biopharmaceutics evaluation of an investigational lipophilic compound. *Pharm. Res.,* 9(1), 87–93.

Chi, E.Y., Krishnan, S., Randolph, T.W., and Carpenter, J.F. (2003). Physical stability of proteins in aqueous solution: Mechanism and driving forces in nonnative protein aggregation. *Pharm. Res.,* 20(9), 1325–1336.

Chiou, W.I. and Riegelman, S. (1971). Increased dissolution rates of water insoluble cardiac glycosides and steroids via solid dispersions in polyethylene glycol 6000. *J. Pharm. Sci.,* 60(10), 1569–1571.

Constantinides, P.P. (1995), Lipid microemulsions for improving drug dissolution and oral absorption, *Pharm. Res.,* 12(11) 1651–1572.

Gershanik, T. and Benita, S. (2000), Self-dispersing lipid formulations for improving oral absorption of lipophilic drugs. *Eur. J. Pharmaceutics and Biopharmaceutics,* 50, 179–188.

Gilbert, P. and Allison, D.G. (1996). Kinetics of heat sterilization, *Eur. J. Parenteral Sci.,* 1, 19–23.

Gopal, E.S.R. (1968). Principles of emulsion formation. In: *Emulsion Sciences,* P. Sherman, ed. Academic Press, London, 1–75.

Groves, F.M. and Groves, M.J. (1991). Dry heat sterilization and depyrogenation. In: *Encyclopedia of Pharmaceutical Technology,* J. Swarbrick and J.C.Boylan, eds. Marcel Dekker, New York, 4, 447–484.

Groves, M.J. (1998). Heat sterilization, In: *Sterilization of Drugs and Devices,* F.M Norhauser and W.P. Olson, eds. Interpharm Press, Buffalo Grove, IL, 5–44.

Groves, M.J. (1978). Spontaneous emulsification. *Chem. Ind.,* 17, June, 417–423.

Groves, M.J. and Mustafa, R.M.A. (1973). Measurement of the 'spontaneity' of self-emulsifiable oils. *J. Pharm. Pharmacol.,* 26, 671–681.

Groves, M.J. and Teng, C.D. (1992). The effect of compaction and moisture on some physical and biological properties of proteins. In: *Stability of Protein Pharmaceuticals, Part A: Chemical and Physical Pathways of Protein Degradation,* T.J. Ahern and M.C. Manning, eds. Plenum Press, New York, 311–359.

Gupta, P. and Garg, S. (2002). Recent advances in dosage forms for dermatological application. *Pharm. Tech.,* 26(3), 144–162.

Guy, R.H. (1998). Iontophoresis: Recent developments. *J. Pharm. Pharmacol.,* 50, 371–374.

Hopp, M.S. (2002). Developing custom adhesive systems for transdermal drug delivery products, *Pharm. Tech.,* 28(3), 30–36.

Iranloye, T.A. and Pilpel, N. (1984). Simple, rapid method for comparing the self-emulsifiability of hydrocarbon oils. *J. Pharm. Sci.,* 73(9), 1267–1270.

Jornitz, M.W. and Meltzer, T.H. (2004). Sterilizing filtrations with microporous membranes. *Pharmacopeial Forum,* 30(5), 1903–1910.

Nazzal, S., Zaghloul, A., and Khan, M.A. (2002). Effect of extragranular microcrystalline cellulose on compaction, surface roughness and in vitro dissolution of a self-nanoemulsified solid dosage form of ubiquinone. *Pharm. Tech.,* 26(4), 86–98.

Novitsky, T.J. (1996). Limulus amebocyte lysate (LAL) assays. In: *Automated Microbial Identification and Quantitation: Technologies for the 2000s,* W.P. Olson, ed. Interpharm Press, Buffalo Grove, IL, 277–298.

Olson, W.P. (1998). Sterility testing. In: *Sterilization of Drugs and Devices,* F.M Norhauser and W.P. Olson, eds. Interpharm Press, Buffalo Grove, IL, 471–486.

Pouton, C.W. (1985). Self-emulsifying drug delivery systems: Assessment of the efficiency of emulsification. *Int. J. Pharmaceutics,* 27, 335–348.

Pouton, C.W. (2000). Lipid formulations for oral administration of drugs: Non-emulsifying and "self-emulsifying" drug delivery systems. *Eur. J. Pharm. Sci.,* Suppl. 2, S93–S98.

Rehman, M., Kippax, P., and York, P. (2003), Particle engineering for improved dispersion in dry powder inhalers. *Pharm. Tech. Europe,* 15(9), 34–39.

Rosano, H.L. and Clausse, M., eds. (1987). *Microemulsion Systems.* Marcel Dekker, New York.

Sutton, S.V.W. and Cundell, A.M. (2004). Microbial identification in the pharmaceutical industry. *Pharmacopeial Forum,* 30(5), 1884–1894.

Stratton, C.C. (1991). Cyclodextrins and biological mmacromolecules. *BioPharm.,* 4(10), 44–51.

Teng, C.D., Zarrintan, M.H., and Groves, M.J. (1991). Water vapor adsorption and desorption isotherms of biologically active proteins. *Pharm. Res.,* 8(2), 191–195.

Thompson, D. and Chaubal, M.V. (2002). Excipient update. *Drug Delivery Technol.,* 2(7), 34–38.

Trotter, M.A., Rodrigues, P.J., and Thoma, L.A. (2002). The usefullness of 0.45 μm-rated filter membranes. *Pharm. Tech.,* 26(4), 60–70.

Wakerly, M.G., Pouton, C.W., Meakin, B.J., and Morton, F.S. (1986). Self-emulsification of vegetable oil-nonionic surfactant mixtures: A proposed mechanism of action. *A.C.S. Symposium,* 311, 242–255.

Walters, K.A. (1989). Penetration enhancers and their use in transdermal therapeutic systems. In: *Transdermal Drug Delivery — Development Issues and Research Initiatives,* J. Hadgraft and R.H. Guy, eds. Marcel Dekker, New York, 197–246.

York, P. and Hanna, M. (1996). Particle engineering by supercritical fluid technologies for powder inhalation drug delivery. *Proc. Conf. Respiratory Drug Delivery V,* Phoenix, AZ, 231–239.

# 8 Proteins as Drug Delivery Systems

*Michael J. Groves, Ph.D., D.Sc.*

## CONTENTS

## INTRODUCTION

Many proteins and protein-like materials are precipitated from aqueous solution by the application of heat or the addition of electrolyte, and this general reaction can be used to prepare solid particles with various particle size ranges. Under the right conditions these particles can be made reproducibly and serve as drug delivery systems in their own right, being especially suitable for the delivery of other proteins or polypeptides.

However, one point needs to be made immediately in that many authors have added their drug *ab initio* and then proceeded to denature or chemically cross-link the carrying system by heat or the addition of aldehydes in order to make it insoluble. They are therefore apparently forgetting that the same consideration applies to any proteinaceous incorporated drug. Cross-linked proteinaceous drug entities tend to be ineffective and it needs to be emphasized that, in most cases, the drug needs to be added to the preformed drug delivery system *after* these have been formed. In the following section gelatin and albumin delivery systems will be discussed in reasonable depth but this consideration needs to be borne in mind throughout.

# COLLAGEN

Collagen is the main constituent of connective tissue in the body and is commercially available when obtained mainly from cattle. The structural properties are described elsewhere in this book but excellent reviews by Geiger and Friess (2002a) are available. Friess has developed this subject and demonstrated how drug delivery systems could be prepared from dried sheets of specially purified collagen by incorporating a drug and punching pellets out of the sheet. Effectively this was the method employed by Friess et al. (1996) when a high molecular weight antineoplastic proteoglycan obtained from *Mycobacterium bovis* (Bacille Calmette-Guérin [BCG] vaccine) was incorporated into collagen sheets and shown to be released in a controlled fashion *in vivo* and *in vitro*.

Geiger and Friess (2002b) have described how collagen solutions can be foamed and dried out to form sponges which can be used to revitalize skin or for hemostasis, wound healing, and as drug delivery systems, especially for local antibiotic applications. The flexible and sterile sheets obtained by this technology would appear to have considerable future application. However, as with gelatin, the only major issue would appear to be the possible contamination with prions responsible for Creutzfeldt-Jakob syndrome although to date this has not been demonstrated to be a realistic concern.

# GELATIN

## The Origins and Constitution of Gelatin

Gelatin is a ubiquitous pharmaceutical adjuvant and can be readily modified to form insoluble macroparticles, i.e., particles larger than ~100 μm in diameter. However, by careful attention to the details of the manufacturing process, microparticles (<1 μm) or even nanoparticles (<<1 μm) can readily be made.

Gelatin is obtained commercially by the controlled hydrolysis of animal collagen tissues. Native collagen consists of rods formed from triple helices that are held together by covalent bonds. During the denaturation process these covalent bonds are progressively destroyed and the helical structures disrupted. The hydrolytic process consists of heating in the presence of either acid or base and this is usually disclosed in the label, e.g., gelatin A or gelatin B. Gelatin solutions in water at temperatures in excess of 37°C are liquids and the individual molecules behave effectively as polymeric coils. When allowed to cool to ambient room temperature, solutions containing more than 0.5% gelatin change to the random chain molecules which themselves revert to triple helix structures. The bulk (>58%) of the gelatin chains consist of glycine, proline, and hydroxyproline that are weakly hydrophobic in nature and it is their periodicity along the chain, combined with the inherent structural flexibility of the glycine molecule, that facilitates the formation of helical structures.

At about neutral pH the gelatin chain is effectively amphoteric and individual groups such as lysine and arginine (~7.5%) are positively charged and others such as glutamic and aspartic acids (~1 1.8%) are negatively charged. These entities allow the formation of centers for electrostatic binding and this process is controlled by the environmental pH.

Light, small angle neutron and x-ray scattering all provide evidence of the length and stiffness of the average gelatin chain as being approximately 20 nm, with a radius of gyration, Re, of ~35 nm. As the concentration is increased two separate regions become evident by the quasi-elastic light scattering techniques. The fast mode is apparently due to cooperative movements of the entangled network of chains and appears to be generally persistent over a range of temperatures and concentrations. The second, slower, mode has been associated with the diffusion coefficient being inversely proportional to the bulk (Newtonian) viscosity of the solution and is considered to be due to the self-diffusion of clusters of chains that effectively form particles with a constant hydrodynamic radius, Re, of ~75 nm. Unlike the properties of the individual chains, the clusters of chains are affected by the molecular weight of the chains themselves, the temperature, and the presence of salts or surfactants such as sodium dodecyl sulfate (SDS). Herning et al. (1991) concluded that clusters were stabilized by hydrophobic interactions between apolar lateral groups of the protein.

Thus, it can be argued that gelatin itself is not a single entity but in many ways it does behave like a model protein. In bulk the material is deficient in some amino acids, having only 18 constituent units, of which the principle constituents are glycine (32–35%), proline (11–13%), alanine (10–11%), hydroxyproline (9–19%), glutamic acid (7–8%), aspartic acid (4–5%), and arginine (5%). The parent collagen, for the most part, consists of three subunits of approximately equal molecular weight (~95 kDa). During the denaturation process to form commercial gelatin, the collagen is broken down to form single chain fragments, the $\alpha$-chains, molecular weight 95 kDa, and $\beta$-fragments, mw ~190 kDa. In addition there are low molecular weight oligomeric fragments and some branched $\gamma$-chains, mw 285 kDa, which actually consist of three $\alpha$-chains joined near their mutual center points.

The main commercial use for gelatin is in the preparation of food, approximately 70% of all production used for this purpose, with an additional 15% going to the pharmaceutical industry, and 10% used for photographic film. The remaining 5% is used in wine making, printing, abrasive manufacturing, and paper production.

Photographic film-making has become a highly specialized use for gelatin and, with this purpose in mind, Lorry and Vedrines (1985) used high performance liquid chromatography (HPLC) and gel electrophoretic techniques to examine the constitutional differences of gelatins from various sources. Skin and bone gelatins were found to be basically similar, the β-chains being substantially absent from skin-derived material. Differences between acid and alkali processed gelatins were noted, the former contained more degradation products. These authors suggested that gelatins with required properties could be obtained by blending low molecular weight (α) or high molecular weight (γ) fractions according to the desired endpoint application.

As a side comment, it might be noted there is little evidence that the gelatin producers would be willing to carry out this blending process for the pharmaceutical industry, in part because there have been very few studies which have helped to define the essential requirements of gelatin for pharmaceutical applications. The main uses for gelatin in pharmacy are the preparation of hard or soft gelatin capsules and coated tablets, and, in Europe, gelatin-based suppositories and ointments. Another economic factor which must have an influence concerns the relatively small requirement of the pharmaceutical industry which would make the returns unrealistic or prohibitively expensive for even large quantities of specialized forms of gelatin.

The purest form of gelatin is probably used in food. Commercially, food-grade material is supplied as sheets of brittle, transparent, and only faintly colored dried solid bearing the marks of the support grid during the final drying process. In use, these sheets are soaked in cold water for a few minutes, followed by gentle heating and stirring to affect solution. Pharmaceutically, gelatin is usually supplied as a coarsely ground powder, but this often contains solid impurities and wherever possible the sheet gelatin should be used.

## PHASE RELATIONSHIPS OF GELATIN SYSTEMS

Although gelatin behaves as a model protein in many respects, it has little or no tertiary structure and is not denatured by heating at temperatures in excess of 100°C. Albumin, on the other hand, is a true protein and is denatured at temperatures of 5–60°C, as well as being precipitated out of aqueous solution by the addition of ethanol or by sodium or ammonium sulfates.

Gelatin is slowly precipitated by the addition of either of these desolvating components, which results in a very complex phase diagram, strongly influenced by temperature. Essentially, gelatin behaves like a mildly hydrophobic protein in dilute aqueous solution and the addition of solutes such as ethanol usually results in two-phase liquid systems, today called coacervates. It is the formation of coacervates that constitutes a key step in the formation of many drug-encapsulated pharmaceutical drug delivery systems. Although known for many years, only recently attempts have been made to determine the effects of temperature and solute concentrations on the transitional boundaries of the various elements. Ternary phase diagrams are often utilized to demonstrate the effect of the addition of a third component on the behavior of another material dissolved in water. Because of the complexity of gelatin-based systems, it should be pointed out that determination of the phase boundaries requires care and interpretive skill so that the published boundaries are rarely precise or exact.

Although the first studies on gelatin go back to the 1930s, a recent systematic evaluation of the effects of ammonium sulfate and ethanol on the solubility or constitution of gelatin B (Bloom 225) has recently been published by Elysée-Collen and Lencki (1996). After the inspection of several hundred gelatin-water combinations

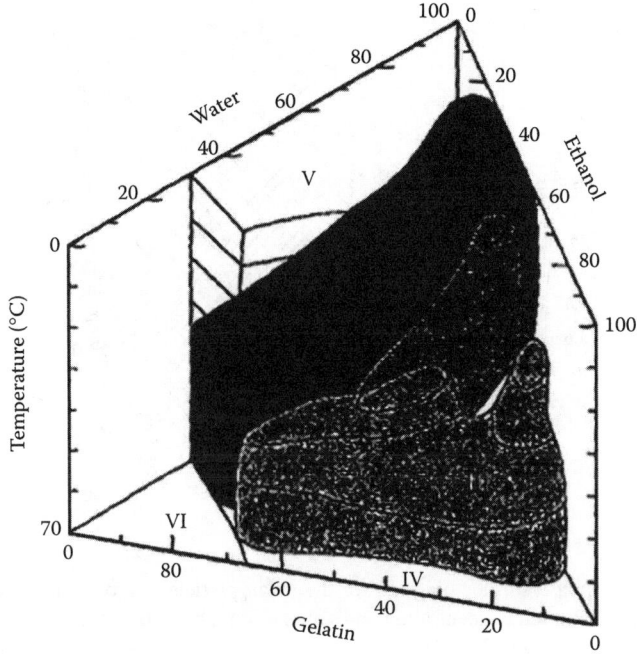

**FIGURE 8.1** Phase diagram for water-gelatin-ethanol as a function of temperature. (Reprinted with permission from Elysée-Collen, B. and Lencki, R. (1996). *J. Agric. Food Chem.*, 44, 1651–1657.)

these authors found that seven distinct morphologies could be distinguished visually, although only six could be found in the presence of ethanol. It must be assumed that these are stable states although this assumption may not be entirely justified. At least 33% water is needed to break apart and hydrate the gelatin to form a solution (Figure 8.1). Interestingly, the addition of ethanol does not appear to change this. This observation suggests that powdered gelatin is itself relatively hydrophobic and does not have a preference for either water or ethanol. Water-structure promoters such as sodium chloride are capable of destabilizing gelatin structures so that almost 50% water is required to form a solution in the presence of 5% sodium chloride. This suggests that both gelatin and salt compete for the same water of hydration.

These authors pointed out that at below 40°C gelatin molecules begin to interact with each other to form linkages stabilized by hydrogen bonds. This accounts for the place of ethanol in encouraging gel formation, morphology I (Figure 8.2). Nevertheless, these authors did not explain the nature of the opaque sols and gels that are so often observed in practice.

Attention may be drawn to the fact that Herning et al. (1991) suggested that clusters formed at concentrations in excess of 1% of high molecular weight gelatins, with hydrodynamic radii of the component particles in the clusters of around 75 nm. However, they also suggested that the radii were around 60 nm when the molecular

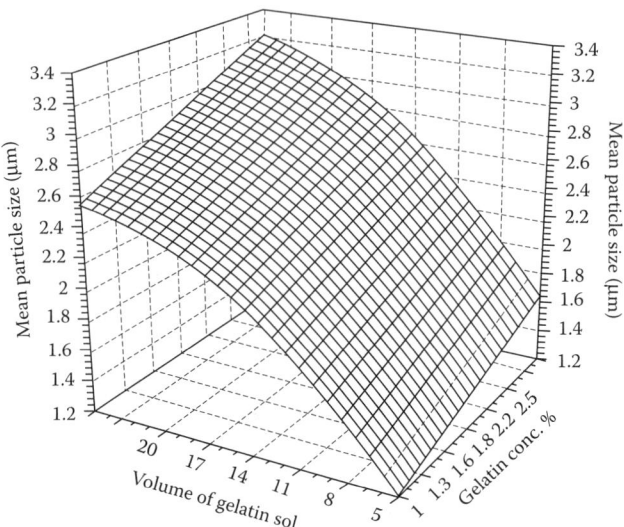

**FIGURE 8.2** Response surface diagram for the mean particle size of gelatin microparticles prepared with variation in concentration and volume of type B, Bloom 225, gelatin solutions. (From Öner, L. and Groves, M.J. (1993a). *Pharm. Res.,* 10(9), 1385–1388. With pemission.)

weight was lower. Farrugia (1998) found that gelatin microparticles had diameters of between 100–200 nm, consistent with the suggestion of Herning et al. that molecular chain clusters of 150 nm diameter were present.

The exact mechanism of gel formation remains unclear at present but probably involves the formation of triple helices at some stage. Certainly, as noted earlier, the solid, nonhydrolyzed, collagen structure consists of triple helices covalently bonded together. These bonds are broken by hydrolysis to allow subsequent reformation of triple helices, facilitated by the presence of large numbers of glycine and proline molecules along the gelatin chain. Since different segments of adjacent chains can be involved, this suggests that the gel consists of linked three-dimensional structures. These structures have been examined by various methods. Conclusions from these researchers indicate that the chains themselves may be 16–20 nm long and a 5% gelatin gel has three-dimensional structures with a mesh size of 5 nm, the exact size being affected by the presence of salts or surfactants.

Most coacervates explored as drug delivery systems consist of particles dispersed in a liquid continuum. The questions as to how these systems are made and how they may be presented as drug delivery systems depends to a large degree on the individual formulator.

## The Preparation of Gelatin Particles

As noted earlier, gelatin can be converted to macro- or microparticulate systems, depending on the conditions of preparation.

## Emulsification Processes

A number of authors have made gelatin microparticles containing relatively stable drugs such as 5-fluorouracil, mitomycin, or adriamycin by an emulsification process in which aqueous solutions of drug and gelatin, with an appropriate emulsifier, have been emulsified in an appropriate inert oil phase. The water-in-oil system is then hardened with an aldehyde, usually glutaraldehyde, and the particles collected by centrifuge or filtration. However, the overall process requires the use of solvents such as chloroform or diethylether to remove the adhering oil phase. Quantities of these environmentally undesirable solvents may be large and this makes it extremely difficult to scale the process up from the laboratory to industrial production, quite apart from heat generation and chemical interactions which may be detrimental to the required properties of the incorporated drug (Öner and Groves 1993c).

## Precipitation Processes

With a view to employing unmodified gelatin particles as immunological drug delivery systems in their own right, the concept of precipitating gelatin into cold, water miscible, solvents such as ethanol was explored by Öner and Groves (1993c). The basic process involved placing 500 mL of a water-miscible solvent such as anhydrous ethanol, acetone, or 2-propanol in a flask suspended over a bath of granules of solid carbon dioxide. The system was slowly stirred until the temperature had dropped to 15°C, at which point various volumes of 1% aqueous gelatin were slowly stirred into the cold solvent. The temporary particles or droplets were held under these conditions for another 15 minutes, and glutaraldehyde solution added to cross-link the dispersed gelatin particles. The system was then placed in a refrigerator overnight or for up to 48 hours and the cross-linking process stopped by adding the contents of the flask to 1500 mL of 5% sodium metabisulfite at 4°C before collecting the particles by filtration. After this the particles were repeatedly washed with 0.01 M phosphate buffer at pH 7.0 before suspending in 5% aqueous mannitol. The system was then lyophilized and stored in the refrigerator.

Two aspects of this simple process need to be noted. The first point is that the cross-linking process takes an appreciably longer time than those described in the literature. Using literature times described for the process resulted in particles that visibly dissolved while being analyzed by Coulter Counter but this issue went away if the cross-linking times were prolonged. The other point is that prolonged washing with water is involved, precluding the prior addition of any drug at the beginning of the process. However, it was shown that the final particles could be conveniently loaded by soaking in an aqueous or ethanolic solvent containing the drug, the solvent being removed by an additional lyophilization process.

The authors applied factorial design to their experiments and were able to demonstrate that the optimal solvent was ethanol, with the smallest particles, ~1 μm diameter, resulting when dilute concentrations of gelatin at acidic pH were utilized (Figure 8.3). Interestingly, the Bloom number or the acid or base-derived gelatin did not make a great deal of difference (Figure 8.2 and Figure 8.3). At high pH the diameter of the resulting particles tended to become bimodal (Figure 8.4) with one

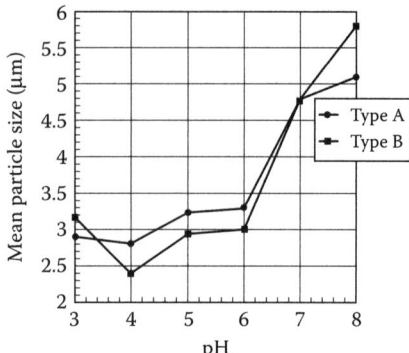

**FIGURE 8.3** The relationship of pH of gelatin solutions and particle size. Type A, Bloom 60; type B, Bloom 225. (From Lou and Groves 1995.)

population around 1 μm and the other around 4–5 μm in diameter (Table 8.1). In addition, the precipitation temperatures had a profound influence on the resulting precipitated particle diameters (Figure 8.7).

In an effort to improve the process Öner and Groves (1993a) evaluated the use of a popular form of nebulizer, the Turbotak atomizer (Turbotak Ltd., Waterloo, Ontario, Canada). The desolvating solvent was again maintained at 15°C but the system was enclosed in order to reduce the volume of solvent required and to minimize the risk of the operators inhaling the generated droplets (Figure 8.7 and Figure 8.8). The

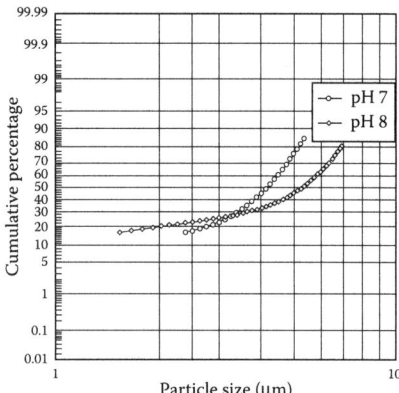

**FIGURE 8.4** Cumulative size distribution of gelatin (lime-cured type B, bloom number 225) microparticles prepared at high pH obtained using a Coulter Counter and plotted as a log-probit function. (From Lou, Y. and Groves, M.J. (1995). *J. Pharm. Pharmacol.,* 47, 97–102. With permission.)

## TABLE 8.1
## Size Characteristics of Gelatin (Type B, Bloom Strength 225) Microparticles as Measured by Coulter Counter

| pH | Mean volume diameter[a] | | | | Type of distribution |
| | Expt 1 | | Expt 2 | | |
| | μm | τ | μm | τ | |
|---|---|---|---|---|---|
| 3 | 3.2 | 1.22 | 3.1 | 1.29 | Unimodal |
| 4 | 2.4 | 1.28 | 2.4 | 1.29 | |
| 5 | 2.9 | 1.24 | 3.0 | 1.20 | |
| 6 | 3.0 | 1.26 | 3.0 | 1.30 | |
| 7 | 1.1 | 1.36 (15%)[b] | 1.1 | 1.27 (17%) | Bimodal |
| | 4.7 | 1.24 (85%) | 4.9 | 1.25 (83%) | |
| 8 | 1.1 | 1.49 (20%) | 1.1 | 1.52 (19%) | |
| | 5.9 | 1.32 (80%) | 5.7 | 1.30 (81%) | |

[a] The mean values and standard deviations (SD) are geometric and geometric standard deviations, respectively, of distributions.
[b] Percentage of total distribution.

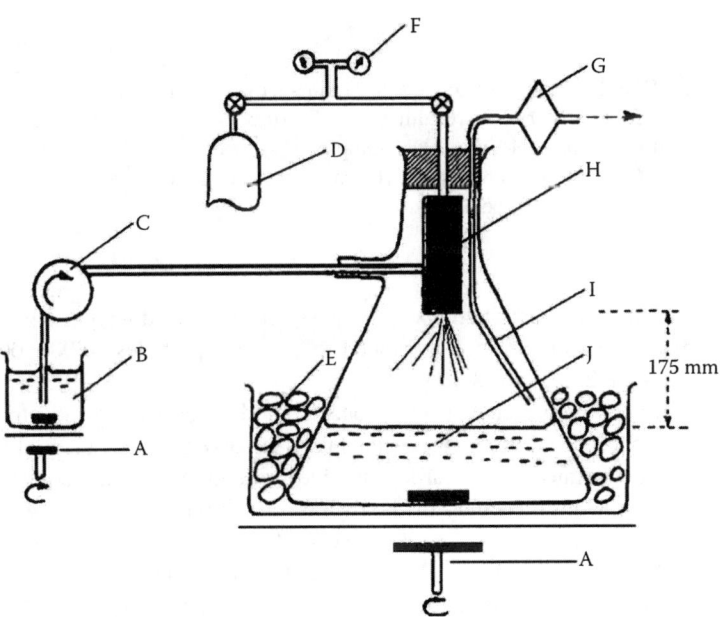

**FIGURE 8.5** Diagram of single-stage carbon dioxide activated spray system. A. Magnetic stirrer; B. hot gelatin solution; C. peristaltic pump; D. carbon dioxide tank; E. solid carbon dioxide pellets; F. gas regulator; G. 0.45-mm membrane filter venting to atmosphere; H. Turbotak; I. vent tube; J. cold stirred anhydrous ethanol. (From Öner, L. and Groves, M.J. (1993a). *Pharm. Res.,* 10(9), 1385–1388. With pemission.)

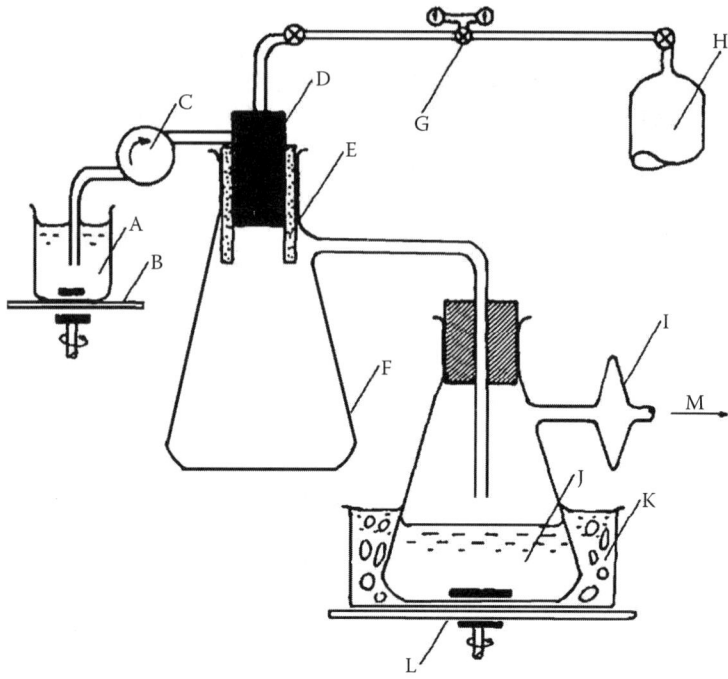

**FIGURE 8.6** Diagram of two-stage carbon dioxide-activated spray system. A. Hot protein solution; B. magnetic stirrer; C. peristaltic pump; D. Turbotak; E. sleeve cut from Masterflex tubing, 1.5 cm below base of Turbotak; F. empty 2-L glass conical flask; G. gas regulator; H. $CO_2$ tank; I. 0-45-μm membrane filter. (From Öner, L. and Groves, M.J. (1993a). *Pharm. Res.*, 10(9), 1385–1388. With pemission.)

direct mixing process described above consistently produced ~1 μm particles, but the closed atomization process produced submicron particles ~800–900 nm in diameter (Table 8.2A, Table 8.2B).

These authors suggested that the process could be scaled up industrially and was attractive because, although the water-miscible solvents have potential for flammability, it was operated under carbon dioxide gas and at low temperatures which considerably improved the safety of the overall operation.

Taking the process to successful delivery of drugs is another issue. Lou and Groves (1995) made gelatin particles of around 2 μm using the open process and incorporated thiotepa, an alkylating agent which does not appear to have been extensively studied from a formulation point of view.

The kinetics of absorption of the drug following administration of the formulation of thiotepa in gelatin particles to rabbits suggested that not only was the blood level elevated compared to a control but the area under the curve (AUC) was essentially more than doubled. Since this represents the amount of drug absorbed the authors suggested that administration of a simple gelatin particulate formulation

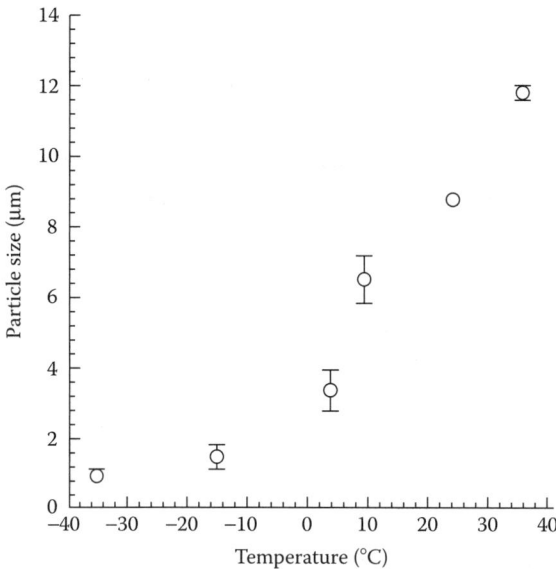

**FIGURE 8.7** Effect of precipitation temperature on the size (dvn) of the resulting gelatin microparticle (lime-cured type B, bloom number 225). Bar = s.e., n = 3. (From Lou, Y. and Groves, M.J. (1995). *J. Pharm. Pharmacol.,* 47, 97–102. With permission.)

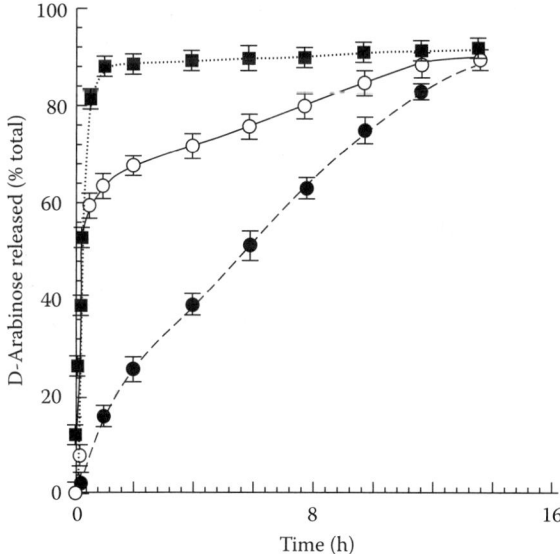

**FIGURE 8.8** Effect of D-arabinose loading on the release profile from gelatin microparticles in 1/15 M phosphate buffer (pH 7.4) at 25°. •- - •, D-Arabinose loading 2.5%, — 5%, 50%. Bar = s.e., n = 6. (From Lou, Y. and Groves, M.J. (1995). *J. Pharm. Pharmacol.,* 47, 97–102. With permission.)

**TABLE 8.2A**
**Size Characteristics of Gelatin Microparticles Prepared Using Chilled Dehydration Techniques**

| | Direct Mixing | | $CO_2$ atomization | |
| --- | --- | --- | --- | --- |
| Lot No. | Mean diameter (μm) | $\tau^a$ | Mean diameter (μm) | $\tau$ |
| 1 | 1.15 | 1.39 | 0.85 | 1.76 |
| 2 | 1.15 | 1.47 | 0.90 | 1.83 |
| 3 | 1.20 | 1.54 | 0.85 | 1.94 |
| 4 | 1.20 | 1.62 | 0.90 | 1.77 |
| 5 | 1.20 | 1.41 | 0.85 | 1.82 |
| 6 | 1.15 | 1.74 | 0.80 | 1.88 |
| Mean | 1.18 | 1.53 | 0.86 | 1.83 |
| $CV^b$ | 2.3% | 8.7% | 4.4% | 3.7% |

[a] Geometric standard deviation (log probit).
[b] Coefficient of variation.
* Significantly different ($p < 0.05$).

of thiotepa would enable the amount of drug administered to be halved, thereby reducing the side effects associated with this otherwise effective but toxic drug. This suggestion has not been followed up clinically.

Another paper by the same authors demonstrated that the adsorption process of D-arabinose, itself highly water soluble, onto preformed gelatin particles could be

**TABLE 8.2B**
**Size Characteristics of Albumin Microparticles Prepared Using Chilled Dehydration Techniques**

| | Direct Mixing | | $CO_2$ atomization | |
| --- | --- | --- | --- | --- |
| Lot No. | Mean diameter (μm) | $\tau^a$ | Mean diameter (μm) | $\tau$ |
| 1 | 3.00 | 2.27 | 1.70 | 1.71 |
| 2 | 3.00 | 1.77 | 1.80 | 1.88 |
| 3 | 2.80 | 1.94 | 1.90 | 1.78 |
| 4 | 2.90 | 2.12 | 1.70 | 1.72 |
| 5 | 3.00 | 1.64 | 1.85 | 1.64 |
| 6 | 3.00 | 1.84 | 1.75 | 1.88 |
| Mean | 2.97 | 1.93 | 1.78 | 1.76 |
| $CV^b$ | 3.4% | 12.0% | 4.6% | 5.1% |

[a] Geometric standard deviation (log probit).
[b] Coefficient of variation.
* Significantly different ($p < 0.05$).

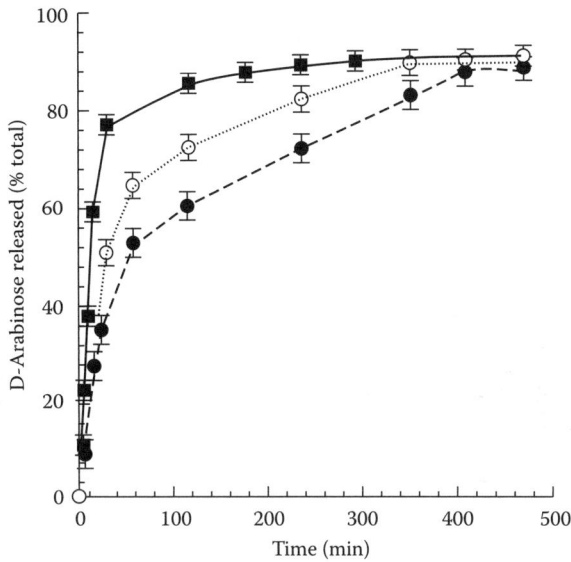

**FIGURE 8.9** Effect of glutaraldehyde concentration on release profile of D-arabinose from gelatin microparticles (D-arabinose loading 16%) in 1/15 M phosphate buffer (pH 7.4) at 25°C. — 4% glutaraldehyde,   6%, •- - - •, 8%. Bar = s.e., n = 6. (From Lou, Y. and Groves, M.J. (1995). *J. Pharm. Pharmacol.*, 47, 97–102. With permission.)

reversed but, effectively, slowed the release process *in vitro* for as long as 14 hours (Figure 8.8 and Figure 8.9). Subtracting the initial amount of D-arabinose released over the first hour of the dissolution process from subsequent results allowed the release rates to be determined. Surprisingly, these rates were effectively steady until the sugar had been stripped off the gelatin particles. The effect was to produce a zero-order release pattern after the initial burst effect. The same effects were observed with a polysaccharide of significantly higher molecular weight, PSI (Table 8.3).

It may be surprising to realize that sugars are adsorbed to a protein-like material although, in practice, this is the principle underlying some chromatographic separation processes used with proteinaceous systems.

What was interesting about this study is that it was possible to demonstrate that the sugar was adsorbed onto the internal gelatin particle matrices as well as the visible external particle surface (Table 8.3). This factor needs to be taken into account in any studies on the use of gelatin particles as drug delivery systems, especially if the initial production process is likely to have any influence on the intrinsic characteristics of the matrix.

## Submicron Diameter Gelatin Particles

Marty and her colleagues (1978) pointed out that colloidal drug delivery systems were well suited for parenteral administration and may have use as sustained release products. Since gelatin is readily available, has low antigenicity, and has been utilized

# TABLE 8.3
## The Effect of Glutaraldehyde Concentration on the Absorption of D-Arabinose and PS1 from the External or Internal Surface of Gelatin

| Glutaraldehyde | External specific surface[a] (cm² g⁻¹ × 10⁴) | Amount adsorbed in external surface (mol g⁻¹ × 10⁻⁶) | | Maximum adsorbed[b] surface[c] (mol g⁻¹ × 10⁻³) | | Total specific surface (cm² g⁻¹ × 10⁻³) | | Amount adsorbed in internal surface (mol g⁻¹ × 10⁻⁵) | | Internal specific surface[d] (cm² g⁻¹ × 10³) | |
|---|---|---|---|---|---|---|---|---|---|---|---|
| | | D-Arabinose | PSI | D-Arabinose | PSI | D-Arabinose | PSI | D-Arabinose | PSI | D-Arabinose | PSI |
| 4% | 3.466 | 10.5 | 0.763 | 51.8 | 0.478 | 1.713 | 2.17 | 50.7 | 0.402 | 16.78 | 1.823 |
| 6% | 1.340 | 4.07 | 0.295 | 47.4 | 0.436 | 1.567 | 1.98 | 46.9 | 0.406 | 15.54 | 1.846 |
| 8% | 0.964 | 2.93 | 0.213 | 39.1 | 0.403 | 1.293 | 1.83 | 38.8 | 0.382 | 12.83 | 1.734 |

[a] Calculated from gelatin particle size and density (based on Coulter Counter measurements and the Hatch-Choate relationship).
[b] Obtained from Langmuir isotherm.
[c] Calculated from Langmuir isotherm and BET equation.
[d] Obtained from total surface substrate less that adsorbed on the external surface.

in parenteral products, Marty et al. were able to describe a method for preparing gelatin nanoparticles. Although this paper has been extensively cited since its publication over 25 years ago, in fact the methodology has proved to be exceptionally difficult to replicate (Farrugia 1998).

## GELATIN PRECIPITATION

The process described by Marty et al. (1978) involves adding ethanol slowly to a 1–3% solution of gelatin in water containing surfactant such as polysorbate 20 and buffering salts at 37°C until the turbidity disappears. At this point a cross-linking agent such as formaldehyde or glutaraldehyde is added to harden the particles. Excess cross-linking agent is then removed and the system is then lyophilized to remove the water and the ethanol. The particles were claimed to be spherical, with diameters of around 200 nm. However, closer examination of their electron photomicrographs suggested that the authors were somewhat optimistic in their estimates of both size and sphericity.

Others have attempted to repeat these experiments. As discussed above, Öner and Groves (1993a) demonstrated that it was possible to prepare micron-sized particles by effectively reversing the process described by Marty et al. (1978) in that warm 1% gelatin solutions were added to stirred dehydrating solvents such as anhydrous ethanol (95% ethanol plus 5% isopropanol). By using solid carbon dioxide around the reaction flask the temperature inside was maintained at 15°C. However, longer periods of contact with glutaraldehyde were required and environment conditions such as pH, volume, and concentration of the initial solution as well as temperature all required careful control (Figure 8.2).

Scanning electron photomicrographs taken by these authors suggested that the cross-linked microparticles were comprised of aggregates of small primary particles of ~200 nm in diameter. This invited comparison with the estimate of sizes obtained by Marty et al. and the measurements of stable gelatin chain clusters by Herning et al. Lou and Groves (1995) demonstrated that the resultant diameter of particles was only modestly affected by the origin of the gelatin but was significantly affected by the pH (Figure 8.4). Moreover, systems with high pH were shown to have a bimodal size distribution, the smaller particles having diameters ~2.5 μm at low pH (<<6.0) were preserved throughout and larger particles of ~5.0 μm started to appear as the pH was increased. The response curve is illustrated in Figure 8.2 for systems made at 15°C.

## OPTIMIZATION OF GELATIN NANOPARTICLE PRODUCTION

As noted earlier, gelatin is a complex material, consisting of various fractions of materials with different molecular weights (Table 8.4). Farrugia (1998) returned to the method described by Marty et al. in 1978 in order to systematize the available information and improve on the methodology which had proved to be somewhat variable.

Earlier literature had demonstrated that gelatin samples could be arbitrarily subdivided into a number of molecular weight classes (Table 8.4), and these could be identified by the careful use of size-exclusion chromatography (Figure 8.10). By adding two more categories, a logical explanation for the structures found based on

### TABLE 8.4
### Molecular Weight Classes Used for the Analysis of the Size
### Exclusion HPLC Chromatograms of Gelatin

| Molecular Weight Class | Molecular Weight Range (kDa) | Approximate Number of β Chains |
|---|---|---|
| Low molecular weight | < 50 | — |
| Sub-alpha | 50–80 | — |
| Alpha (α) | 80–125 | 1 |
| Beta (β) | 125–225 | 2 |
| Gamma (γ) | 225–340 | 3 |
| Epsilon (ε) | 340–700 | 6 (2γ) |
| Zeta (ζ) | 700–1000 | 9 (3 γ) |
| Delta (δ) | 1000–1800 | 12 (4 γ) |
| Microgel | >1800 | >12 |

*Source:* From Farrugia (1998).

a single unit (the α-fraction) could be made. A process was devised based on the need to eliminate the high molecular weight—but essentially amorphous—microgel phase. The separation of different molecular weight fractions from various sources of gelatin proved to be unrealistic but molecular weight profiles could be obtained for each material and this was valuable information. As noted earlier it is unlikely that gelatin manufacturers would be interested in taking this work to the next logical stage, but at least methodology is now available for characterizing mixed samples and matching performance with identity.

Thermal degradation studies demonstrated that deterioration of the high molecular weight fractions occurred at temperatures in excess of 50°C, with virtually no degradation being found at 37°C or room temperature. This pretreatment therefore offers another way of removing or minimizing the quantities of higher molecular weight fractions present, depending on the pH of the medium. It should also be noted that these studies were carried out in dilute (0.1%) solutions of gelatin in order to avoid the other molecular interactions that occur at higher concentrations. As might be anticipated, the closer the pH came to the isoelectric point of the gelatin sample, the lower was the net particle charge. This favored the formation of noncovalent interactions required for the formation of gelatin aggregation. However, the systems were also shown to be extremely sensitive to environmental temperature, which clearly requires careful control in order to obtain reproducible aggregate sizes.

## THE TOXICITY OF GELATIN

Although widely perceived as being innocuous, studies on the intravenous use of gelatin as a blood expander showed a number of unexpected toxicities in clinical practice. For this reason, injected gelatin microparticles might be anticipated to suffer from the same effects and should not be regarded as being completely innocuous.

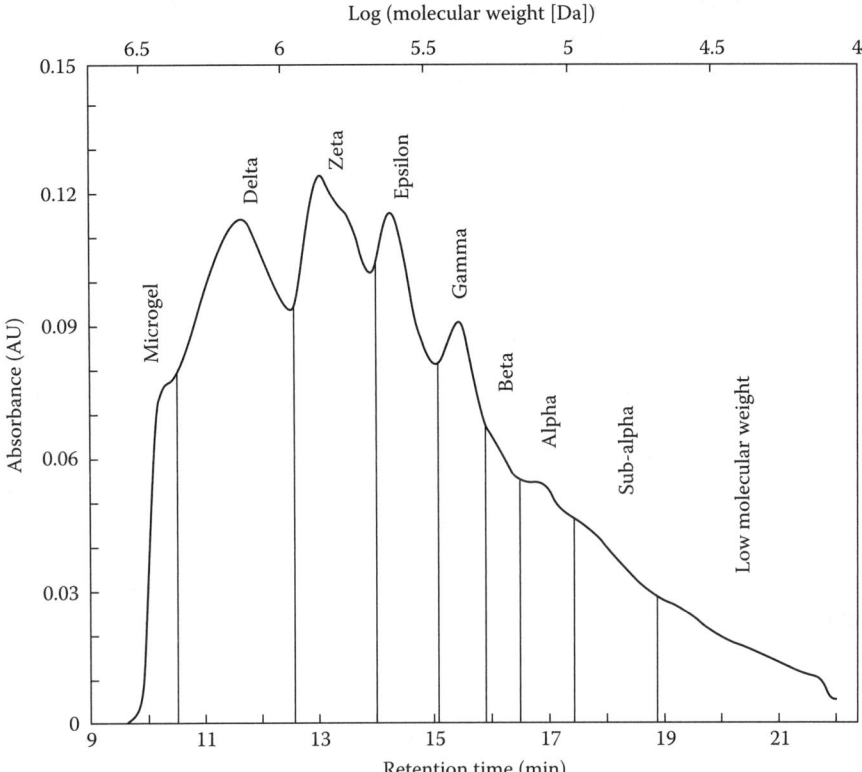

**FIGURE 8.10** HPLC size exclusion chromatogram of B225 gelatin obtained using the Bio-Sil SEC 400-5 and SEC 250-5 columns. (From Farrugia, C.A. (1998). The formulation of gelatin nanoparticles and their effect on melanoma growth in vivo. Ph.D. thesis, University of Illinois at Chicago. With permission.)

For example, both collagen and gelatin adhere to fibronectin-enriched surfaces and although this reaction is valuable for targeting particulate drug delivery systems to tumor surfaces, the reaction is also important in the healthy body. The gelatin-based plasma substitutes completely depleted plasma fibronectin, producing undesirable effects on the plasma clotting cascade. This effect would be anticipated to be less severe with gelatin microparticles than with gelatin in solution.

Two other issues need to be raised here although, again, they are more theoretical than realistic. The first is the possible presence of prions, the causative organism responsible for bovine spongiform encephalopathy (BSE) or "mad cow disease." A "new variant" of the so-called Creutzfeldt-Jakob prion is now known to have jumped species, causing deaths in humans. At present this is mainly a European issue and various remedies have been proposed, including the banning of all leather goods made from cattle skins. The British now only slaughter young cattle for food and experience is beginning to indicate that the disease is under control for the most part. As yet it has not materialized in the United States and it might be noted that most collagen for gelatin production comes from India these days, so the issue is somewhat remote from reality. Other, nonbovine sources of gelatin could be used such as porcine but this is not acceptable in some cultures. Prions have not to date been found in gelatin samples and the issue is substantially remote, no matter the origin of the gelatin.

The second issue is the slow release in the body of the cross-linking aldehydic reagents used to cross-link and stabilize the gelatin microparticles. This will occur slowly *in vivo* as the gelatin matrix itself dissociates and goes into solution. However, the effect of low level, slow, localized, release of glutaraldehyde over a period of time has not been evaluated clinically and the issue remains open for discussion.

## STERILIZATION OF GELATIN PRODUCTS

As noted, the effect of heat on gelatin solutions is enough to cause the deterioration of the highest molecular weight fractions and this usually manifests as a reduction in the viscosity of the solution. However, after heating, gelatin has a tendency to revert to structures found in the original collagen so the heated gelatin breaks down initially but, on cooling, tends to recover some of its original properties. Undoubtedly, some irreversible degradation must occur and, if feasible, aseptic filtration of dilute aqueous solutions should be used to sterilize gelatin intended for parenteral administration.

## TISSUE TARGETING WITH GELATIN MICROPARTICLES

For the most part, tissue targeting depends on the interaction between surface fibronectin and gelatin at some site in the body or the free fibronectin in the plasma. Fibronectin is a ubiquitous, asymmetric glycoprotein or, more accurately, a family of at least 100 related glycoproteins with structural homology (Figure 8.11). They each have two similar but not identical polypeptide chains joined near their C-termini held together with disulfide bridges. This family of proteins have high molecular weights (440 kDa) and are normal constituents of blood plasma besides being widely dispersed on connective tissues, blood vessel walls and basement membranes. Excreted onto the surface of a cell, the fibronectin functions as a cell-to-cell or cell-basement membrane adhesive. Free fibronectin in the plasma serves as the main opsonic protein, modulating the functional capacity of the reticuloendothelial system. In addition, fibronectin is believed to be the main factor involved when tumor cells metastasize from their original site to some other site in the body. This is important since it is well recognized that the high mortality rates following the diagnosis and surgical removal of many cancers is due to tumor cells becoming detached during the surgery and finding other sites in the body to develop. Only about half of the patients diagnosed and treated for their cancers survive beyond the initial 5-year period following surgery, chemotherapy, or radiotherapy, and this is inevitably due to metastases developing elsewhere.

Because of the complexity of its function in the biological environment, fibronectin has been shown to have selected areas or domains along its chain which can be associated to specific binding functions. For example, as shown in Figure 8.11, there are at least two domains with direct cellular binding activity, another domain binds

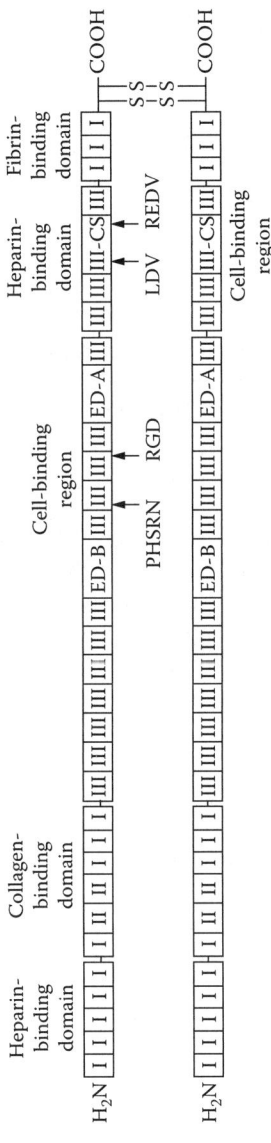

**FIGURE 8.11** Model of the structure of a fibronectin dimer, showing molecular composition of the various domains, and major amino acid sequences involved in cell binding. (From dos Santos (2005). Ph.D. thesis, University of Lisbon.)

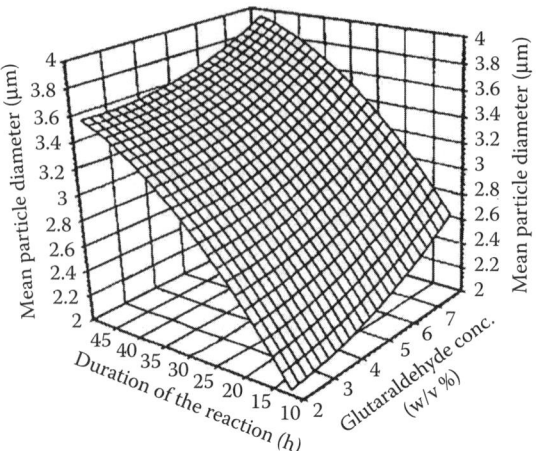

**FIGURE 8.12** Response-surface diagram for the mean particle diameter of albumin microparticles prepared by variation of glutaraldehyde concentration and duration of cross-linking. (From
Öner, L. and Groves, M.J. (1993b). *J. Pharm. Pharmacol.*, 45, 866–870. With permission.)

to DNA and yet others to fibrin and to heparin. However, one domain has been
identified as being specifically responsible for binding to collagen or, in its degraded
form, gelatin.

As a sidebar, it should be noted that there is a domain on the fibronectin molecule
which binds to the simple peptide RGD (arginine-glycine-aspartic acid) and this has
been explored as a means of interfering with the tumor adhesion process using
longer, more stable, polypeptide sequences and other analogues or derivatives
(Humphries et al. 1987).

In the clinical treatment of superficial bladder cancer with viable cells of an
attenuated *Mycobacterium bovis* vaccine, BCG (see Chapter 12), it became evident
that the bacterial cells adhered to the tumor cells through fibronectin produced at or
near the tumors. A fibronectin receptor on the bacterial cell wall has been identified
as an Antigen 86 complex homologue. Outside the scope of this present review, the
interested reader is referred to publications noted in the bibliography. However, by
analogy with the known reaction between collagen and gelatin, Lou and her colleagues (1995) suggested that the nonviable gelatin microparticles could be substituted for the viable organism, which is usually directly responsible for most clinical
side effects observed during treatment of bladder cancer with BCG.

This supposition was essentially confirmed and Lou et al. were able to demonstrate that high and low molecular weight antineoplastics could be loaded onto
gelatin particles following lyophilization.

Farrugia and Groves (1999b) demonstrated *in vitro* and *in vivo* that gelatin
microparticles, on their own, without any incorporated drugs, would block the
dissemination of melanoma cells. This is not surprising in view of the known ability

of gelatin to interact directly with fibronectin. The clinical potential of this concept has not been evaluated.

# ALBUMIN MICROPARTICLES

## INTRODUCTION

Albumin is ubiquitous in mammals: it comprises roughly half of the total plasma proteins, ranging from 6–9 g/L in humans. It is a single chain protein with a molecular weight of ~67 kDa and consisting of about 400 amino acid moieties. About half of these are hydrophilic and the remainder hydrophobic in character. Present in such large quantities, it might be anticipated that this protein serves more than one function in the body. For example, it is responsible for maintaining the osmotic pressure in blood, and therefore the blood volume, increases the buffer capacity of blood and functions as a major reservoir for amino acids as well as being involved in the binding and transport of smaller molecules.

Albumin is a major transport facilitator of hydrophobic compounds which would otherwise disrupt cellular membranes. These compounds include free fatty acids and bilirubin as well as hormones such as cortisol, aldosterone, and thyroxine when these materials have exceeded the capacity of proteins normally associated with them. Albumin also binds ions, including toxic heavy metals and metals such as copper and zinc which are essential for normal physiological functioning but may be toxic in quantities in excess of their binding capacity for their carrier proteins. Binding of protons is the basis for the buffering capacity of albumin.

Albumin will not cross membranes, mainly because of its size, and exerts an osmotic pressure counterbalanced by a counterflow of water into the circulation. If the plasma concentration of albumin is increased this will cause dehydration of tissues such as in lungs and cause an increase in blood volume. If the body is dehydrated due to a reduction of fluid intake this results in a reduction in the volume of blood and a decrease in the amount of blood albumin.

Use of albumin as an amino acid reservoir is analogous to the use of glycogen for storing sugars. Starvation will result in serum albumin being broken down to allow the synthesis of other proteins. In addition, the amino acids can function as an energy source by being converted into glucose in the liver.

## ALBUMIN MICROPARTICLES

Commercially albumins are obtained from avian eggs, fractionated blood products of human or bovine origin, or from plant sources such as soy bean.

The use of albumin microparticles as a drug delivery system was first suggested by Kramer (1974) and several methods for their production were subsequently developed (Gupta and Haung 1989). Most methods involved the application of emulsification methodology and factors involved in this process have been evaluated by a number of authors. However, studies of the *in vitro* disintegration process of protein microspheres, induced by the presence of protease enzymes in the environment, are limited (El-Samaligy and Rohdewald 1983).

## Emulsification Processes

Being a globular protein, containing hydrophobic entities that are retained in the center of the molecule, and hydrophilic entities tending to point into the surrounding aqueous medium, albumin is destabilized by heat. This process is readily illustrated with egg albumin, known commercially as *albumen*. The mixture of albumins, mucoids, and other materials starts to precipitate when heated at about 55°C and the process is irreversibly complete in the case of egg albumen at about 61°C which goes from a clear limpid liquid to a white, rubbery solid. It might be noted that, at the lower end of the temperature range, the process is reversible but somewhat slow. Denaturation is also induced by vigorous shaking, electrical currents, and various chemicals such as acids or bases, heavy metal salts, and alcohols.

Micron-range insoluble particles are therefore readily made from albumins by an emulsification process in which the protein is dissolved in water and emulsified with a surfactant in oil. The protein is denatured by heating the system above its denaturation point or with chemical cross-linking agents (e.g., glutaraldehyde) and collected by removal of the oil using solvents (e.g., petroleum ether, chloroform, or diethylether). These solvents are environmentally undesirable which makes the process unattractive for large-scale manufacture and the resulting microparticles often contain surfactants in them which can interfere with the resultant properties as drug delivery systems.

## Atomization Processes

As described earlier, Groves and his colleagues prepared gelatin microparticles by adding or spraying aqueous solutions of gelatin into cold anhydrous alcohol (Figure 8.7 and Figure 8.8). Öner and Groves (1993b,c) applied substantially the same process to the preparation of albumin particles and applied factorial analysis to optimize the conditions for the preparation of micron-ranged materials (Figure 8.13).

In addition, the enzymic disintegration of the prepared albumin microparticles was systematically evaluated by carrying out a dissolution process in the presence of trypsin.

The response surface (Figure 8.13) suggested that the cross-linking process in this case was essentially dynamic; particles becoming larger with time of exposure and concentration of glutaraldehyde. This was completely unlike the situation observed with gelatin where, once formed, the particles retained their original size. This may be due to the higher molecular weight of the globular albumin molecule which is more sensitive to structural changes induced by the cross-linking agent. In addition, as noted, the gelatin particles tended to precipitate in units of about 200 nm diameter or aggregates of that initial size.

Moreover, as noted, gelatin particles tended to redissolve unless they had been exposed to the glutaraldehyde for a sufficiently long period of time. Albumin particles, on the other hand, formed very rapidly at low concentrations of reagent and low temperatures and it appears that they grow in size as the attached water molecules are removed by the excess of anhydrous alcohol. It is likely that the denatured, insoluble, protein aggregates grow in size with time during the addition of the

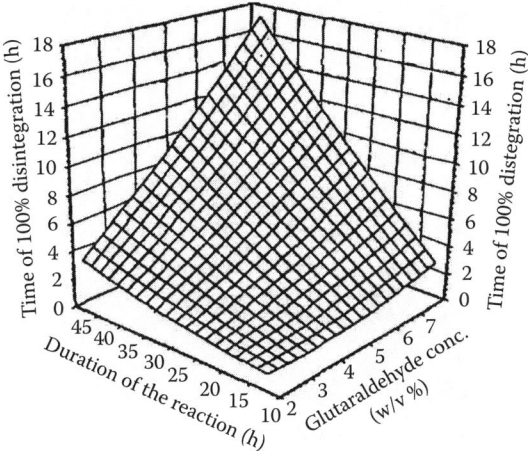

**FIGURE 8.13** Response-surface diagram for the time of 100% *in vitro* enzymatic disintegration of albumin microparticles prepared by variation of glutaraldehyde concentration and duration of cross-linking. (From Öner, L. and Groves, M.J. (1993b). *J. Pharm. Pharmacol.*, 45, 866–870. With permission.)

glutaraldehyde. Ideally, therefore, it should be possible to prepare albumin particles of any predetermined size by very careful attention to the conditions of preparation.

This observation needs to be compared to the few literature reports on the underlying factors that control the preparation of the albumin particles by the emulsification process. For example, it has been widely reported that parameters such as the variability in stirring rates and temperature had a significant influence on the size of the resulting beads and it has been concluded that the main process variables were controlled by the oil phase of the emulsion.

## Properties of the Particles

There have been relatively few studies of the properties of the albumin beads or particles. Gupta and Haung (1989) demonstrated that beads made by the emulsification process at temperatures over the range of 105–150°C varied in the release of incorporated doxorubicin, rates decreasing with an increase of the denaturation temperature. This indicated that the beads themselves were becoming hard or more dense.

Öner and Groves (1993c), on the other hand, repeated the experiment and were able to derive two response surfaces that correlated the production parameters of the chilled cross-linking process with the disintegration process in the presence of enzyme (Figure 8.13 and Figure 8.14). Although the relationship between the time for complete enzymic disintegration and particle diameter was not linear, there were strong indications that particles made under different condition did have different properties. This was considered to be related to the cross-linking activity which influenced the formation of an insoluble matrix as well as the actual nature of the matrix itself. Interestingly,

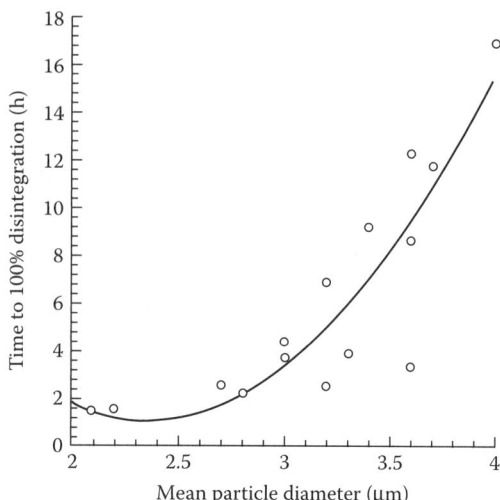

**FIGURE 8.14** The relationship between the mean particle diameter and the time of 100% *in vitro* enzymatic disintegration for albumin microparticles for all conditions under which the microparticles were made. (From Öner, L. and Groves, M.J. (1993b). *J. Pharm. Pharmacol.*, 45, 866–870. With permission.)

these authors did not see evidence of the swelling reported for cases in microparticles made by the emulsification process. This may have been due to the lyophilization stage which is an integral part of the chilled cross-linking process. The chilled cross-linking process, apart from requiring relatively large volumes of ethanol, has many features that make it attractive for scale-up.

## REFERENCES AND FURTHER READING

Ames, W. (1947). Heat degradation of gelatin. *J. Soc. Chem. Ind.*, 66, 279–284.

El-Samaligy, M. and Rohdewald, P. (1983). Reconstituted collagen nanoparticles. A novel drug carrier delivery system, *J. Pharm. Pharmacol.*, 35, 537–539.

Elysée-Collen, B. and Lencki, R. (1996). Protein ternary phase diagrams. I. Effect of ethanol, ammonium sulfate and temperature on the phase behaviour of type B gelatin. *J. Agric. Food Chem.*, 44, 1651–1657.

Farrugia, C.A. (1998). The formulation of gelatin nanoparticles and their effect on melanoma growth in vivo. Ph.D. thesis, University of Illinois at Chicago.

Farrugia, C.A. and Groves, M.J. (1999a). Gelatin behaviour in dilute aqueous solution: designing a nanoparticulate formulation. *J. Pharm. Pharmacol.*, 51, 643–649.

Farrugia, C.A and Groves, M.J. (1999b). The activity of unloaded gelatin nanoparticles on murine melanoma B16-FO growth in vivo. *Anticancer Res.*, 19, 1027–1032.

Friess, W., Zhou, W, and Groves, M.J. (1996). In vivo activity of collagen matrices containing PS1, an antineoplastic glycan, against murine sarcoma cells. *Pharm. Sci.*, 2, 121–124.

Geiger, M. and Freiss, W. (2002a). The use of collagen as a biomaterial. *Pharm. Tech. Europe*, 14(2), 40–48.

Geiger, M. and Freiss, W. (2002b). Collagen sponge implants. *Pharm. Tech. Europe,* 14(4), 58–66.

Gupta, P.K. and Haung, C.T. (1989). Albumin microspheres II. Applications in drug delivery. *J. Microencap.,* 6, 463–472.

Herning, T., Djabourov, M., Leblond, J., and Takerkart, G. (1991). Conformation of gelatin chains in aqueous solutions: 2. A quasi-elastic light scattering study. *Polymer,* 32, 3211–3217.

Humphries, K., Komoriya, A., Akiyama, S., Olden, K., and Yamada, K. (1987). Identification of two distinct regions of the type III connecting segment of human plasma fibronectin that promote cell type-specific adhesion. *J. Biol. Chem.,* 262, 6886–6892.

Hynes, R. (1990). Structure of fibronectins. In: *Fibronectins,* R. Hynes, ed. Springer-Verlag, New York, 113–175.

Jones, R. (1987). Gelatin: structure and manufacture. In: *Hard Capsules: Development and Technology,* K. Ridgeway, ed. Pharmaceutical Press, London, 13–30, 31–48.

Kramer, P.A. (1974). Albumin microspheres as vehicles for achieving specificity in drug delivery. *J. Pharm. Sci.,* 63, 1646–1647.

Lewis, D.A., Field, W.N., Hayes, K., and Alpar, H.O. (1992). The use of albumin microspheres in the treatment of carrageen-induced inflammation in the rat. *J. Pharm. Pharmacol.,* 44, 271–274.

Lorry, D. and Vedrines, M. (1985). Determination of molecular weight distribution of gelatins by H.P.S.E.C. In *Photographic Gelatin,* H. Ammann-Brass and J. Pouradier, eds. Proc. 4th IAG Conf., Fribourg, Germany, 1983, 35–54.

Lou, Y. and Groves, M.J. (1995). The use of gelatin microparticles to delay the release of readily water-soluble materials. *J. Pharm. Pharmacol.,* 47, 97–102.

Lou, Y., Olson, W.P., Tian, X., Klegerman, M., and Groves, M.J. (1995). Interaction between fibronectin-bearing surfaces and Bacillus Calmette-Guérin (BCG) or gelatin micro-particles. *J. Pharm. Pharmacol.,* 47, 177–181.

McNally, E.J., ed. (2000). *Protein Formulation and Delivery.* Marcel Dekker, New York.

Marty, J., Oppenheim, R., and Speiser, P. (1978). Nanoparticles—a new colloidal drug delivery system. *Pharm. Act.. Helv.,* 53, 17–23.

Öner, L. and Groves, M.J. (1993a). Preparation of small gelatin and albumin microparticles by a carbon dioxide atomization process. *Pharm. Res.,* 10(9), 1385–1388.

Öner, L. and Groves, M.J. (1993b). Properties of human albumin microparticles prepared by a chilled cross-linking technique. *J. Pharm. Pharmacol.,* 45, 866–870.

Öner, L. and Groves, M.J. (1993c). Optimization of conditions for preparing 2- to 5-micron range gelatin microparticles by using chilled dehydration agents. *Pharm. Res.,* 10, 6212–626.

Stainsby, G. (1977). The physical chemistry of gelatin. In: *The Science and Technology of Gelatin,* A. Ward and A. Courts, eds. Academic Press, London, 109–137.

Wang, Y.J. and Pearlman, R., eds. (1993). *Stability and Characterization of Protein and Peptide Drugs.* Plenum Press, New York.

# 9 Proteins and Phospholipids

*Michael J. Groves, Ph.D., D.Sc.*

## CONTENTS

## INTRODUCTION

This chapter is effectively an extension of the formulation chapter (Chapter 8) because the use of phospholipids drug delivery systems has become important over the past decade and deserves to be treated as a separate subject.

## STRUCTURAL PROPERTIES OF PHOSPHOLIPIDS

Phospholipids are widely found in nature and constitute, for example, the wall structure of most living cells in the plant and animal kingdoms. Structurally they are complex and the subtleties of the main structural elements have been elucidated only within the past few years.

As the name suggests, phospholipids contain both phosphorus and lipids and, structurally are based on lipid esters of glycerol (Figure 9.1). The basic structure can be substituted widely (Figure 9.2). In nature the lipid side chains vary in length from about $C_{12}$ to $C_{20}$ , with varying degrees of substitution and saturation. This results in the various compounds having a broad variety of properties, for example,

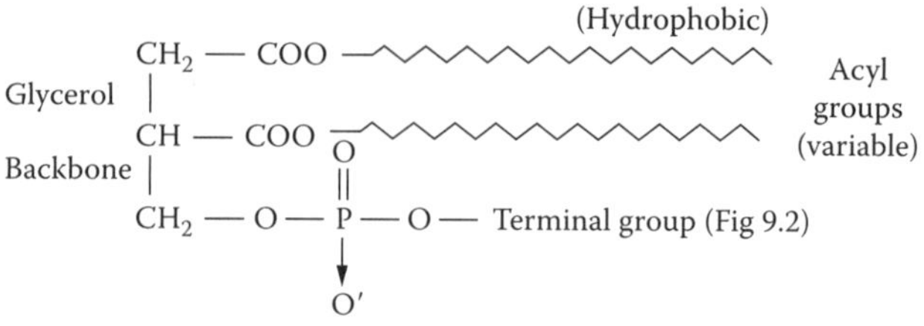

FIGURE 9.1 Phospholipid general structure.

a range of aqueous solubilities. However, as a generalization it may be noted that phospholipids are not very soluble in either water or oil and it is this feature that has resulted in their interesting physical properties.

It is possible to manipulate the structure chemically to provide artificial substitutes, the most familiar of which will be the pegylated derivatives in which polyethylene derivatives are bonded to the basic structure of, for example, phosphatidylethanolamine (Figure 9.3).

As might be anticipated, on their own these materials undergo complex decomposition reactions, illustrated in the case of phospholipids used to stabilize injectable lipid emulsions. As discussed later, these systems are usually steam-sterilized so that they readily hydrolyze both in the presence of heat during the sterilization process and on subsequent storage (Figure 9.4). Oxidation reactions also occur in the presence

| Name of group | Structure | Phospholipid name and abbreviation |
|---|---|---|
| CHOLINE | $-CH_2CH_2N(CH_3)_3^+$ | Phosphatidylcholine (PC) |
| ETHANOLAMINE | $-CH_2CH_2NH_3^+$ | Phosphatidylethanolamine (PE) |
| SERINE | $-CH_2CH(NH_3^+)COO^-$ | Phosphatidylserine (PS) |
| GLYCEROL | $-CH_2CH(OH)CH_2OH$ | Phosphatidylglycerol (PG) |
| INOSITOL | 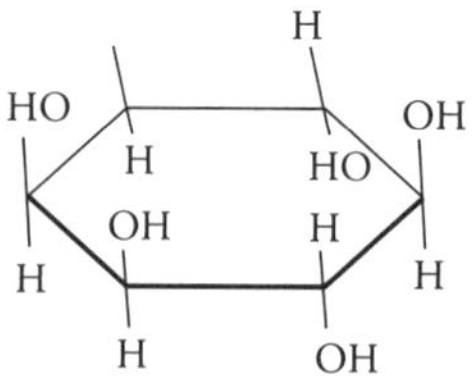 | Phosphatidylinositol (PI) |
| N-GLUTARYL-ETHANOLAMINE | $-CH_2CH_2NHCO(CH_2)_3$ $-COO^-$ | N-glutaryl-phosphatidyl -ethanolamine (N-glut. PE)(artificial) |

FIGURE 9.2 Basic structures of natural substituted phospholipids.

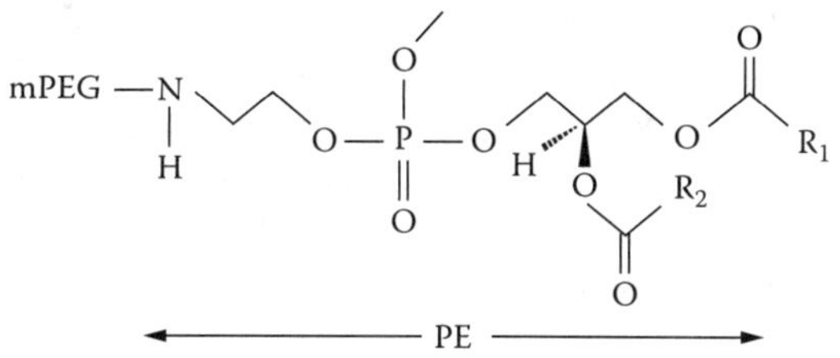

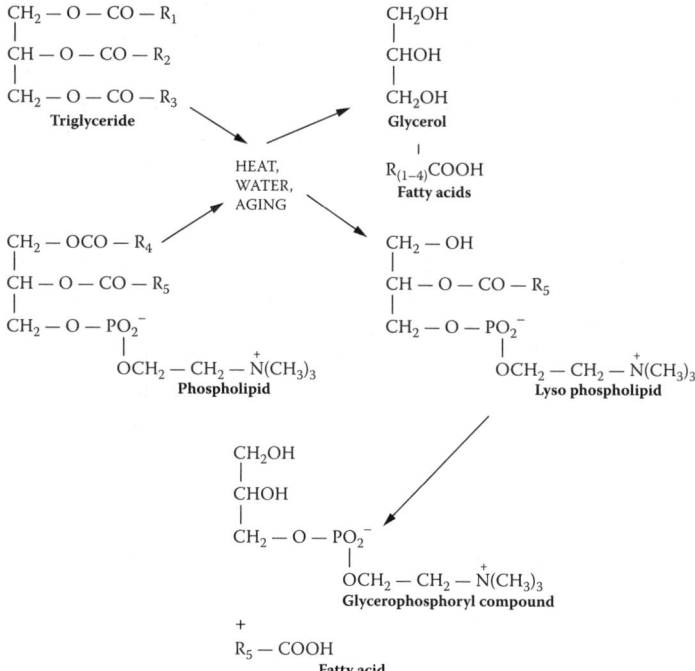

**FIGURE 9.3** Attachment of methoxy polyethylene glycol to PE.

of air or oxygen; important considerations from the packaging point of view for any formulation.

Collectively, as found in nature, phospholipids are usually mixtures of different acyl groupings and terminal head groups. These mixtures can be extracted using solvents from, for example, plants (e.g., soybeans) or eggs and constitute the material usually known in commerce as *lecithin*. Unfortunately there has been a tendency to call dipalmitoylphosphatidycholine, PC, one of the major constituents of most natural mixtures, *lecithin*, and it may be better to avoid applying this name to a purified derivative.

**FIGURE 9.4** Schematic showing the hydrolytic degradation of components of a phospholipids-stabilized triglyceride emulsion. (From Klegerman and Groves, 1992.)

As noted, phospholipids as a class tend to be only partially soluble in oil or in water. In the presence of water, for example, dispersed phospholipids tend to form characteristic structures under predetermined conditions. Examples of these would be the linear pairing, inside and out, of cellular bilayers and, as an extension, the unique hollow spherical structures of free phospholipids called liposomes which have proved to be valuable as drug delivery systems.

## INJECTABLE LIPID EMULSIONS

Injectable lipid emulsions are used to provide parenteral nutrition and their use can be traced back to the 1920s. However, because they are particulate systems by their very nature, administration of emulsions into the blood system must be viewed with care, requiring precautions and special requirements. Indeed, until the 1950s it was not realized that one essential requirement for injectable emulsions was that the droplet diameter must be below 1 μm in diameter. Otherwise there is always a finite risk of blocking the smaller blood vessels.

Another associated issue was the possibility of inactivating the LRES (lymphoreticuloendothelial system). By analogy with other injectable systems, it could also be deduced that the injectable emulsion system needed to be sterile and apyrogenic and free of acute or chronic toxicities from components or their associated degradation products. It also followed that the injectable system required to be *stable*, although how stability was to be determined and, more to the point, measured, has remained an issue to the present day. This is mainly because emulsions are thermodynamically unstable although their *stability* can be extended by formulation. As a result emulsion products are now available that are submicron in diameter, sterile, and stable for several years after preparation. In major part this has been due to the use of phospholipids as stabilizers and emulsifiers, in particular the mixed products identified as the lecithin of commerce.

The administration of metabolizable vegetable oils as concentrated sources of nutrition has proved to be valuable for patients who are debilitated and who are unable to take nourishment orally. In addition, oils such as soy bean oil provide a source of essential fatty acids which can be rapidly depleted in a patient after starvation for only a few days. Wretlind and his colleagues devised the phospholipids-stabilized soy oil emulsion now marketed as Intralipid™ (Pharmacia, now Pfizer, New York, NY) in Sweden during the 1960s and this product has been modified to carry oil-soluble drugs such as diazepam. In Europe this is marketed as Diazemuls™ and it may be anticipated that other drugs may be presented in the same or similar vehicles.

## FORMULATION AND PREPARATION

As noted, the key to successful preparation of an injectable emulsion product proved to be the selection of a suitable commercial source of lecithin, in this case a purified material obtained from the yolk of hen eggs. Lecithin can also be obtained from a number of natural sources such as soy beans and comprises a number of different phospholipids. These include phosphatidylcholine (PC) and phosphatidylethanolamine

## TABLE 9.1
## Approximate Constitution of Purified Lecithin Used to Stabilize Injectable Emulsions

Major Components

- Phosphatidylcholine (PC) ~65–70% w/w
- Phosphatidylethanolamine (PE) ~20–25% w/w

Minor Components

- Lysophosphatidylcholine (lyso-PC) <4% w/w
- Sphingomyelin (SP)
- Phosphatidic acid (PA)
- Phosphatidylinositol (PI)
- Phosphatidylserine (PS)
- Cholesterol
- DL-tocopherol (vitamin E)

(PE), usually present in the ratio of approximately 3:1 and making up about 90% of the total weight of the lecithin phospholipids (Table 9.1). It is known that the two main phospholipids account for most of the stabilization and emulsification activity of the lecithin, but it is thought that minor components such as sphingomyelin and phosphatidic acid also play some as yet undefined role in the process. It might be emphasized here that the natural mixture of components is more effective at stabilizing emulsions than any of the major components in either purified or synthetic form, alone or in artificial admixtures.

The presence of α-tocopherol, together with smaller quantities of β- and γ-tocopherols, should not be surprising since these are natural antioxidants found in both egg yolk and vegetable oils and are usually regarded as useful contaminants. However, some manufacturers are known to be adding up to 0.2% of the α-tocopherol as an antioxidant.

On this point, phospholipids as a class are readily oxidized by air and hydrolyzed in the presence of water. It follows that injectable emulsions should be handled with care in order to prevent degradation.

Hydrolysis that occurs to some degree on autoclaving and storage results in the production of corresponding lyso-compounds, phospholipids without one of the acyl chains. These materials are called lyso- because, on their own, lyso-compounds produce lysis of red blood cells and are generally regarded as toxic. Analytically it is not difficult to demonstrate that there are significant quantities of these compounds present in injectable emulsions but there have never been any clinical reports of toxicity. The probability is that these materials form some type of association complex with the other phospholipids present and are not available on their own to exert any effect which would result in toxicity.

From a formulator's perspective, the clinical emulsions contain usually 10% w/v oil, usually soy oil or, occasionally, safflower oil and other metabolizable oils. The oil concentration is also found at the 20% or 30% levels but in almost all cases a level of 1.2% lecithin is sufficient to stabilize the emulsion system. A point not usually appreciated is that most commercial vegetable oils are hydrogenated because they are used in the preparation of margarine. However, for injection, the oil must be nonhydrogenated and any batches of incoming oil for manufacturing injectable emulsions must be checked for the presence of hydrogenated material. This is especially significant because tanks used for the bulk transportation of the oil may not necessarily have been adequately cleaned out between one lot of hydrogenated oil normally carried in the tank and the incoming lot of untreated oil.

Glycerol is also added to the emulsion to raise the osmotic pressure since electrolytes tend to destabilize the system.

These emulsions can be prepared by mixing either an oil dispersion of lecithin with glycerol followed by some of the water or, conversely, by dispersing the lecithin in most of the water followed by the glycerol and, at the end, the oil. In either case some complex phase changes may occur during the mixing process. After mixing the coarse emulsion is passed repeatedly through a homogenizer of the Manton-Gaulin type until the droplet size is consistently below 500 nm in diameter. This machine is also widely employed to homogenize milk products so the associated technology is well understood.

The completed emulsion is then filled into large-volume glass containers and sealed with oil-resistant composition stoppers. These containers are then autoclaved, in some cases using shaking or rotating autoclaves to minimize any thermal damage to the product as a whole.

The fact that these phospholipids-stabilized emulsions are sterilized by heat may be surprising since most emulsions stabilized by almost any other surfactant is readily destabilized by heat. Indeed, the fact that the droplet size of phospholipids-stabilized emulsions actually decreases on the application of thermal stress is probably due to the behavior of the phospholipids which move from the aqueous phase to the oil phase, especially to the interfacial mesophase, during the heating process.

Chemical degradation of phospholipids results in the production of lyso-compounds but also free fatty acids (Figure 9.4). The pH of a heated emulsion will tend to fall as more fatty acids are produced and the hydrolysis rate slows down until a pH of around 4 is reached, at which point it starts to increase again. Accordingly, the initial emulsion is adjusted to pH 9 by the addition of small quantities of sodium hydroxide, and this has the effect of prolonging the shelf life of the product.

From a pharmaceutical perspective, phospholipids-stabilized emulsions are remarkable. For example, they are relatively stable, with shelf lives of 18 months to 2 years being obtained after the initial heat sterilization. They resist the increased shear rates as the bottles are transported from producer to user and they can tolerate the addition of a wide variety of monovalent electrolytes for at least short periods prior to administration. However, they cannot resist freezing and changes in droplet size following exposure to freeze-thaw cycles can be used as a measure of the stability of the emulsion system. Most injectable emulsions are sensitive to multivalent cations such as calcium or magnesium salts, which rapidly flocculate the phospholipids-stabilized systems.

# THE PHYSICAL STATE OF PHOSPHOLIPIDS-
# STABILIZED EMULSIONS

The fact that phospholipids-stabilized emulsions behave differently from what might be called classical emulsions suggests that they are structurally different and it is this feature that allows them to resist environmental stresses. The likelihood is that the phospholipids, otherwise substantially insoluble in water or oil, exist at the interphase between the oil and water. This interphase has been termed a mesophase and in this case has a structure that gives it a separate identity. Although difficult to prove because they constitute only a small fraction of the overall constitution, the probability is that the mesophase consists of intimate mixtures of phospholipid, the free fatty acids formed by hydrolysis of the phospholipids, oil, and water. This close association may also vary in constitution as one approaches the triglyceride oil droplet from the external water phase, being richer in water on the outside and in oil on the inside of the mesophase layer. However, there is a strong possibility that some type of liquid crystal is to be found in the mesophase and this may account for the evident resistance of the droplets to coalescence under various environmental stresses. In addition, it is known that the cold dispersion of oil and phospholipid in water prior to application of heat during the sterilization process is constitutionally different to the cooled, heater-treated system. PE, for example, moves out of the aqueous phase to the oil phase on heating and remains there after cooling, indicating an interfacial reaction which would affect the constitution and therefore behavior of the mesophase (Herman and Groves 1993).

Another complication is the fact that the phospholipids are not soluble in water, forming in many cases liposomal structures, but they are soluble in electrolyte solutions. Thus, the mere addition of an electrolyte to a phospholipids-stabilized emulsion may physically affect the structure of the mesophase and, hence, the stability of the whole system.

## INJECTABLE EMULSIONS AS DRUG DELIVERY SYSTEMS

Injectable emulsion formulations have an advantage as drug delivery systems since hydrophobic (lipophilic) drugs can be incorporated into the oil phase of the emulsion. How it is incorporated is another matter since the drug may simply be dissolved in the oil at the beginning of the manufacturing process and, if it is stable, carried through the final heat-sterilization process. Alternatively, if the drug is labile, it may be incorporated passively by simply adding the drug or a solution of the drug in, for example, ethanol to the final, sterile, oil/water emulsion and allowing the compound to diffuse into the oil phase.

Proteins for the most part are water soluble but some are not and most have domains along the molecule that differ in their hydrophobicities. There is therefore a strong possibility that some proteins or polypeptides, when added to a sterile emulsion, would distribute themselves into the mesophase layer of the emulsion droplets and remain associated with these droplets on delivery.

Biologically active materials such as prostaglandin $E_1$ have been reported to benefit from being incorporated into phospholipids-stabilized emulsions (Teagarden et al. 1988, 1989). In this case, the kinetics of degradation indicated that this strongly surface-active drug was located in the droplet mesophase. The degradation was influenced by the presence of a charged interfacial group as well as the overall pH of the emulsion system. Both penclomedine and diazepam are substantially water-insoluble drugs and have been incorporated into phospholipid-stabilized emulsions. Muramyl dipeptide and some of its analogues derived from mycobacteria have also been incorporated into phospholipid-stabilized squalane emulsions. These were extensively tested for immunological activity against a number of human tumors, although extended clinical trials indicated that there was little or no effect on established tumors in humans.

A number of other emulsion formulations have been tried as drug delivery systems on an experimental basis. No doubt these studies will continue into the future because emulsions are attractive as drug delivery systems and have been thoroughly studied by a number of researchers.

## LIPOSOMES

Liposomes are characteristic hollow spherical aggregates or vesicles that form spontaneously when phospholipids are dispersed in water. This is a function of their low solubilities in both oil and water and results from the hydrophobic nature of the twin acyl tails and the strongly hydrophilic polar head group which are the main characteristics of phospholipids (Figure 9.1). As a result, phospholipid molecules, when dispersed in water form double layers where the phospholipids align themselves, tail-to-tail and head-to-head (Figure 9.5)

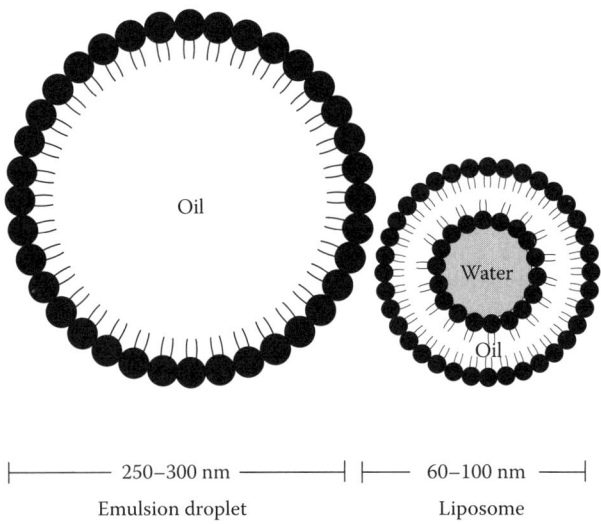

|————— 250–300 nm —————| |——— 60–100 nm ———|

Emulsion droplet                              Liposome

**FIGURE 9.5** Schematic representation of phospholipids distribution in lipid emulsions and liposomes. Both drug carrier systems show similar surface. The main differences between liposomes and lipid droplets are related to the mean diameter and nature of the inner contents. (From dos Santos, 2005.)

Phospholipids, when dispersed in water, form spherical vesicular structures, an observation first made by Alex Bangham and Robert Horne in 1952. An interesting and humorous account of the early work on liposomes has been published by Bangham as an introduction to the book by Ostro (1983) which is recommended reading for the interested student.

At low concentrations, a hollow vesicle results with usually just one double layer and, as the concentration is increased, the number of double layers can increase in a transition from unilamellar vesicles to multilamellar structures. Since the hydroplilic head groups are exposed on the inside as well as the outside of the vesicular structure this provides an opportunity to entrap hydrophilic *guest* drug molecules both inside the center of the vesicle and, if multilamellar, between the phospholipid bilayers as well. On the other hand, hydrophobic molecules can become incorporated in the hydrophobic regions of the bilayers where the hydrophobic tails overlap.

One approach to entrapping guest molecules is to use a different ionic environment inside the hydrophilic portion of liposomes. This is used with, for example, doxorubicin, where the drug is precipitated inside the central liposomal core containing sulfate ions. The precipitation reaction between the charged drug molecules and the counter-ion is controlled by pH and enables the drug to be trapped inside the carrier and, subsequently, released slowly when injected into the body.

Cholesterol is often incorporated in to liposome formulations because it is capable of moving into the hydrophobic regions of the bilayers and has the effect of stabilizing these structures. Single component phospholipid bilayers can exist in the "gel" state or as liquid crystalline "fluid" structures. The transition temperature between these two states, $T_c$ is a measurement of the melting point which is influenced by the nature of the polar head group, the degree of unsaturation, the length of the acyl groups in the phospholipids, and the presence of additives such as cholesterol. A bipolar layer in the gel state tends to be more rigid and less permeable to entrapped guest molecules than the fluid state.

However, the stability of unmodified liposomes is generally not good because of the inherent properties of the phospholipids themselves. Any individual phospholipid molecule is capable of moving along the surface in which it was originally incorporated and, indeed, more slowly, across into the adjacent surface. This molecular fluidity of the bilayers allows incorporated guest molecules to escape into the external environment. Ultimately this *leakage* removes any advantage that the liposomal vesicle may have as a drug delivery system since, after initial entrapment, the system, as a whole, moves to an equilibrium state in which the guest molecule is likely to be equally distributed between the interior and exterior (aqueous) environments. Formulations with cholesterol and other components are made in order to increase the rigidity of the bilayers, encouraging the formation of a gel state, thereby reducing the leakage rate.

This leakage can also be reduced by chemical modification of the external surface of the bilayers in order to introduce polymeric components into the interfacial structure (see the section on surface modified liposomes).

In principle, a stabilized liposome might be valuable as a passive drug targeting system since the site of disposition in the body is significantly influenced by the diameter of the vesicle. For example, small liposomes below ~50 nm diameter can pass through the capillary wall surrounding a tumor, allowing suitably surface-modified vesicles to specifically target a tumor surface. Larger liposomes, >10 μm in diameter, could be trapped in the alveolar lung tissues and passively release incorporated drug. The issue with passive targeting is that it is not necessarily precise enough in its targeting ability to hit a specific target and this may result in collateral damage to healthy tissues.

Unfortunately, unmodified liposomes are rapidly coated with globular apoproteins found in serum. This is also true of injectable lipid emulsions but here this reaction is desirable since the oils are administered for nutritional purposes and it is an advantage for them to be rapidly metabolized. In the case of liposomes, passive destruction by the lymphoreticular endothelial system (LRES) is not necessarily desirable, especially if the system is targeted to a specific surface such as a tumor which may be remote from the injection site.

## Charged Liposomes

Liposomes prepared from the usual mixed phospholipids found in commerce tend to be electrostatically neutral or carry a slight electronegative charge. Most cellular surfaces in the body also tend to be negatively charged. For this reason there might be advantages to using cationic liposomes to target surfaces of opposite charge. These systems have been used to interact with and deliver DNA sequences as genes. Cationic liposomes can be prepared by incorporating positively charged lipids such as DOTMA (N-[1-(2,3 dioleyloxy)propyl]-N,N,N-trimethylammonium chloride) in equal weight with the liposome-forming dioleylphosphatidylcholine, otherwise known as DOPE. Other positively charged derivatives such as cholesterol esters or stearyl amine can also be incorporated to provide positively charged drug delivery systems. The process is discussed by Crommelin and Schreier (1994 (Kreuter)) in an excellent review of the state of the art of liposomes prior to the introduction of sterically hindered liposome surfaces.

Electronegative liposomal structures can readily be formed by the addition of phosphatidylglycerol or phosphatidic acid or diacetylphosphate. Dadey (in press) has recently introduced the concept of sulfating the terminal hydroxyl groups of phosphatidylinositol, thereby producing a strongly negatively charged phospholipid which may prove to have profound physiological effects *in vivo*. These modifications should also increase the stability of the bilayers, thereby increasing the overall stability of the liposomal structure.

## Surface-Modified Liposomes

Although the concept of using liposomes as drug delivery vehicles was initially very attractive, there was not a great deal of success in bringing the ideas to the market place. In part this was due to the questionable stability of liposomes when injected into the body. As noted above, the lipid surfaces become rapidly coated with the apoproteins found in serum and this is the first stage of the short degradation process following injection.

Recently a delayed release type of liposome has been developed (Allen et al.) in which some of the external phosphatidylethanolamine (PE) has been replaced by

PE to which is covalently bonded polyethylene glycol (PEG). The net effect of this is to cover the external liposome surface with PEG, forming a convoluted but essentially hydrophilic surface which resists the involvement of apoproteins on injection (loc. cit.). Bangham rather whimsically, described the effect as effectively camouflaging the surface so that it appeared to be water-like to any nearby protein molecules. These liposomes have been called "Stealth" liposomes for this reason although this is a registered trademark and "sterically hindered" liposomes might be a better term to use in general practice and is more accurate.

Clinically these systems are proving to be more successful as drug delivery devices than unmodified liposomes. For the most part this may be attributed to the improved stability and reduced interaction with serum apoproteins when injected. Other forms of sterically hindered liposomes are being explored such as PE associated with chitosan although these have not been tested clinically. However, this development has been extended to incorporate other entities into the liposomal structure which facilitates targeting of liposomes to structures such as tumor cells. This takes the targeting concept beyond the passive size approach in which small liposomes can move through blood vessel walls and larger liposomes can be trapped in blood vessels in, for example, the lungs.

## Direct Interactions between Proteins and Liposomes

It should be noted that liposomes in effect mimic cell walls and proteins are to be found in nature associated with cell walls. However, the association may differ physical according to the conformation for the protein under consideration. For example some proteins fold so that they are exposing hydrophobic regions which fit into the hydrophobic regions of the liposomal structure (or cell wall). The net effect is that the transbilayer protein exposes its hydrophilic regions both inside and outside of the liposomal structure to the water surrounding the structure (Figure 9.6).

The most likely type of interaction between a phospholipid surface layer and a protein would be entirely hydrophilic in its nature. Here oppositely charged groups interact electrostatically. Zwitterionic phospholipids will not react in significant amounts and the vesicle has to be composed of phospholipids containing quantities of, for example, phosphatidylglycerol or phosphatidylserine to ensure that the liposome carries a significant electronegative charge. The interaction is largely controlled by the environmental pH which influences the ionization of surface groups on the protein and phospholipids. Unless there is a region of marked hydrophobicity along the length of the protein molecule, the protein is most likely to lie along the phospholipid surface on the inside or outside of the vesicle. The protein is readily transported attached to the vesicle although it may not be protected from the action of environmental proteases which can attack the external surface material. However, if the protein is encapsulated inside the vesicle this issue is considerably reduced and liposomes have been used as drug delivery vehicles for vaccine antigens. Care needs to be taken to ensure that a native protein does not change its conformation at a pH selected to enhance the adsorption of protein to the vesicle.

On the other hand, hydrophobic interactions can also play a part in the interactions between phospholipids and proteins or drugs. If the reacting entity has a high

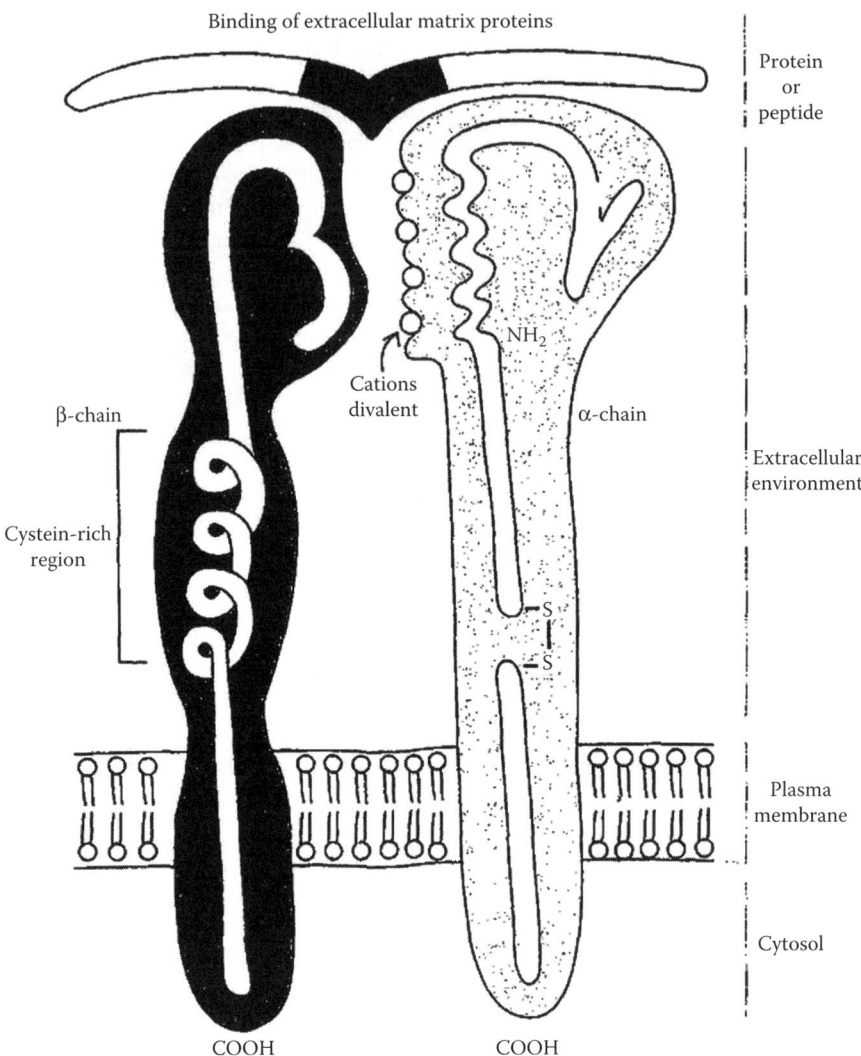

**FIGURE 9.6** Structure features of integrin receptors. Peptides or proteins containing the tripeptide sequence of RGD are recognized by several integrins present in the tumor cells. However, the protein "tails" pass through the plasma membrane with the hydrophobic domains and are locked into place. (From dos Santos, 2005)

water/octanol coefficient it will tend to form a stable association with long chain lipids such as those found in phospholipids. This reaction can be stabilized by appropriate pH buffer conditions but when introduced into the body, where conditions could be described as being "sink," that is, there is a relatively large volume of solvent available for solution, the bound entity tends to be rapidly released and becomes associated with the lipoproteins of the serum. Under these conditions the lipid system is acting as a carrier for the drug but, effectively, is not a delivery system in the sense that the drug's pharmacokinetics remain unchanged. An unpublished example of this type of reaction is the lipid/drug formulation of verteporfin (Visudyne, QLT, Vancouver, BC, Canada).

As noted, it is possible to design a liposome in which the protein is entrapped in the vesicular structure without necessarily being adsorbed to the surface. If the vesicle is also designed to target a biological surface or entity then it is important to destabilize the liposomal bilayers at that target surface. The reason here is that the protein will not be released by the leakage mechanism associated with smaller guest molecules under the same conditions and it is necessary to ensure that the large molecule is released rapidly and intact.

Proteins attached to the outer surface of a liposomal vesicle may become valuable if the protein is an antigen. There seems to be general agreement that such systems result in enhancement of the antigen activity, the phospholipids acting as adjuvants for the antigen. An adjuvant is here defined as any substance which enhances the immunological response to that particular antigen. This type of system was tried clinically for muramyl tripeptide encapsulated in a liposomal formulation although the trials were abandoned when little enhancement could be demonstrated in humans. This disappointment has tended to discourage further clinical trials of other systems although, on a much smaller scale, a number of systems have been demonstrated to produce significant enhancement of antigen activity in mice.

Finally, covalent bonding between phospholipids incorporated in liposomal vesicles and various proteins has been explored and may provide an avenue for future development of the concept and design of targeting liposomes.

Factors influencing the properties of liposomes are summarized in Table 9.2. Lipid-based carriers which include liposomes are classified in Table 9.3.

---

**TABLE 9.2**
**Factors that Affect the Properties of Liposomes**

Number of lamellae in liposome (SUV, MLV)
Average diameter of vesicle
Identity of surface phospholipid head groups
Presence of bound structural modifiers such as PEGs
Surface charge
Phase transition temperature of bilayers
Presence of cholesterol or cholesterol esters in formulation
Presence of other lipids to enhance required properties
Manufacturing conditions
Hydrolysis or oxidative stress factors on stability
Presence or absence of additives to enhance stability (antioxidants, lyophilization aids)
Packaging
Head-space gases

---

**TABLE 9.3**
**Classification Scheme for Lipid-Based Drug Carriers**

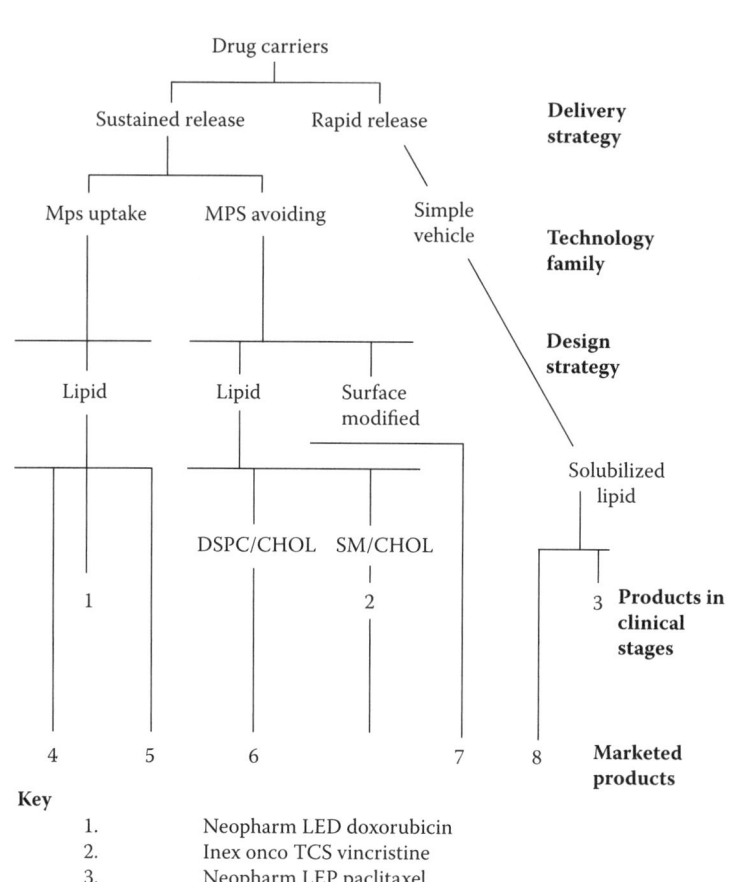

Key

| | |
|---|---|
| 1. | Neopharm LED doxorubicin |
| 2. | Inex onco TCS vincristine |
| 3. | Neopharm LEP paclitaxel |
| 4. | Elan inyocet doxorubicin |
| 5. | Gilead ambisome amphoteracin B |
| 6. | Gilead daunXome daunorubicin |
| 7. | Alza doxil doxorubicin |
| 8. | QLT visudyn verteporfin |

## PEPTIDES AND PHOSPHOLIPIDS*

An interesting sidelight on the dilemma of drug delivery has recently been published which may represent a solution to this ever present problem. It will be recalled that with most modern drugs, especially those of natural origin there is

* This section written with material supplied by Professors Hayat Önyüksel and Israel Rubinstein, University of Illinois at Chicago.

an issue since only small quantities of the biologically active material are required at the site of activity although relatively massive amounts of the physiologically active substance may need to be administered orally or parenterally. The result is that activity may be manifest at other sites throughout the body, an activity usually termed *toxicity*.

Önyüksel, Rubinstein and their coworkers have recently published a series of papers on the formulation and biological activity of the short-chain vasoactive intestinal peptide, VIP. This peptide, as well as secretin and the pituitary adenylate cyclase-activating peptide, ($PACAP_{1-38}$), are all widely distributed amphipathic mammalian neuropeptides that exert diverse biological effects in target tissues located some distance from their site(s) of release. Furthermore, the biological half lives of these exogenously administered peptides in blood are extremely short — often less than one minute, and this paradox is difficult to explain.

VIP itself is a ubiquitous 28-amino acid neuropeptide which displays a wide range of biological activities. It is found in the central and peripheral nervous systems, functioning as a non-adrenergic, non-cholinergic neurotransmitter and neuromodulator. It is also present in pervascular nerves and functions to relax smooth muscle. In addition, significant quantities of VIP are to be found in almost every other body tissue. Secretin, on the other hand, is a 27-amino acid hormone with substantial sequence homology with VIP. It is known to be involved with the secretion of pancreatic exocrine factors, including insulin release and growth hormone releasing factor, but is also capable of inhibiting gastrin secretion and gastrin release. Like VIP, it also produces a potent, albeit transient, vasodilatation.

VIP or its analogues have been demonstrated to strongly associate with natural phospholipids or with polyethylene glycol-grafted stabilized phospholipids. This observation goes some of the way to explain the biological paradox noted earlier. Interactions between phospholipids on one hand and the peptides themselves on the other have been shown to increase the peptide stability *in vitro* and amplify the biological activity *in vivo*. This activity is presumed to enable the stabilized peptides to be transported from their site of synthesis to distant sites of application within the body. Intrinsically the peptides themselves have some curious properties such as the ability to lower interfacial tension and form micellar structures just like a synthetic surfactant. They also directly interact with phospholipid membranes. These peptides are generally long enough to fold back upon themselves and in water will generally form random coils. These coils, in the presence of some organic solvents, detergents or anionic lipids, will fold to form α-helical structures. VIP binds reversibly to liposomes at neutral pH and in this configuration is resistant to proteolysis by trypsin. This interaction and stabilization of the α-helix by phospholipid is believed to occur *in vivo* but, more to the point, the complex of VIP with phospholipids is stabilized in blood for at least eight days at 37°C. The half-life of liposomal VIP in blood was at least eightfold longer than the VIP alone so obviously a significant stabilization mechanism is at work here.

That these mechanisms may operate in the body has been mentioned and certainly could account for the paradox noted earlier. However, from a practical standpoint, delivery of VIP as a drug in sterically stabilized phospholipid liposomes would be very attractive and may find future application.

# COCHLEAL PHOSPHOLIPID STRUCTURES

In 1975 Papahadjopoulos and his colleagues observed that some phospholipids precipitated out of liposomal suspensions in the presence of multivalent cations in the form of lipid cylinders or cochlear structures. These cochleates were shown to have unique multilayered structures formed by the internal collapse of liposomes. They consisted of lipid sheets, rolled up into spiral or stacked sheets but, unlike the original liposomes, there was no internal space in between the sheets. Formed mainly from the anionic phosphatidylserine (Fig. 9.1), the addition of cationic calcium ions caused them to bind to the oppositely charged head groups and the spherical structures collapsed to roll or stack into essentially solid lipid structures. However, although there is no trapped water in these cochlear structures, any drug or guest molecule incorporated in the original liposomes will become trapped in between the adjacent lipid sheets, becoming protected from the surrounding aqueous medium.

This principle has been extended to DNA-protein complexes or genes which can be entrapped and delivered to target cells, the major advantage of the systems being that they tend to be able to penetrate the cell walls, thereby enabling external materials such as genes to penetrate to the cell nucleus.

This hypothetical transfer process has in fact been observed for human hemopoietic stem cells and may provide a means of treating or preventing a number of diseases such as a variety of anemia's, immune deficiencies and cancers. Some of these possibilities have recently been discussed by Delmarre et al. (2004).

Cochlear formulations are interesting in other ways since they offer a real prospect of being able to deliver nuclear materials through cell walls without the issues associated with the use of viable vaccines. As noted, this may be especially true because not only do the phospholipid structures pass through the phospholipid cell walls but incorporated materials are protected from the surrounding environment such as enzymes that would otherwise destroy them.

Cochleates are readily lyophilized to free flowing powders that can be incorporated into capsules for oral administration or resuspended in aqueous vehicles for injection. Since they are generally made from phosphatidylserine, itself mainly obtained from soy beans, the claim had been made that they are non-toxic. However, cytoplasmic transfer of incorporated material does undoubtedly occur and this may affect physiological or toxicological profiles of incorporated drugs. Such claims as a generality may be acceptable but when it comes to specific cases, some initial caution would seem to be justified.

# REFERENCES AND FURTHER READING

Allen, T.M., Hansen, C., Martin, F., Redemann, C. and Yau-Young, A. (1991), Liposomes containing synthetic lipid derivatives of poly(ethyleneglycol) show prolonged circulation half-lives *in vivo. Biochim. Biophys. Acta,* 1066, 29–36.

Ashok, B., Rubinstein, I., Tsueshita, T. and Önyüksel, H. (2004), Effect of molecular mass and PEG chain length on the vasoreactivity of VIP and PACAP$_{1-36}$ in PEGylated phospholipid micelles, *Peptides,* 25, 1253–1258.

Benita, S. (Ed.) (1998), *Submicron Emulsions in Drug Targeting and Delivery,* Harwood Academic Publishers, Amsterdam.

Crommelin, D.J.A. and Schreier, H. (1994), Liposomes, In Kreuter, J. (Ed.) *Colloidal Drug Delivery Systems.* Marcel Dekker, New York, pp. 73–190.

Dadey, E.J. (2003), unpublished.

Delmarre, D., Lu, R., Tatton, N., Krause-Elsmore, S, Gould-Fogerite, S., and Mannino, R.J. (2004), *Drug Delivery Technol.,* 4(1), 64–69.

Dos Santos, A.M. (2005), PhD Thesis, University of Lisban, Portugal.

Ghandi, S., Rubinstein, I., Tsueshita, T., and Önyüksel, H. (2002) Secretin self-assemblies and interacts spontaneously with phospholipids *in vitro, Peptides,* 23, 201–204.

Gordon, R.D. and Peterson, T.A. (2003), Four myths about transdermal drug delivery, *Drug Delivery Technol.,* 3(4), 44–50.

Gould-Fogerite, S., Mannino, R.J., and Margolis, D. (2003), Cochleate delivery vehicles: Applications to genetherapy, *Drug Delivery Technol.,* 3(2), 40–47.

Groves, M.J. and Herman, C.J. (1993), The redistribution of bulk aqueous phase phospholipids during thermal stressing of phospholipid-stabilized emulsions, *J. Pharm. Pharmacol.,* 45, 592–596.

Groves, M.J. (1999), Parenteral drug delivery systems, In Mathiowitz, E. (Ed.), *Encyclopedia of Controlled Drug Delivery,* Vol. 3, John Wiley & Sons, New York, pp. 743–777.

Hanin, I. and Pepeu, G. (Eds.) (1990), *Phospholipids: Biochemical, Pharmaceutical and Analytical Considerations,* Plenum Press, New York.

Herman, C.J. and Groves, M.J. (1992), Hydrolysis kinetics of phospholipids in thermally stressed intravenous lipid emulsion formulations, *J. Pharm. Pharmacol.,* 44, 539–542.

Klegerman, M.E., and Groves, M.J. (1992), *Pharmaceutical Biotechnology: Fundamentals and Essentials,* Interpharm Press, Buffalo Grove, IL.

Kreuter, J., (Ed.) (1994), *Colloidal Drug Delivery Systems,* Marcel Dekker, New York.

Lasic, D. and Martin, F. (Eds.), (1995), *Stealth® Liposomes,* CRC Press, Boca Raton.

Önyüksel, H., Ashok, B., Dagar, S., Sethi, V. and Rubinstein, I. (2003), Interaction of VIP with gel-phase phospholipid bilayers: Implications for vasoreactivity, *Peptides,* 24, 281–286.

Önyüksel, H., Ikezaki, H., Patel, M., and Rubinstein, I. (1999), A novel formulation of VIP in sterically stabilized micelles amplifies vasodilation *in vivo, Pharm. Res.,* 16, 155–160.

Ostro, M.J. (Ed.), (1983), *Liposomes,* Marcel Dekker, New York.

Teagarden, D.L., Anderson, B.D. and Petiz, J.W., (1988), Determination of the pH-dependent phase distribution of prostaglandin E in a lipid emulsion by ultrafiltration, *Pharm. Res.,* 5(8), 482–487.

Teagarden, D.L., Anderson, B.D. and Petiz, J.W., (1989), Dehydration kinetics of prostaglandin E in a lipid emulsion, *Pharm. Res.,* 6(3), 210–215.

# 10 Pulmonary Drug Delivery Systems for Biomacromolecules

*Camellia Zamiri, Ph.D. and Richard A. Gemeinhart, Ph.D.*

## CONTENTS

## THE PULMONARY ROUTE OF DRUG DELIVERY

Demand for improved methods for the delivery of biopharmaceutical agents initiated by the Human Genome Project (Olson 2002) has resulted in the development of numerous technologies. Although considerable effort has been spent on finding ways to deliver the biomacromolecules by the convenient gastrointestinal, nasal, buccal, and transdermal routes, these body surfaces are fundamentally impermeable and have restricted surface area, limited residence time, or other factors that make the efficiency of penetration very low to all but the most potent small peptides. Of the small peptides absorbed through these routes, many have specific uptake mechanisms (Russelljones 1996), thus, the potential for widespread biomacromolecular drug delivery via these routes is difficult.

Two distinct areas of application of the pulmonary route for systemic delivery are inhalation delivery for situations where (1) rapid absorption of drugs is required than can be achieved by other noninvasive routes, especially by oral administration; and (2) absorption by other noninvasive routes is quite low (Gonda 2000). Attempts to deliver macromolecules through the lung for systemic effect have achieved mixed success. To date, the most successful example of delivery of biologically derived macromolecules is insulin. This work has been so successful that an inhaled formulation has reached phase III clinical trials (Laube et al. 1998). Other large biologically active proteins have also been delivered via the airways either to experimental animals or to humans. Proteins include recombinant human granulocyte colony stimulating factor (rhG-CSF) (Niven et al. 1993, 1995; Machida et al. 1996), interferon-$\alpha$ (INF-$\alpha$) (Giosue et al. 1996), interleukin-2 (IL-2) (Huland et al. 1992), and human growth hormone (hGH) (Patton et al. 1989–1990; Colthorpe et al. 1995; Wall and Smith 1997).

We present the current status of delivery of biotechnology-derived products by the pulmonary route of administration. Examples are included to demonstrate the usefulness of this approach for systemic delivery of therapeutic peptides, proteins and poly(nucleic acids) (PNAs). When possible, factors with a significant effect on observed performance results are discussed. This review will not be exhaustive, and further reading is suggested. The ideas and examples presented are intended to give a point of reference for the reader to gain an understanding of the possibilities of pulmonary drug delivery and to initiate the search for information.

Pulmonary drug delivery is one of the more underutilized and potentially beneficial routes of drug delivery available. Pulmonary drug delivery has been in use since the early 1950s, with many drugs absorbed by inhalation, although technological problems have prevented more drugs being administered via this route. Technology problems associated with pulmonary drug delivery are being overcome with advances in design. Delivery of a biologically derived agent by the pulmonary route requires an understanding of the anatomy and physiology of the lungs to determine the methods that can be used to administer a drug as well as potential problems that may be encountered.

## ANATOMY AND PHYSIOLOGY OF THE RESPIRATORY SYSTEM

To administer a drug using adsorption in the lungs, the drug molecules must first arrive in the lungs. For healthy humans, there are two routes that lead into the lungs. The first leads through the nose and the second through the mouth; the nose and mouth along with the pharynx and associated structures are referred to as the upper respiratory system. The upper respiratory system leads directly into the lower respiratory system (Figure 10.1): larynx, trachea, bronchi, and lungs (Tortora and Grabowski 1993). Traditionally, pulmonary administration of drugs avoids the nose and nasal cavity by using the oral cavity; however, this is not always the case. The main reason for using oral and not nasal administration is that the nasal cavity creates a very turbulent environment that may cause an aerosolized particle to contact the

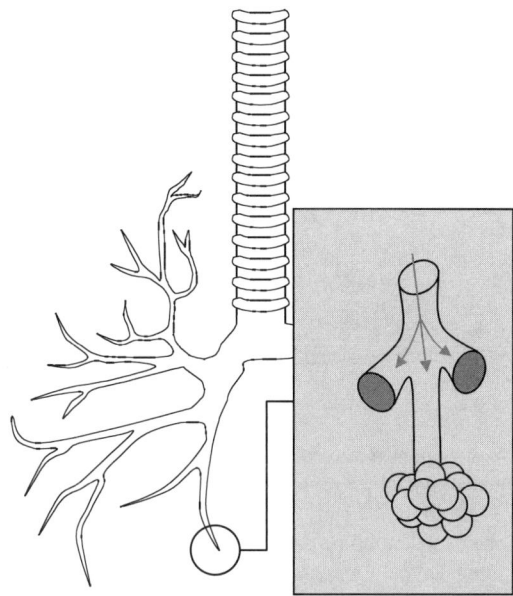

**FIGURE 10.1** Schematic representation of the lower respiratory system. The trachea leads into the bronchioles that branch and finally reach the alveolar sacs only on one side of the bronchial tree. *Inset* represents the alveolar tubules and alveolar sacs present at the terminal portions of the bronchial tree.

nasal mucus and epithelium This disadvantage for pulmonary administration is actually capitalized upon when using nasal delivery.

By avoiding the turbulent and restrictive environment of the nasal passage, upper respiratory deposition of aerosol particles for pulmonary delivery is very low (Cheng et al. 1999). Air that reaches the larynx from the nasal passage or from the oral cavity proceeds along the same route through the trachea to the lungs. The trachea splits into the two primary bronchi, left and right; the primary bronchi split into secondary bronchi, tertiary bronchi, and bronchioles. Secondary bronchi each lead to a lobe of the lung: superior, middle, and inferior lobes on the right and superior and inferior on the left. The bronchioles lead into terminal bronchioles that then lead into respiratory bronchioles. This network of branching bronchi and bronchioles is similar to the branching of a tree, thus it is referred to as the bronchial tree.

Analogous to the leaves of a tree are the alveolar sacs of the lungs; leaves convert sunlight to energy while the alveolar sacs allow gas exchange for the future production of energy for the organism (Patton 1996). It is at this point in the bronchial tree that most absorption of a drug takes place, although a molecule may not undergo transport exclusively at this point. This is not unexpected, as the alveoli are designed for optimal transport of molecules into the bloodstream. Only two cellular layers are present that allow molecular transport (Figure 10.2). Along with the alveolar epithelium and capillary endothelium, there is a bilayer of basement membrane,

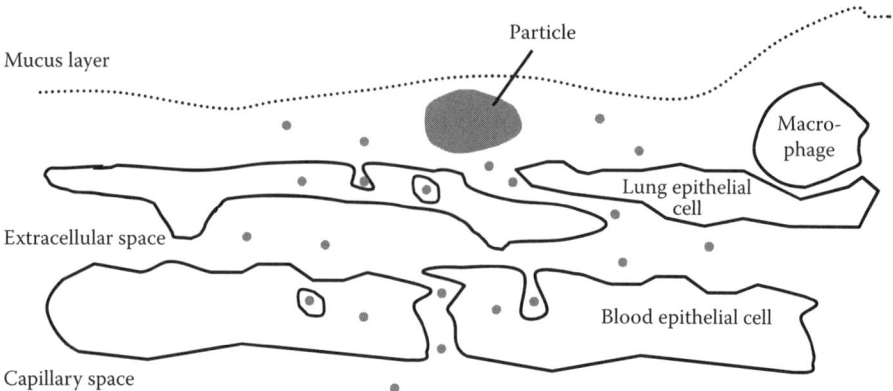

**FIGURE 10.2** Schematic representation of alveolar cells and possible mechanism of transport of molecules from the alveolar space into the circulation. Particles will release molecules of interest (*gray circles*) into the mucus in which the particle is embedded. The molecule can either be lost in the mucus, taken up by alveolar macrophages by phagocytosis or diffusion, taken up by alveolar epithelial cells by passive or active transport, or bypass the alveolar cells via paracellular transport depending upon the properties of the drug. Once a molecule has reached the extracellular space, the same mechanisms are possible for transport from the extracellular space into the blood. Molecules in the extracellular space may also reach to circulation via the lymph.

which consists of alveolar epithelial basement membrane and capillary basement membrane. This distance, approximately 0.5 μm, is one of the most narrow barriers for molecular transport into the body.

For this reason, the lungs represent an attractive route for drug delivery mainly due to the high surface area for absorption: 100–140 $m^2$ in humans (Altiere and Thompson 1996). Not only is the area for absorption high, but extensive vascularization of the alveolar epithelium allows efficient molecular transport. More important for small molecules than for biomacromolecules, the vasculature of the lung does not lead directly into the liver. Vessels leave the lungs and reenter the heart for distribution to the remainder of the body. Most biomacromolecules do not undergo the so-called first-pass effect, but many small biomolecule-derived molecules will have higher apparent absorption simply due to the lack of immediate hepatic clearance or metabolism.

Despite the fact that most of the alveolar surface is composed of alveolar epithelium, three primary types of cells are present in the alveoli: type I alveolar cells, type II alveolar cells, and alveolar macrophages. Type I alveolar cells are also referred to as squamous pulmonary epithelial cells and are the continuous lining of the alveolar sac. Type II alveolar cells are also referred to as septal cells. Type II alveolar cells secrete the alveolar fluid that is necessary to keep the surface moist and to maintain surface tension of the alveolar fluid; surface tension is necessary to keep the alveoli from collapsing. Alveolar fluid is a suitable environment for proteins when compared to the low pH and high protease levels associated with the intestine and oral drug delivery, although some proteolytic activity does exist (Yang et al. 2000). Transport of molecules across the mucus of the intestine may even be slower than the transport in the surfactant layer of the lungs. Lung surfactant is particularly devoid of protein content. Many suggestions have been proposed for this lack of protein in the fluid of the lung, but cellular uptake/renewal and fluid flow/regeneration have been cited as potential reasons. If cellular uptake/renewal is the main mechanism, then molecular delivery following pulmonary administration is very promising. A hypothetical model for absorption of macromolecules across alveolar type I cells is shown in Figure 10.2 (Patton 1996).

## LIPID-BASED PULMONARY DELIVERY

Liposomes and micelles are lipid vesicles composed of self-assembled amphiphilic molecules. Amphiphiles with nonpolar tails (i.e., hydrophobic chains) self-assemble into lipid bilayers, and when appropriate conditions are present, a spherical bilayer is formed. The nonpolar interior of the bilayer is shielded by the surface polar heads and an aqueous environment is contained in the interior of the sphere (Figure 10.3A). Micelles are small vesicles composed of a shell of lipid; the interior of the micelle is the hydrophobic tails of the lipid molecules (Figure 10.3B). Liposomes have been the primary form of lipid-based delivery system because they contain an aqueous interior phase that can be loaded with biomacromolecules. The ability to prepare liposomes and micelles from compounds analogous to pulmonary surfactant is frequently quoted as a major advantage of liposomes over other colloidal carrier systems.

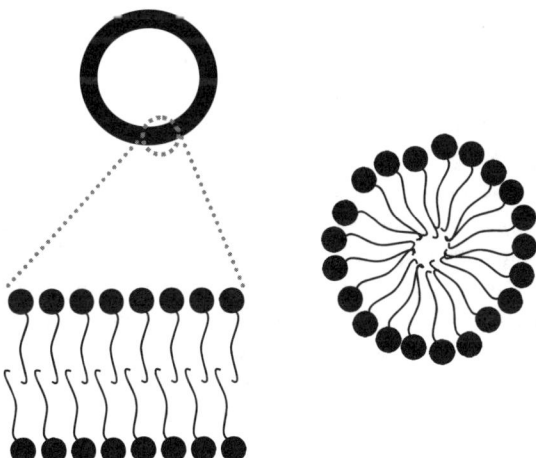

**FIGURE 10.3** Schematic presentation of lipid based drug delivery systems. Micelles (right) are composed of a solid lipid core with the polar heads exposed to the aqueous environment. Liposomes (left) are particles with a lipid bilayer surrounding an aqueous core. Drug can be encapsulated in the hydrophobic regions of the lipid particle, in the aqueous environment of the liposome, or adsorbed to the surface of the lipid particle.

Most materials used to produce liposomes are derived from natural materials, thus are thought to be safe when administered. Generally, however, phospholipids administered in liposomal form are cleared from the lungs more slowly than comparable doses of lung surfactant (Oguchi et al. 1985). Many macromolecules have been incorporated into liposomes in order to improve their pulmonary delivery. Some lipid-entrapped macromolecules have been tested in animal models and human volunteers to determine efficacy (Kellaway and Farr 1990).

Insulin is the protein that has been most investigated for pulmonary administration. Insulin levels are not maintained in diabetic patients, and precise control over blood glucose levels is needed. Insulin is a small protein, 5.8 kDa, which is composed of two chains that are covalently linked by an interchain disulfide bond. Currently, insulin is administered by injection, several times a day for many diabetics. The ability to deliver insulin via a noninvasive route would free diabetics from inconvenient, invasive insulin delivery methods and possibly eliminate secondary problems associated with diabetes, such as diabetic retinopathy.

Liposomes were formed from 1,2-dipalmitoylphosphatidylcholine (DPPC) and cholesterol (Chol) and the effect of liposomal entrapment on pulmonary absorption of insulin was related to oligomerization of insulin (Liu et al. 1993). Instillation of both dimeric and hexameric insulin produced equivalent duration of hypoglycemic response. However, the initial response from the hexameric form was slightly slower than that from dimeric insulin, probably due to lower permeability across alveolar epithelium of the hexameric form caused by larger molecular size. The intratracheal administration of liposomal insulin enhanced pulmonary absorption and resulted in an absolute bioavailability of 30.3%. Nevertheless, a similar extent of absorption and hypoglycemic effects was obtained from a physical mixture of insulin and blank liposomes and from liposomal insulin. This suggests a specific interaction of the phospholipid with the surfactant layer or even with the alveolar membrane.

Charge on the surface of a liposome can be controlled by the addition of specific molecules to the liposome composition; charge can influence both the interaction of the lipid with the lung and with the molecule being delivered. Physical mixture of insulin with positively charged liposomes was observed to show a strong hypoglycemic effect. Negatively charged and neutral liposome physical mixtures with insulin showed a lower hypoglycemic effect, but still had a significant response when compared to insulin alone (Figure 10.4.) (Li and Mitra 1996). Greater bioavailability of insulin observed using positively charged liposomes is suggested to be due to the membrane destabilizing effect of stearylamine. Stearylamine is one of the more toxic components that have been traditionally added to liposomes as a positive charge-inducing agent due to its basic nature. Unfortunately, the more toxic charge-inducing agents are the most effective at transferring proteins. This fact is not unexpected since the mechanism that is expected to allow molecular transport is the destabilization of cell membranes; toxicity from lipids is also typically by a destabilization mechanism.

Phospholipids, such as DPPC, act as absorption enhancers in the lung. A significantly higher reduction in blood glucose levels was observed with a DPPC-insulin physical mixture compared to liposome-insulin following intratracheal instillation into rats (Figure 10.5) (Mitra et al. 2001). In this study, insulin alone, 1 U/kg, resulted

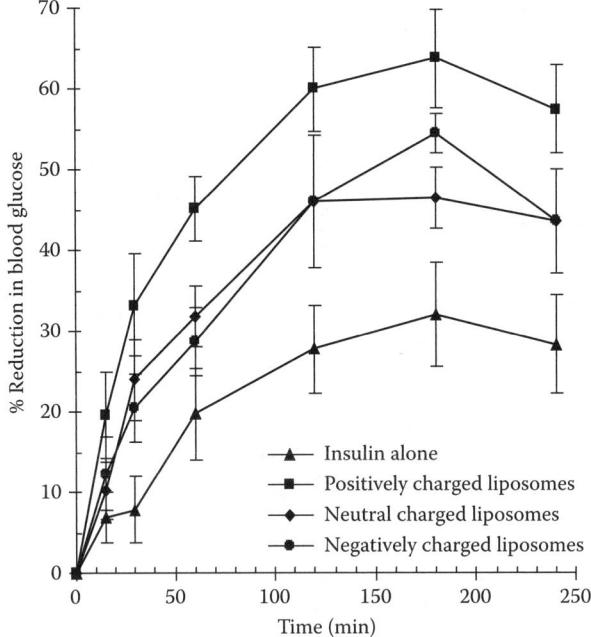

**FIGURE 10.4** Reduction in blood glucose following pulmonary administration of 1 U/kg insulin (▲), insulin encapsulated in positively charged liposomes (■), insulin encapsulated in negatively charged liposomes (●), insulin encapsulated in neutral liposomes (◆). (Adapted from Li and Mitra 1996).

in a 32% reduction in blood glucose after 3 hours, while an insulin-liposome mixture resulted in a 50% reduction in blood glucose. The highest hypoglycemic effect was obtained with a DPPC-insulin physical mixture, generating a 73% reduction. This suggests that the physical mixture of lipid and protein may actually be a more efficient transport mechanism, potentially due to entrapment of insulin in the liposome that cannot escape. This was just one case of increased delivery by physical mixture with lipid, but the idea that entrapment may not be necessary cannot be overlooked when delivering biomacromolecules.

Not only will the charge of a lipid and the composition of lipids affect the delivery of biomacromolecules, but the size of the liposome may alter the transport. Mixtures of insulin with three different diameter (1.98 μm, 0.4 μm, and 0.1 μm) neutral liposomes (DPPC: Chol) resulted in similar overall hypoglycemic effects to insulin alone. Contrary to this finding is the fact that pulmonary absorption of liposomal [3H] terbutaline, a small molecule, has been reported to be dependent on both composition and size of the liposomes used (Abra et al. 1990). Differences in the absorption mechanism may be the explanation for this contradictory evidence; further studies are needed to clarify this and other uncertainties about the uptake mechanism of macromolecules (Patton 1996).

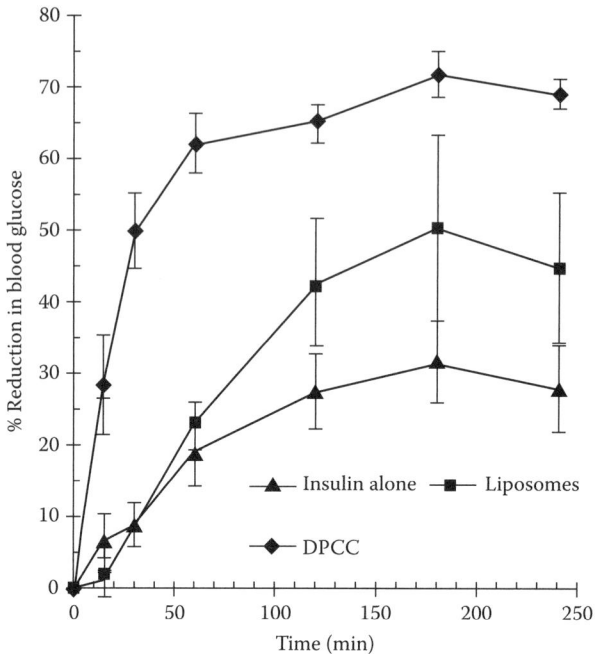

**FIGURE 10.5** Reduction in blood glucose following pulmonary administration of 1 U/kg insulin (▲), insulin encapsulated in liposomes (■), insulin physical mixture with DPCC particles (♦). (Adapted from Mitra et al. 2001).

Proteins are transported across the pulmonary epithelium effectively, and poly(nucleic acids), such as DNA, are also transported. Nucleic acids can be efficiently expressed in the lungs by intravenous injection of cationic lipid and plasmid DNA complexes (Liu et al. 1995, 1997). The safety of intrapulmonary cationic lipid delivery in healthy volunteers has been investigated (Chadwick et al. 1997). A single application of aerosol formulation of a cationic lipid [$N^4$-spermine cholesterylcarbamate (GL-67), dioleoyl phosphatidyl-ethanolamine (DOPE), dimystoyl phosphatidyl-ethanonamine conjugated with poly(ethylene glycol) (DMPE-PEG$_{5000}$)] did not result in clinically detectable changes when given by nebulization into the lungs of healthy volunteers and provides an indication of a lipid dose tolerated in man. Cationic lipids are necessary to transfer anionic PNAs into cells as they bind to the anionic DNA macromolecules; anionic lipids repel anionic PNAs.

The therapeutic efficacy of either systemic or local pulmonary delivery of the IFN-γ gene was evaluated in a murine allergen–induced airway hyperresponsiveness (AHR) model (Dow et al. 1999) and it was found that a high efficiency of gene transfer could be achieved. Intratracheal administered cationic liposomes were prepared from a mixture of 1,2-diacylglycero-3-ethylphosphocholine (EDMPC) and cholesterol. Intravenous injections were prepared from 1,2-dioleyl-3-trimethylammonium propane (DOTAP) and cholesterol and compared with pulmonary administered

IFN-γ. Although the reason for using different cationic lipids in different routes was not mentioned, probably the size of liposomes would have had an impact on choosing alternate cationic lipids for administration via different routes. Intravenous IFN-γ gene delivery significantly inhibited development of AHR and decreased serum IgE levels, compared to control mice. Intratracheal IFN-γ gene delivery also significantly inhibited AHR and airway eosinophilia, but did not affect serum IgE levels. Treatment with IFN-γ was much less effective than IFN-γ gene delivery by either route. This is one example of an effective gene therapy that illustrates the potential of pulmonary lipid-based PNA delivery. Of major importance is the fact that careful selection of lipid composition is required when considering both protein and PNA delivery.

Pulmonary administration of PNAs has great potential for the same reasons that pulmonary protein and peptide delivery have been successful. Predominantly, the distance for transport and ease of administration of agents are the advantages of pulmonary delivery, but the formulation of labile molecules for eventual pulmonary administration as lipid-based aerosols may be problematic.

## SOLID COLLOIDAL PARTICLES

Microparticles (MP) are small colloidal particles, with a diameter less than 1000 μm, and generally composed of a solid matrix in which a drug is dispersed and/or dissolved. Microparticles have been produced from a wide variety of biodegradable materials of both natural (e.g., chitosan, gelatin, albumin) and synthetic origin (e.g., polyesters of lactic acid, glycolic acid, and ε-caprolactone). Microparticles can be prepared in different sizes and many hydrophilic and hydrophobic drugs have been entrapped or incorporated with relatively high efficiency and different drug release rates can be achieved. In comparison to liposomes, MP may be more stable both *in vitro* and *in vivo*, therefore producing a slower release rate and a longer duration of action of the incorporated agent. Higher stability of MP may provide an opportunity for optimization of duration of pulmonary release and action. The following sections describe several of the most promising materials used to produce small colloidal particles and how they have been used for pulmonary drug delivery of biotechnology-derived products.

## POLY(DL-LACTIDE-*CO*-GLYCOLIDE)

Poly(DL-lactide-*co*-glycolide) (PLGA) is a synthetic polymer that has been successfully applied in the medical field. PLGA is composed of lactic acid and glycolic acid monomers, which are natural molecules in the body. The only problem that has been associated with PLGA polymers is the fact that as the polymer degrades by water hydrolysis in physiologic conditions, acid is formed as the lactic and glycolic acid is liberated. Consequently, degradation of the incorporated molecules is possible, so care must be taken during formulation. To counteract this effect, buffering molecules are at times added to the formulation to keep the environment neutral and prevent acid catalyzed degradation of the incorporated molecules.

Rifampicin-loaded PLGA microparticles (RifMP) were studied for treatment of pulmonary tuberculosis (Suarez et al. 2001a,b). A guinea pig infection model has been adopted as a post-treatment screening method for antimicrobial effect. Rifampicin, alone or encapsulated in microparticles was delivered by insufflation or nebulization. There was a dose-dependent relationship between insufflated RifMP and burden of bacteria in the lungs. In addition, guinea pigs treated with RifMP had a smaller number of viable bacteria, reduced inflammation, and lung damage than control (lactose and saline), PLGA, or RIF treated animals. Nebulization was more efficient in reducing the number of viable microorganisms in the lungs at equivalent doses of RifMP than was insufflated. This is due to the deeper penetration of the nebulized MP when compared to insufflated MP. Results with a small molecule in a locally administered model suggested that PLGA microparticles should in fact be nebulized or aerosolized in order to be active in the lungs. It was still necessary to show that a biotechnology-derived agent could be administered by this method.

Poly(DL-lactide-co-glycolide) MP have successfully been used to deliver insulin via the pulmonary route (Kawashima et al. 1999). Insulin loading was much improved compared to the methods previously described (Masinde and Hickey 1993); incorporation of hydrophilic molecules, and specifically proteins, is typically quite low for PLGA MP. Eighty-five percent of the drug was released from the MP at the initial burst *in vitro*, followed by prolonged release of the remaining drug for a few hours. After nebulization, a prolonged hypoglycemic effect was achieved (48 h) compared to the nebulized aqueous solution of insulin (6 h); however, only 2.6% of the nebulized mist was delivered into guinea pig lungs during a 20 minute nebulization.

One important factor that has to be considered for all types of pulmonary delivery: particles deposit in the lungs based on their aerodynamic diameter (Equation 1). For a spherical particle, the aerodynamic diameter ($d_{aer}$) is equal to the product of actual diameter ($d$) times the square root of particle density ($\rho$) (Gonda 1992).

$$d_{aer} = d \cdot \sqrt{\rho} \qquad (1)$$

Particles of aerodynamic diameter greater than 5 µm deposit primarily in the upper airways or mouth and throat region, while a significant percentage of those less than 1 µm do not deposit but are exhaled (Darquenne et al. 1997). Therefore, for optimum deposition in the alveolar region and systemic delivery, particles have to be between 1 and 5 µm.

In an attempt to increase the amount of particles retained in the lungs, large porous particles with low density ($\rho < 0.1$ g/cm²) have been designed (Edwards et al. 1997). The particles were composed of 50% lactide and 50% glycolide. Porous and nonporous particles loaded with testosterone were aerosolized into a cascade impactor system from a dry powder inhaler (DPI) and the respirable fraction was measured. Nonporous particles ($d$ = 3.5 µm, $\rho$ = 0.8 g/cm³) exhibited a respirable fraction of 20.5 ± 3.5%, whereas 50 ± 10% of porous particles ($d$ = 8.5 µm, $\rho$ = 0.1 g/cm³) were respirable, even though the aerodynamic diameter of the two particle types were nearly identical. Porous particles as a consequence of their large size and low mass density can aerosolize from a DPI more efficiently than smaller, nonporous particles, resulting in higher respirable fraction. Insulin encapsulated in porous ($d = 6.8$ μm, $d_{aer} = 2.15$ μm) and nonporous ($d = 4.4$ μm, $d_{aer} = 2.15$ μm) PLGA particles and administered in rats by force ventilation resulted in activity of insulin in both cases. For large porous particles, insulin bioavailability relative to SC injection was 87.5%, whereas the small nonporous particles yielded a relative bioavailability of 12% after inhalation. This dramatic increase in insulin activity should be achievable for other biologically derived molecules. One particular aspect of increased insulin bioavailability that must be examined is the fact that large particles can avoid phagocytic clearance from the lungs until they have delivered their therapeutic dose. For nonporous particles, $30 \pm 3\%$ of phagocytic cells contained particles immediately after inhalation, compared to $8 \pm 2\%$ for large porous particles. This again is a dramatic difference in phagocytosis. The combined effect of greater respirable fraction and lower phagocytosis should greatly improve the efficiency of pulmonary biomolecular delivery.

Although PLGA microparticles, in general, offer several advantages including providing sustained delivery of drug from a biocompatible and nontoxic polymer, some limitations as a carrier for drugs in the lungs do exist. First, small PLGA microparticles degrade over a period of weeks to months, but typically deliver drugs for a shorter period of time (Batycky et al. 1997; Edwards et al. 1997). Therefore, unwanted buildup of polymer in the lungs upon repeat administration may occur. Also, the hydrophobic surface of PLGA MP can cause rapid opsonization (protein adsorption), resulting in a rapid clearance by alveolar phagocytic cells (Tabata and Ikada 1988). However, this problem was solved by large porous PLGA microparticles with low mass density (Edwards et al. 1997). Finally, bulk degradation of PLGA microparticles creates an acid core, which can damage pH-sensitive drugs such as peptides and proteins (Mader et al. 1996). Depending on the compositions of the MP and the desired dosing regimen, PLGA MP may be an optimal delivery vehicle, but for frequent administrations, high molecular weight PLGA may not be an optimal choice.

## POLYCYANOACRYLATES

An alternative to PLGA for the production of MP is poly(alkyl cyanoacrylates) (PCNA). PCNAs are a type of polymer that is produced under mild conditions by anionic polymerization. PCNAs have low toxicity and are used in medical applications, including surgical adhesives. These polymers also degrade rapidly in the body when compared to PLGAs, decreasing the residence time for the particles after drug delivery has been completed. These advantages over PLGAs have been exploited for delivery of insulin.

Prolonged hypoglycemic effect of insulin was reported after using poly(butyl cyanoacrylate) microparticles with a mean diameter of 254.7 nm (Zhang et al. 2001). Insulin-loaded poly(butyl cyanoacrylate) microparticles were prepared by emulsion polymerization in the presence of insulin. Insulin-loaded microparticles were administered intratracheally to normal rats. The duration of glucose levels below 80% of baseline was maintained for a longer period when insulin was administered in microparticles than when insulin was administered as a solution at all doses. Relative pharmacological bioavailability of the insulin loaded in microparticles was 57.2% compared to the same formulation by subcutaneous administration. Nevertheless, insulin solution resulted in decreased glucose levels compared to insulin-loaded microparticles after single intratracheal administration. The equivalent initial decrease in glucose levels is explained by immediate permeability of the lung epithelium to insulin, but the prolonged glucose level decrease suggests that PCNA microparticles can be used for extended delivery of biologically derived molecules.

## GELATIN

Gelatin (see also Chapter 8, p. 216 ff) is a naturally derived polymer that has been used to produce microparticles for pulmonary delivery. The natural derivation and biodegradability of gelatin have pushed it to the forefront as a material for producing pulmonary drug delivery systems. Gelatin must, however, be crosslinked by synthetic means to produce MP. Typically, gluteraldehyde is used to crosslink gelatin into a solid matrix. Gluteraldehyde is a toxic molecule and can retain activity in the MP. For this reason, care must be taken to fully remove unreacted gluteraldehyde prior to administration. Gluteraldehyde, after reacting with gelatin, may revert to form reactive gluteraldehyde over time; even when care was taken to remove or react all gluteraldehyde, free, reactive gluteraldehyde may be present. Gluteraldehyde may also react with the molecule being incorporated, thus decreasing the activity of the molecule or changing the activity altogether. For this reason, biomolecules must be added to the MP after the particle has been formed. Typically, loading efficiency is quite low and the release period is not typically more than a few hours. For this reason, residual gelatin MP may be problematic as described with PLGA MP. Control of gluteraldehyde crosslinking can be used to decrease the time for complete degradation, but not completely.

Gelatin MP loaded with salmon calcitonin have been investigated (Morimoto et al. 2000). Positively and negatively charged gelatin microparticles with different sizes were prepared. *In vitro* release studies showed that the release of salmon calcitonin from negatively charged gelatin microparticles was slower than from positively charged gelatin microparticles indicating a potential ionic interaction with the MP. Cumulative calcitonin release from negatively charged gelatin microparticles reached approximately 40% after 2 hours and then leveled off compared to 85% in positively charged microparticles. In addition, the release profiles were not influenced by particle size. Pulmonary absorption of salmon calcitonin from positively and negatively charged gelatin microparticles were estimated by administering the formulation intratracheally (via tracheotomy) in rats. Results indicated that salmon calcitonin in positively and negatively charged gelatin microparticles reduced plasma calcium levels greater than salmon calcitonin in PBS (pH 7) (Figure 10.6). Administration of salmon calcitonin in positively charged gelatin microparticles with smaller size particles led to higher pharmacological availability.

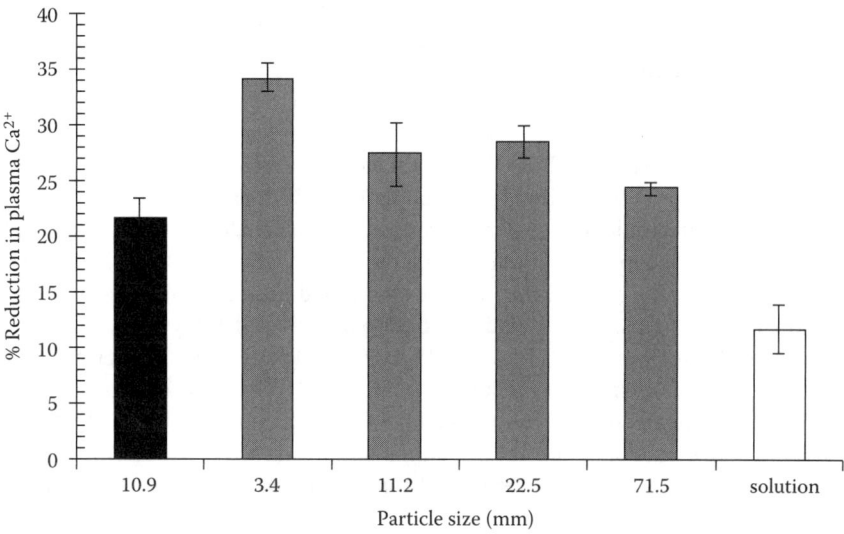

**FIGURE 10.6** Hypercalcemic effect in rats following administration of 3 U/kg of salmon calcitonin in solution (*white fill*) and negatively charged (*black fill*) or positively charged (*gray fill*) gelatin microspheres of indicated size. The time to reach maximum effectiveness ranged from 0.5 to 1.5 hours. (Adapted from Morimoto et al. 2000.)

## POLY(ETHER-ANHYDRIDES)

A relatively new class of polymers, polyether-anhydrides (PEAs), for pulmonary delivery has been reported (Fu et al. 2002). PEAs are hydrolyzed in a fashion similar to PLGAs, but due to the lability of anhydride bonds, PEAs have significantly faster degradation rates than PLGAs. By controlling the ratio of anhydride to ether bond, the degradation rate can be tuned. These polymers are typically composed of the monomers sebacic acid (SA) and PEG. Microparticles are prepared from poly(SA-co-PEG) using a double-emulsion solvent-evaporation method. Large, low-density particles with aerodynamic diameters between 1.9 μm (30% PEG) and 3.7 μm (0% PEG) have been produced. Various amounts of PEG (5–50% by mass) were incorporated into the backbone of new polymers. The variations in PEG allowed tuning of particle surface properties for potentially enhanced aerosolization efficiency. The PEG also decreased particle clearance rate by phagocytosis in the deep lung; steric stabilization was proposed as the mechanism of phagocytosis control. Drug delivery from these particles have not been examined *in vivo*, however, cell culture models have shown that the particles do affect the transepithelial resistance in some models, but not others (Fiegel et al. 2003). Because of the biocompatibility and degradation rates of these polymers, these materials may, in fact, be excellent for use as pulmonary drug carriers.

## DIKETOPIPERAZINE DERIVATIVES

A composition based on diketopiperazine derivatives (3,6-bis (N-fumaryl-N-(n-butyl) amino-2, 5-diketopiperazine) has been investigated as a pulmonary drug delivery system, termed Technospheres™ (Pharmaceutical Discovery Corp., Elmsford, NY) (Pohl et al. 2000; Steiner et al. 2002). The diketopiperazine derivatives self-assemble into microparticles at low pH with a mean diameter of approximately 2 μm. During self-assembly, diketopiperazine derivatives microencapsulate peptides present in the solution. Insulin incorporated in diketopiperazine derivatives (TI) was administered to five healthy humans by the use of a capsule-based inhaler with a passive powder deagglomeration mechanism. Relative and absolute bioavailability of TI in the first 3 hours (0–180 min) were 26 ± 12% and 15 ± 5%, and for 6 hours (0–360 min) 16 ± 8% and 16 ± 6%, respectively (Steiner et al. 2002). The time to peak action for glucose infusion rates was shorter with both IV (14 ± 6 min) injection and inhalation (39 ± 36 min), as compared to SC administration (163 ± 25 min). This rapid absorption of insulin would be beneficial for diabetic patients who need to rapidly affect their glucose levels.

By the use of a breath-powered unit dose dry powder inhaler, which was adapted to the physical properties of TI, relative bioavailability was 50% for the first 3 hours and 30% over the entire 6-hour period in 12 healthy volunteers (Pfutzner et al. 2002). However, although the studies demonstrated pulmonary administration of TI has the advantages of fast onset of action, short duration of action, and lower variability over the SC injections of insulin; no attempt has been made to compare pulmonary administration of insulin alone with the same inhaler device. This method of encapsulating biomacromolecules has some advantages and must be considered when electing to deliver a molecule.

## POLYETHYLENEIMINE (PEI)-DNA COMPLEXES

Aerosol delivery of genes has been drawing more attention during recent years due to a targeted and noninvasive approach to the treatment of a wide range of pulmonary disorders. Polycations have been reported to be effective vectors for transfecting cells *in vitro* and *in vivo* (Boussif et al. 1995, 1996). PEI-DNA complexes have been administered through the pulmonary route via aerosol as a means of gene therapy and genetic immunization (Densmore et al. 2000). *In vivo* transfection efficiencies of liposome-DNA and PEI-DNA complexes after aerosol and nasal delivery were compared, and the results indicated PEI was a 10-fold better vector for expression in the lungs than bis-guanidinium-tren-cholesterol (BGTC): DOPE and compared even more favorably to other cationic lipid formulations. Aerosol delivery resulted in about 20-fold greater transfection efficiency in the lungs than in the nose for PEI-based formulations. Genetic immunizations by aerosol and instillation administration of PEI-cytomegalovirus promoter-human growth hormone (pCMV-hGH) plasmid vaccines has been studied in mice with IM injection of naked DNA as a positive control (Figure 10.7). Antibody induction by genetic immunization with PEI-pCMV-hGH in animals exposed to the PEI-pCMV-hGH IM injection showed the levels of the antibody response persisted through 8 weeks from a single administration of the

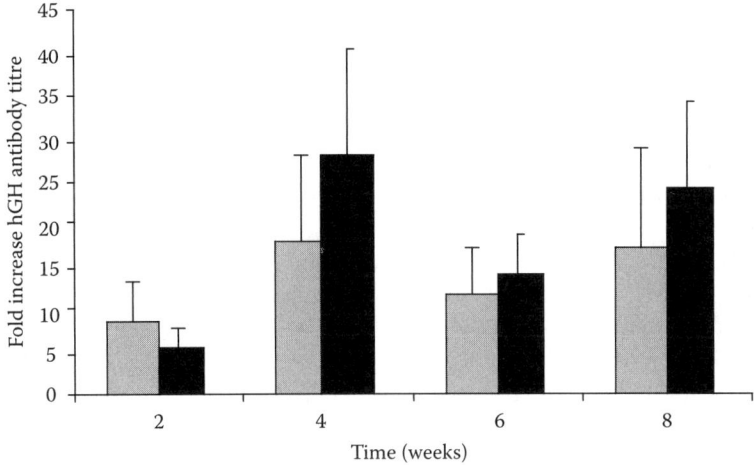

**FIGURE 10.7** Antibody to hGH produced in mice following PEI-pCMV-hGH plasmid immunization via IM (*black fill*) and aerosol (*gray fill*) administration detected 2, 4, 5, and 8 weeks following induction. Fold increase is the number of times greater than untreated control animals. (Adapted from Densmore et al. 2000.)

plasmid-PEI preparation and were essentially equivalent to those achieved with IM injection. This suggests that PEI-DNA complexes could be used for pulmonary administration of genes. Poly(ethyleneimine) is more efficient than liposomes at transferring the DNA to the cells, but may be more toxic. Alternative polymers are being investigated by many groups to create a less toxic polymeric carrier for gene transfer.

## POLY(ETHYLENE GLYCOL) CONJUGATES

Poly(ethylene glycol) (PEG)-protein conjugates have become of interest in medical applications due to the ability of PEG to protect a protein from degradation and recognition. PEG is covalently linked to the protein through amino groups, carboxylic groups, or thiol groups depending on the biomacromolecule. Care must be taken as modification of a protein at or near the active site will decrease the biological activity of the protein; although in some cases the slight decrease in activity is insignificant when compared with the protection from degradation and clearance. Even modification at a site far from the active site has devastating consequences on the activity of many proteins, so care must be taken and each individual protein must be examined.

The pulmonary delivery of rhG-CSF (18.8 kDa) and two PEG conjugates of rhG-CSF (P1, 81.5 kDa and P2, 146.8 kDa) was investigated in rats (Niven et al. 1994). Comparison of white blood cell responses after IT instillation of 500 µg/kg P1 and P2 and rhG-CSF alone, demonstrated a greater response for more substituted PEG-rhG-CSF than rhG-CSF alone. The plasma concentration vs. time curve showed a $T_{max}$ at $160 \pm 65$ minute for rhG-CSF versus $347 \pm 183$ minute and $210 \pm 83$ minute for P1 and P2, respectively. However, lower bioavailability was obtained for PEG-rhG-CSF. All bioavailability values (500 µg/kg and 50 µg/kg) were less than 30%, and those of both PEG proteins were less than 5%. There was an apparent inverse relationship between the extent of absorption and the size of the protein. However, the absorption of rhG-CSF at the two doses was markedly different (3%, 50 µg/kg vs. 27.4%, 500 µg/kg) suggesting absorption may be influenced by other clearance mechanisms. Even though the absorption of PEG-rhGH-CSF is low, biologic activity is present. Again, this is a positive sign for pulmonary biomolecular delivery. This field is in its infancy of development and more research is needed to determine the exact mechanisms of biomolecular transport in the alveolar epithelium. However, it is clear that by use of appropriate modification or encapsulation, the uptake of biomolecules can be increased in the lung.

## FACTORS AFFECTING PULMONARY DOSING

There are some factors that must be considered when comparing different formulations with respect to efficiency and bioavailability, and the results have to be interpreted very carefully. There are differences between inhaled aerosols and instilled liquids. Comparison of the pharmacokinetics following IV insulin administration with the pharmacokinetics following IT and aerosolized insulin administration in rabbits showed improved peripheral lung deposition after aerosol administration (Colthorpe et al. 1992). The penetration index (peripheral vs. central deposition) was significantly higher for the aerosol treatment ($1.52 \pm 0.36$) compared to IT treatment ($0.32 \pm 0.08$). Aerosol formulation administration produced a bioavailability of nearly 10-fold greater than produced by IT instillation (Table 10.1). It follows that more test macromolecules will be trapped in the mucus layer of the conducting airways of the central rather than peripheral deposition pattern and therefore will be cleared faster by mucociliary clearance.

**TABLE 10.1**
**Insulin Pharmacokinetic Parameters by Different Routes of Administration (mean™ SD, n = 4)**

|                    | IV          | IT Instillate | Aerosol                          |
| ------------------ | ----------- | ------------- | -------------------------------- |
| Dose               | (0.1 IU/kg) | (5.0 IU/kg)   | (300 IU/ml nebulized for 4 min)  |
| $t_{max}$ (min)    | —           | $11.3 \pm 4.8$ | $12.5 \pm 2.9$                  |
| $t_{1/2}$ (min)    | 3.0         | 49            | 69                               |
| F (%)[a]           | 100         | $5.6 \pm 3.3$[b] | $57.2 \pm 28.5$[b]            |

[a]F-values are calculated relative to 100% after IV dosing.
[b]Significantly different ($P < 0.05$)
Adapted from Colthorpe et al. 1992.

In general, IT instillation is an easy, rapid, and relatively inexpensive method for testing pulmonary drug delivery. It requires a small amount of drug for efficient administration and provides precise and noninvasive dosing. Intratracheal administration is best used to determine whether a test macromolecule is absorbed from the lung (El Jamal et al. 1996). In contrast, aerosol delivery is expensive, inefficient, technically challenging, and difficult to precisely quantify the delivered dose in animals (Brown and Schanker 1983). However, this technique has the advantages of being physiologically relevant and providing a more even distribution of test molecules in the lung than intratracheal administration (El Jamal et al. 1996). Even when the appropriate method of drug installation is chosen, care must be taken to appropriately choose an animal model.

Species differences in lung structure and function illustrate that caution must be exercised in extrapolating to the human lung (Hickey and Garcia-Contreras 2001). Compared to other species (rats, dogs, and baboons), human lungs were found to contain a greater number of macrophages, alveolar type II, endothelial, and interstitial cells (Table 10.2) (Crapo et al. 1983). The thickness of interstitium and

## TABLE 10.2
## Morphometric Parameters in the Alveolar Region of Normal Mammalian Lungs

| | Sprague Dawley Rat | Dog | Baboon | Human |
|---|---|---|---|---|
| Body weight, kg | $0.36 \pm 0.01$ | $16 \pm 3$ | $29 \pm 3$ | $79 \pm 4$ |
| Lung volume, ml | $10.55 \pm 0.37$ | $1322 \pm 64$ | $2393 \pm 100$ | $4341 \pm 284$ |
| Surface area,[a] m²/both lungs | | | | |
| Alveolar epithelium | | | | |
| Type I | $0.387 \pm 0.025$ | $51.0 \pm 1.0$ | $47.7 \pm 7.7$ | $89.0 \pm 8.0$ |
| Type II | $0.015 \pm 0.002$ | $1.0 \pm 0.2$ | $1.9 \pm 0.3$ | $7.0 \pm 1.0$ |
| Capillary endothelium | $0.452 \pm 0.035$ | $57.0 \pm 2.0$ | $38.6 \pm 9.5$ | $91.0 \pm 9.0$ |
| Tissue thickness, μm | | | | |
| Harmonic mean, air-to-plasma | $0.405 \pm 0.017$ | $0.450 \pm 0.007$ | $0.674 \pm 0.055$ | $0.745 \pm 0.059$ |
| **Total Lung Cells, %** | | | | |
| Alveolar type I | $8.9 \pm 0.9$ | $12.5 \pm 1.7$ | $11.8 \pm 0.6$ | $8.3 \pm 0.6$ |
| Alveolar type II | $14.2 \pm 0.7$ | $11.8 \pm 0.6$ | $7.7 \pm 1.0$ | $15.9 \pm 0.8$ |
| Endothelial | $42.2 \pm 1.1$ | $45.7 \pm 0.8$ | $36.3 \pm 2.4$ | $30.2 \pm 2.4$ |
| Interstitial | $27.7 \pm 1.8$ | $26.6 \pm 0.7$ | $41.8 \pm 2.7$ | $36.1 \pm 1.0$ |
| Macrophage | $3.0 \pm 0.3$ | $3.4 \pm 0.6$ | $2.3 \pm 0.7$ | $9.4 \pm 2.2$ |
| **Alveolar Surface Coverage, %** | | | | |
| Alveolar type I | $96.2 \pm 0.5$ | $97.3 \pm 0.4$ | $96.0 \pm 0.6$ | $92.9 \pm 1.0$ |
| Alveolar type II | $3.8 \pm 0.5$ | $7.1 \pm 1.0$ | $4.0 \pm 0.6$ | $7.1 \pm 1.0$ |

[a] Type I surface area (SA) is the surface area of the basement membrane under type I cells; type II SA is the air surface of type II cells excluding the extra SA contributed by microvilli; endothelial SA is the luminal surface of the endothelial cells.

Adapted from Crapo et al. 1983.

pulmonary capillary endothelia were also significantly greater in human lungs than in lower primates and mammals. However, despite these differences, an overall similarity in the characteristics of individual lung cells was observed (Crapo et al. 1983).

Intracellular pH in alveolar epithelial cells may also vary among species. Several different values for baseline steady state intercellular pH have been published for alveolar pneumocytes (Lubman and Crandall 1992). The baseline intracellular pH of isolated alveolar type II cells has been reported to be between 7.07 (Nord et al. 1987) and 7.36 (Sano et al. 1988) for rats and 7.22 for rabbits (Finkelstein and Brandes 1987). Secretions stored in mucous tubules of glands in the respiratory tract vary among species. Mucous tubules in the mouse form mainly sialomucin and to a smaller extent sulfomucin. In rat tracheolaryngeal glands, the mucus tubules, which predominate, produce and store sulfomucin in abundance. Human respiratory tract mucous tubules, on the other hand, show roughly equal proportions of sulfated and of nonsulfated, sialylated mucosubstances through these structures (Spicer et al. 1983; Hickey and Garcia-Contreras 2001). Cytochrome P-450 monoxygenase activity has been demonstrated in the mouse, rat, hamster, rabbit, and pig. No data are available regarding the activity in other species such as humans, guinea pigs, dogs, cats, and sheep (Plopper 1983; Hickey and Garcia-Contreras 2001). These differences between animal species and humans makes the interpretation of results more complicated, so choice of species for animal studies must be carefully chosen with no specific animal suggested to date for a generic test subject. But species-to-species differences in lung properties are not the only factor that must be examined; there are various types of inhalers, the selection of which greatly influences absorption.

The pulmonary deposition of an aerosol depends on (Matthys and Kohler 1985)

- Type of aerosol generator (physical principle, outlet geometry, and outlet speed)
- Properties of the particles (size, density, shape, solid, fluid, hygroscopicity, and electric charge)
- Route and mode of application (by mouth, nose, tracheostoma, bolus, continuous)
- Breathing maneuver (tidal volume, frequency) and duration
- Morphometry of the upper and lower airways (degree of obstruction)

Aerosols can be generated by three main drug delivery systems: nebulizers, pressurized metered dose inhaler (pMDI), and dry powder inhaler (DPI).

Nebulizers are usually prescribed to patients who are unable to operate other inhalation devices, for example because of poor hand–lung coordination (Nikander 1997). Use of nebulizers have some limitations including relatively long treatment times and lack of portability. Nebulizers use aqueous solutions of drugs, so drug instability in aqueous solutions via hydrolysis would preclude the use of nebulizers. Many colloidal particles can be used to circumvent aqueous degradation, and thus could be used in nebulizers. In addition, the process of nebulization exerts high shear stress on the compounds, which can lead to protein denaturation. Furthermore, the droplets produced by nebulizers are rather heterogeneous, which results in poor drug delivery to the lower respiratory tract (Agu et al. 2001).

MDIs utilize propellants (chloroflurocarbons and increasingly, hydrofluroalkanes) to atomize drug solutions resulting in a more uniform spray and improved effectiveness compared to nebulizers (Agu et al. 2001). Easy handling and convenience to the patient are other advantages of these devices; however, some proteins and peptides are susceptible to denaturation when they come into contact with propellants. The large air–liquid interface that is constantly being generated during aerosolization may also cause denaturation of macromolecules (Banga 1995). The high exit velocity of drug aerosol can lead to high levels of oropharangeal impaction, and the need for users to coordinate the pMDI valve actuation with their breathing maneuver make MDIs a difficult apparatus for some users (Smart 2002). Improvements on the design of MDIs may make this type of device more highly applicable.

DPIs represent a significant advance in pulmonary delivery technology. They are potentially suitable for delivering a wider range of drugs than MDIs, including biopharmaceutically derived therapies for systemic applications such as peptides and proteins. DPIs have a key advantage over pMDIs because they are breath-actuated and therefore do not require patient coordination. Since the production of an aerosol with a DPI is dependent on the breath of the user, less shear stress and no propellants are needed for production of the aerosol. Many modifications of DPIs are currently available with more being designed.

Even when the appropriate inhaler is chosen, the influence of the disease state cannot be ignored. Disease states can influence the dimension and properties of the airways and hence the disposition of any inhaled drug. Thus, great care must be taken when extrapolating the findings based on intratracheal administration to different animal species in order to predict deposition profiles after inhalation of aerosol formulations by patients suffering from airway disease. DPIs are not appropriate in many diseases when the ability to have sufficient airflow is hindered. Since many diseases that we would like to treat via pulmonary administration of biomolecules cause a decrease in airflow, we must be careful in the decision of which type of inhalation mechanism to choose.

## CONCLUSIONS

In this chapter, different delivery systems such as liposomes and microparticles for pulmonary administration of macromolecules were reviewed. However, comparison of various delivery systems is problematic due to different aerosol devices, methods of administration (inhalation vs. IT) and animal models. The past 10 years have seen dramatic changes in delivery of drugs by inhalation and there is good evidence to show that pulmonary administration of drugs with poor oral bioavailability can be very effective in achieving significant plasma concentrations. Few adverse effects have been noted in the clinical trials conducted to date, and the toxicology studies have, in general, produced only modest lung effects. The limited data available also suggest if SC delivery has been well tolerated immunologically, there is a reasonable expectation that inhalation delivery may also be well tolerated (Wolff 1998). This may be of particular benefit for many biologically derived substances, which are currently given parenterally, without the associated complications and discomfort.

The field of pulmonary administration of biomacromolecules is just now in its infancy, and much promise is held in this field. By carefully examining the methods and types of delivery that are possible we may be able to better design easy to use and inexpensive methods for delivering biotechnology-derived products. This chapter was neither all-inclusive nor comprehensive, but was intended to be a starting point for the examination of pulmonary administration of biotechnology derived drugs.

## REFERENCES AND FURTHER READING

Abra, R.M., Mihalko, P.J., and Schreier, H. (1990). The effect of lipid composition upon the encapsulation and in vitro leakage of metaproternol sulfate from 0.2 μm diameter, extruded, multilamellar liposomes. *J. Control. Release,* 14, 71–78.

Agu, R.U., Ugwoke, M.I., Armand, M., Kinget, R., and Verbeke, N. (2001). The lung as a route for systemic delivery of therapeutic proteins and peptides. *Resp. Res.,* 2, 198–209.

Altiere, R.J. and Thompson, D.C. (1996). Physiology and pharmacology of the airways. In: Hickey, A., ed. *Inhalation Aerosols.* Marcel Dekker, New York, 233–272.

Banga, A.K. (1995). *Therapeutic Peptides and Proteins: Formulation, Processing and Delivery Systems.* Technomic, Lancaster, PA.

Batycky, R.P., Hanes, J., Langer, R., and Edwards, D.A. (1997). A theoretical model of erosion and macromolecular drug release from biodegrading microspheres. *J. Pharmaceut. Sci.,* 86, 1464–1477.

Boussif, O., Lezoualc'h, F., Zanta, M.A., et al. (1995). A versatile vector for gene and oligonucleotide transfer into cells in culture and *in vivo*: Polyethyleneimine. *Proc. Nat. Acad. Sci. U S A.,* 92, 7297–301.

Boussif, O., Zanta, M.A., and Behr, J.P. (1996). Optimized galenics improve *in vitro* gene transfer with cationic molecules up to 1000-fold. *Gene Ther.,* 3, 1074–1080.

Brown, R.A.J. and Schanker, L.S. (1983). Absorption of aerosolized drugs from the rat lung. *Drug Metab. Dispos.,* 11, 355–360.

Chadwick, S.L., Kingston, H.D., Stern, M., et al. (1997). Safety of a single aerosol administration of escalating doses of the cationic lipid GL-67/DOPE/DMPE-PEG(5000) formulation to the lungs of normal volunteers. *Gene Ther.,* 4, 937–942.

Cheng, Y.S., Zhou, Y., and Chen, B.T. (1999). Particle deposition in a cast of human oral airways. *Aerosol Sci. Technol.,* 31, 286–300.

Colthorpe, P., Farr, S.J., Taylor, G., Smith, I.J., and Wyatt, D. (1992). Pharmacokinetics of pulmonary-delivered insulin: Comparison of intratracheal and aerosol administration to the rabbit. *Pharmaceut. Res.,* 9, 764–768.

Colthorpe, P., Farr, S.J., Smith, I.J., Wyatt, D., and Taylor, G. (1995). The influence of regional deposition on the pharmacokinetics of pulmonary-delivered human growth hormone in rabbits. *Pharmaceut. Res.,* 12, 356–359.

Crapo, J.D., Young, S.L., Fram, E.K., Pikerton, K.E., Barry, B.E., and Crapo, R.O. (1983). Morphometric characteristics of cells in the alveolar region of mammalian lungs. *A. Rev. Resp. Dis.,* 128, s43–s46.

Darquenne, C., Brand, P., Heyder, J., and Paiva, M. (1997). Aerosol dispersion in human lung: comparison between numerical simulations and experiments for bolus tests. *J. Allied Physiol.,* 83, 966–974.

Densmore, C.L., Orson, F.M., Xu, B., et al. (2000). Aerosol delivery of robust polyethylene-imine-DNA complexes for gene therapy and genetic immunization. *Mol. Ther.,* 1, 180–188.

Dow, S.W., Schwarze, J., Heath, T.D., Potter, T.A., and Gelfand, E.W. (1999). Systemic and local interferon gamma gene delivery to the lungs for treatment of allergen-induced airway hyperresponsiveness in mice. *Hum. Gene Ther.*, 10, 1905–1914.

Edwards, D.A., Hanes, J., Caponetti, G., et al. (1997). Large porous particles for pulmonary drug delivery. *Science*, 276, 1868–1871.

El Jamal, M., Nagarajan, S., and Patton, J.S. (1996). *In situ* and *in vivo* methods for pulmonary delivery. *Pharm. Biothechnol.*, 8, 361–374.

Fiegel, J., Ehrhardt, C., Schaefer, U.F., Lehr, C.M., and Hanes, J. (2003). Large porous particle impingement on lung epithelial cell monolayers—Toward improved particle characterization in the lung. *Pharmaceut. Res.*, 20, 788–796.

Finkelstein, J.N. and Brandes, M.E. (1987). Activation of $Na^+-H^+$ exchange in type II pneumocytes by stimulators of surfactant secretion (Abstract). *Am. Rev. Resp. Dis.*, 135, A63.

Fu, J., Fiegel, J., Krauland, E., and Hanes, J. (2002). New polymeric carriers for controlled drug delivery following inhalation or injection. *Biomaterials*, 23, 4425–4433.

Giosue, S., Casarini, M., Ameglio, F., Alemanno, L., Saltini, C., and Bisetti, A. (1996). Minimal dose of aerosolized interferon-alpha in human subjects: Biological consequences and side- effects. *Eur. Resp. J.*, 9, 42–46.

Gonda, I. (1992). Targeting by deposition. In: Hickey, A.J., ed. *Pharmaceutical Inhalation Aerosol Technology.* Marcel Dekker, New York, 61–82.

Gonda, I. (2000). The ascent of pulmonary drug delivery. *J. Pharmaceut. Sci.*, 89, 940–945.

Hickey, A. J. and Garcia-Contreras, L. (2001). Immunological and toxicological implications of short-term studies in animals of pharmaceutical aerosol delivery to the lungs: Relevance to humans. *Crit.l Rev. Ther. Drug Carrier Syst.*, 18, 387–431.

Huland, E., Huland, H., and Heinzer, H. (1992). Interleukin-2 by inhalation; Local therapy for metastatic renal cell carcinoma. *J. Urology*, 147, 344–348.

Kawashima, Y., Yamamoto, H., Takeuchi, H., Fujioka, S., and Hino, T. (1999). Pulmonary delivery of insulin with nebulized DL-lactide/glycolide copolymer (PLGA) nanospheres to prolong hypoglycemic effect. *J. Controlled Release*, 62, 279–287.

Kellaway, I.W. and Farr, S.J. (1990). Liposomes as drug delivery systems to the lung. *Adv. Drug Delivery Rev.*, 5, 149–161.

Laube, B.L., Benedict, G.W., and Dobs, A.S. (1998). The lung as an alternative route of delivery for insulin in controlling postprandial glucose levels in patients with diabetes. *Chest*, 114, 1734–1739.

Liu, F.Y., Shao, Z.Z., Kildsig, D.O., and Mitra, A.K. (1993). Pulmonary delivery of free and liposomal insulin. *Pharmaceutical Research*, 10, 228–232.

Liu, Y., Liggitt, D., Tu, G., Zhong, W., Gnesler, K., and Debs, R. (1995). Cationic liposome-mediated intravenous gene delivery in mice. *J. Biologic. Chem.*, 27, 24864–24870.

Liu, Y., Mounkes, L.C., Liggitt, H.D., et al. (1997). Factors influencing the efficiency of cationic liposome-mediated intravenous gene delivery. *Nat. Biotechnol.*, 15, 167–173.

Li, Y.P. and Mitra, A.K., (1996). Effects of phospholipid chain length, concentration, charge, and vesicle size on pulmonary insulin absorption. *Pharmaceut. Res.*, 13, 76–79.

Lubman, R.L. and Crandall, E.D. (1992). Regulation of intracellular pH in alveolar epithelial cells. *American Journal of Physiology*, 262(1 Pt 1), L1–L14.

Machida, M., Hayashi, M., and Awazu, S. (1996). Pulmonary absorption of recombinant human granulocyte colony- stimulating factor (rhG-CSF) after intratracheal administration to rats. *Biologic. Pharmaceut. Bull.*, 19, 259–262.

Mader, K., Gallez, B., Liu, K.J., and Swartz, H.M. (1996). Non-invasive in vivo characterization of release processes in biodegradable polymers by low-frequency electron paramagnetic resonance spectroscopy. *Biomaterials*, 17, 457–461.

Masinde, L. and Hickey, A.J. (1993). Aerosolized aqueous suspensions of poly(*L*- lactic acid) microspheres. *Int. J. Pharmaceut.,* 100, 123–131.

Matthys, H. and Kohler, D. (1985). Pulmonary deposition of aerosols by different mechanical devices. *Respiration,* 48, 269–276.

Mitra, R., Pezron, I., Li, Y. P., and Mitra, A.K. (2001). Enhanced pulmonary delivery of insulin by lung lavage fluid and phospholipids. *Int. J. Pharmaceut.,* 217, 25–31.

Morimoto, K., Katsumata, H., Yabuta, T., et al. (2000). Gelatin microspheres as a pulmonary delivery system: Evaluation of salmon calcitonin absorption. *J. Pharm. Pharmacol.,* 52, 611–617.

Nikander, K. (1997). Adaptive aerosol delivery: the principles. *Eur. Resp. Rev.,* 7, 385–387.

Niven, R.W., Lott, F.D., and Cribbs, J.M. (1993). Pulmonary absorption of recombinant methionyl human granulocyte-colony-stimulating factor (R-Hug-Csf) after intratracheal instillation to the hamster. *Pharmaceut. Res.,* 10, 1604–1610.

Niven, R.W., Lott, F.D., Ip, A.Y., and Cribbs, J.M. (1994). Pulmonary delivery of powders and solutions containing recombinant human granulocyte-colony-stimulating factor (rhg-csf) to the rabbit. *Pharmaceut. Res.,* 11, 1101–1109.

Niven, R.W., Whitcomb, K.L., Shaner, L., Ip, A.Y., and Kinstler, O. B. (1995). The pulmonary absorption of aerosolized and intratracheally instilled rhg-csf and monopegylated rhg-csf. *Pharmaceut. Res.,* 12, 1343–1349.

Nord, E.P., Brown, S.E.S., and Crandall, E.D. (1987). Characterization of $Na^+$-$H^+$ antiport in type II alveolar epithelial cells. *Am. J. Physiol. (Cell Physiol.,* 21), 252, C490–C498.

Oguchi, K., Ikegami, M., Jacobs, H., and Jobe, A. (1985). Clearance of large amounts of natural surfactants and liposomes of dipalmitoylphosphatidylcholine from the lungs of rabbits. *Exp. Lung Res.,* 9, 221–235.

Olson, M.V. (2002). The human genome project: A player's perspective. *J. Mol. Biol.,* 319, 931–942.

Patton, J.S. (1996). Mechanisms of macromolecule absorption by the lungs. *Adv. Drug Delivery Rev.,* 19, 3–36.

Patton, J.S., Mccabe, J.G., Hansen, S.E., and Daugherty, A.L. (1989–1990). Absorption of human growth hormone from the rat lung. *Biotechnol. Therap.,* 1, 213–228.

Pfutzner, A., Mann, A.E., and Steiner, S.S. (2002). Technosphere™ /insulin-A new approach for effective delivery of human insulin via the pulmonary route. *Diabetes Technol. Therap.,* 4, 589–94.

Ploer, C.G. (1983). Comparative morphologic features of bronchiolar epithelial cells: The Clara cell. *Am. Rev.Resp. Dis.,* 128, s37–s41.

Pohl, R., Muggenburg, B.A., Wilson, B.R., Woods, R.J., Burell, B.E., and Steiner, S.S. (2000). A dog model as predictor of the temporal properties of pulmonary Technosphere/insulin in humans. *Resp. Drug Delivery,* VII, 463–465.

Russelljones, G. (1996). Utilization of the natural mechanism for vitamin B-12 uptake for the oral delivery of therapeutics. *Eur. J. Pharmaceut. Biopharmaceut.,* 42, 241–249.

Sano, K., Cott, G., Voelker, D., and Mason, R. (1988). The $Na^+$/$H^+$ antiporter in rat alveolar type II cells and its role in stimulated sufactant secretion. *Biochem. Biophys. Acta,* 939, 449–458.

Smart, J.R., (2002). A brief overview of novel liquid-based inhalation technologies. *Drug Delivery Syst. Sci.,* 2, 67–71.

Spicer, S.S., Schulte, B.A., and Thomopoulos, G.N. (1983). Histochemical properties of the respiratory tract epithelium in different species. *Am. Rev. Resp. Dis.,* 128, s20–s26.

Steiner, S., Pfutzner, A., Wilson, B.R., Harzer, O., Heinemann, L., and Rave, K. (2002). Technosphere (TM)/Insulin - proof of concept study with a new insulin formulation for pulmonary delivery. *Exp. Clin. Endocrinol. & Diabetes,* 110, 17–21.

Suarez, S., O'hara, P., Kazantseva, M., et al. (2001a). Respirable PLGA microspheres containing rifampicin for the treatment of tuberculosis: Screening in an infectious disease model. *Pharmaceut. Res.,* 18, 1315–1319.

Suarez, S., O'Hara, P., Kazantseva, M., et al. (2001b). Airways delivery of rifampicin microparticles for the treatment of tuberculosis. *J. Antimicrobial Chemother.,* 48, 431–434.

Tabata, Y. and Ikada, Y. (1988). Effect of the size and surface-charge of polymer microspheres on their phagocytosis by macrophage. *Biomaterials,* 9, 356–362.

Tortora, G.J. and Grabowski, S.R. (1993). *Principles of Anatomy and Physiology.* HarperCollins College Publishers, New York.

Wall, D.A. and Smith, P.L. (1997). Inhalation therapy for growth hormone deficiency. In: L.A. Adjei and P.K. Gupta, eds. *Inhalation Delivery of Therapeutic Peptides and Proteins.* Marcel Dekker, New York, 453–469.

Wolff, R.K. (1998). Safety of inhaled proteins for therapeutic use. *J. Aerosol Medicine-Deposition Clearance and Effects in the Lung,* 11, 197–219.

Yang, X.D., Ma, J.K.A., Malanga, C.J., and Rojanasakul, Y. (2000). Characterization of proteolytic activities of pulmonary alveolar epithelium. *Int. J. Pharmaceutics,* 195, 93–101.

Zhang, Q., Shen, Z.C., and Nagai, T. (2001). Prolonged hypoglycemic effect of insulin-loaded polybutylcyanoacrylate nanoparticles after pulmonary administration to normal rats. *Int. J. Pharmaceutics,* 218, 75–80.

# 11 Polymeric Systems for Oral Protein and Peptide Delivery

*Richard A. Gemeinhart, Ph.D.*

## CONTENTS

## INTRODUCTION

In recent years many peptide and protein drugs have been identified for use in humans (Table 11.1). With rapid progress in human genome analysis and as the structures of more proteins are deciphered, better mimetic peptides will be designed that have higher activity, better stability, and improved physical properties when compared to the original protein. Many biomacromolecules and their synthetic analogs are very potent. Correspondingly, the therapeutic index for the protein or peptide may be quite narrow. Since most biomacromolecules have short *in vivo* half-lives, repeated injections would be necessary for the drug to be continually active.

**TABLE 11.1**
**Protein and Peptide Drugs Currently in Use or Being**
**Investigated that Could Be Delivered via the Oral Route**

| Protein or Peptide | Disease |
|---|---|
| Insulin | Diabetes |
| Calcitonin | Osteoporosis |
| Erythropoietin | Anemia |
| Interleukin-2 | Renal carcinoma |
| Interferon | Multiple sclerosis |
| Hepatitis B subunit | Hepatitis B vaccine |
| Human growth hormone | Growth disorders |
| Tissue plasminogen activator | Heart attack |
| β-Cerebrosidase | Enzyme disorder |
| Deoxyribonuclease | Cystic fibrosis |

New biomacromolecules are not the only recent development. Production methods for biomacromolecules have become more reasonable. Modern cell culture methods allow the production of gram quantities of proteins from bacterial, yeast, or mammalian cells (Hauser and Wagner 1997). Efficient methods are now available to recover the proteins from their producing cells (Seetharam and Sharma 1991). The breakthroughs in protein production allow for more expensive methods to be investigated for the delivery of the protein; as the price of the protein drops, the price of the delivery device can go up while the cost of the final dose and profit for the company remains the same. The combination of new active proteins and the lower cost of the proteins has promoted increased research in the area of protein delivery.

Most biomacromolecules are delivered by injection, but many patients will not submit to the repeated injections that are needed to maintain sufficient levels of protein in the bloodstream. Because of this, alternative methods to deliver biomacromolecules are desired. Pulmonary delivery is promising for certain agents (see Chapter 10), however, many patients are not willing to use inhalation therapy and tend to reduce their dosage because they are not happy using this type of dosing. Transdermal delivery is also possible for delivery of some proteins, but usually not without the use of dermal abrasion or iontophoretic mechanisms. Since the oral route of delivery is the most accepted by patients, many investigators have focused on identifying possible methods for oral peptide delivery. New methods for the delivery of biomacromolecules must be economically feasible, as healthcare costs are currently such a concern. Many of the oral protein and peptide delivery systems to date use polymers.

Polymers are macromolecules composed of specific repeating units. The properties of the polymer, such as viscosity of a solution, elasticity, and solid strength, are determined by the number of repeating units, or monomers, and ultimately the radius of the polymer. The properties of a polymer can be predicted based upon theoretical calculations (Flory 1953). The diversity and predictability of properties are the reason that polymers are so useful in controlled drug delivery. Careful choice of the polymer leads to dosage forms that can deliver an active agent with reproducibility and accuracy. In this chapter, polymers useful in delivery of biomacromolecules are described. The delivery mechanism of the dosage forms containing this type of polymer is explained and future usefulness of the polymer is discussed.

# POLYMERS USED FOR CONTROLLED DRUG DELIVERY

The classification of polymers for oral drug delivery can be done by using various means. To make this discipline readily accessible to the novice reader, the hydrophobic-hydrophilic nature of the polymer was chosen to group polymers since the mechanism of biomacromolecule release from most hydrophobic polymeric devices is similar; the mechanism of release from most hydrophilic polymeric devices also have similar mechanisms. Hydrophobic polymers are described first, followed by hydrophilic polymers.

## HYDROPHOBIC POLYMERS

Hydrophobic polymers are often used to deliver biomacromolecules regardless of the route of administration. The rapid transit time of approximately 8 hours limits the time of a device in the gastrointestinal (GI) system, consequently the mechanisms possible for oral drug release are limited. The predominant method of release from hydrophobic polymers has been degradation, or biodegradation, of a polymeric matrix by hydrolysis (Figure 11.1). In fact, all of the hydrophobic polymers described in this chapter for use as oral protein or peptide delivery are hydrolytically unstable.

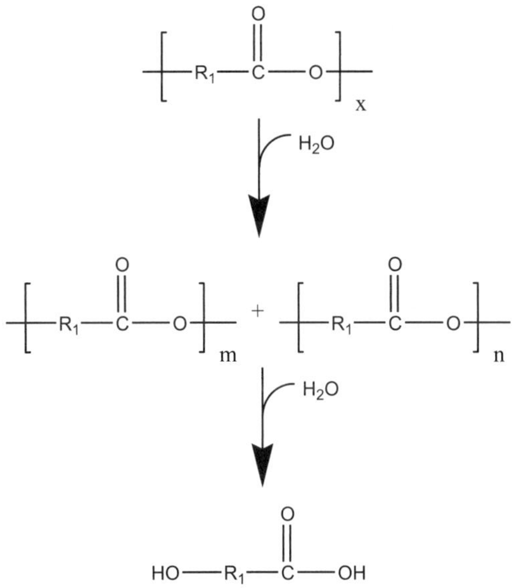

**FIGURE 11.1** Chemical scheme for degradation of poly(esters). Water reacts with the hydrolytic unstable ester bonds and finally produces monomers or short oligomers of the monomer.

The hydrophobic polymers are described in the following sections beginning with the most frequently described hydrophobic polymers used for oral protein and peptide delivery. Some of these polymers have been used for oral peptide and protein delivery, while others have not. Those polymers that have not been used to date for protein or peptide delivery have the potential for future use in devices for oral peptide and protein delivery and should not be overlooked. Each has been used *in vitro* or in animal studies that suggest that the polymer could be used for oral protein or peptide delivery.

## Poly(esters)

Poly(esters) (Table 11.2) are the first class of polymers discussed, as they are the most widely investigated of all of the polymer families for oral protein delivery. Poly(esters) used for oral drug delivery have primarily been biodegradable polymers (Figure 11.1). Biodegradation is the primary delivery mechanism for poly(ester) polymers used for protein and peptide delivery. The degradation properties of poly(esters) are dependent on the monomers used to produce the poly(ester). Several poly(esters) are discussed in detail in the following sections.

**TABLE 11.2**
**Structures of Poly(esters) Used in Oral Protein and Peptide Delivery**

| Polymer | Structure |
|---------|-----------|
| Poly(lactic acid) | |
| Poly(glycolic acid) | |
| Poly(lactic acid-*co*-glycolic acid) | |
| Poly(caprolactone) | |
| Poly(β-hydroxybutyric acid) | |
| Poly(β-hydroxyvaleric acid) | |

## Poly(lactic acid-co-glycolic acid)

The most investigated poly(esters) for oral peptide delivery have been poly(lactic acid-co-glycolic acid) (PLGA) copolymers (Spenlehauer et al. 1989; Wang et al. 1990; Odonnell and McGinity 1997). In actuality, PLGA itself is a group of copolymers with distinctly different properties depending on the composition. PLGA is a combination of D-lactic acid, L-lactic acid, and glycolic acid. Typically, PLGA polymers are referred to by the percent lactic acid to glycolic acid content that is present. A PLGA with 25% D-lactic acid, 25% L-lactic acid, and 50% glycolic acid would be described as 50:50 poly(D,L-lactic acid-*co*-glycolic acid) or 50:50 poly(D,L-lactide-*co*-glycolide). The physical and chemical properties of PLGA vary greatly with poly(glycolic acid) being poorly soluble in most solvents and poly(D,L-lactic acid) being soluble in many organic solvents.

The molecular weight and polydispersity of the polymer is an important factor in determining the properties of PLGAs. The starting molecular weight is important because as the PLGA is degraded, the delivery rate changes with a resultant decrease in molecular weight. When the molecular weight of the polymers is sufficiently low, the remaining chains become soluble in water. These small molecular weight oligomers and lactic and glycolic acid monomers have a plasticizing effect on the higher molecular weight portions of the device (Grizzi et al. 1995). The increased water content of the polymeric device contributes to the degradation process that is observed with poly(esters) and may even account for the diffusion-like delivery of some hydrophilic molecules (Figure 11.2A). The bulk degradation properties of PLGA have frequently been cited as a disadvantage. In bulk degradation of the matrix, the drug molecule may diffuse from the interior and exterior of the matrix; however, in surface erosion, which will be discussed in detail later, the drug will only be released from the surface of the matrix as it dissolves or degrades (Figure 11.2B).

PLGAs are one of the most biocompatible polymers currently used. This is due to the fact that the degradation products, lactic acid and glycolic acid, are produced naturally in the body. PLGA devices have been shown since the late 1960s to be an acceptable material for implantation and have been utilized since that time as surgical sutures (Cutright et al. 1971; Frazza and Schmitt 1971). Inflammatory responses occur with these materials due to causes directly related to the degradation mechanism. As the polymer degrades, an acidic environment is produced surrounding and within the polymer matrix (Fu et al. 2000). As the backbone of the polymer is hydrolyzed, more carboxylic acid groups are produced in the matrix, decreasing the pH of the surrounding fluid. It has been proposed that basic salts can be used to control the local pH of the microparticles (Agrawal and Athanasiou 1997). Control over the internal pH of the microparticles may produce a better system for delivery of peptides as the decreased pH of the current microparticles may contribute to low activity of delivered peptide. Despite the decrease in pH, many biomacromolecules can be delivered using PLGAs in the delivery device.

Due to the short transit time of materials taken orally, large PLGA matrices cannot deliver their entire contents prior to being eliminated. For this reason, microparticles have been the dosage form of choice for not only PLGAs, but also the majority of the hydrophobic polymer based systems. Microparticles of PLGA deliver their contents over a period from hours to years, depending on their composition

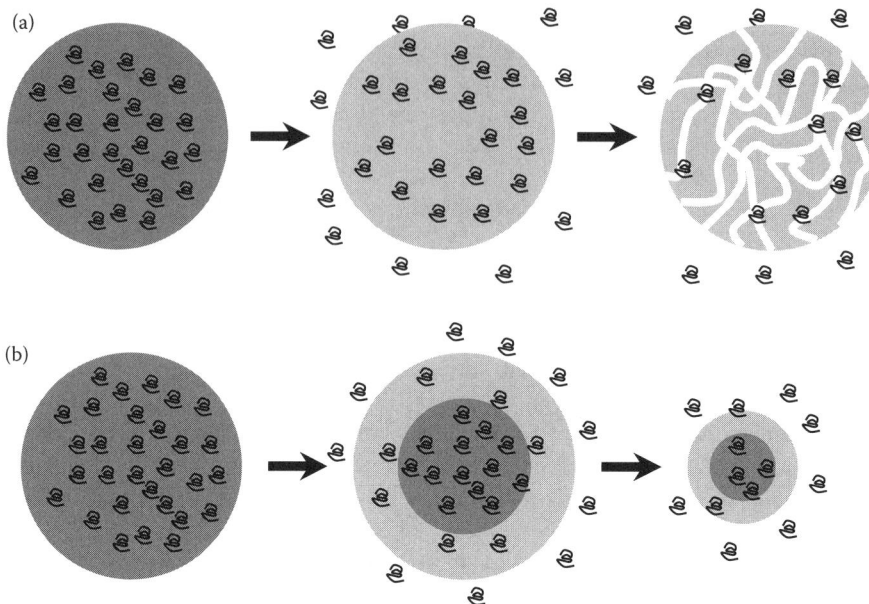

**FIGURE 11.2** Degradation of polymer matrices by (A) bulk degradation or (B) surface erosion. In bulk degradation, the degradation occurs as fluid enters the entire matrix and the encorporated molecule can diffuse from the degrading matrix. During surface erosion, only the biomacromolecules in the degrading layer are released; therefore, there is drug being released as long as the matrix is present in the body.

and site of deposition, but there is a typically large burst of delivery that will occur in the time that the microparticles are retained in the gastrointestinal tract. The time course of delivery is not always a problem, however, as some microparticles are actually absorbed by the gastrointestinal border cells (Figure 11.3). Those particles that are absorbed deliver their contents while the particles degrade in the body. This is typically in the lymphatic tissue of the intestine. To increase the uptake of microparticles by the M-cells in the Peyer's patches of the intestine, nonspecific bioadhesives and cell-specific molecules are being placed on the surface of the PLGA microparticle (Gabor and Wirth 2003). Uptake of microparticles is quite low, so a large excess of microparticles is needed to achieve sufficient drug transport. Bioadhesive and cell-specific molecules increase the contact time between the cells and microparticles, thus increasing the cellular uptake of the particles.

The water-in-oil-in-water (w/o/w) emulsion method (Figure 11.4) is the predominant method used for encapsulation of biomacromolecules in these microparticles. Protein solution forms the internal water phase of the w/o/w emulsion. Loading efficiency of the microparticles has not been optimal using water or buffer as an internal phase, so water is sometimes substituted with polymeric liquids, such as low molecular weight polyethylene glycol. The primary emulsion is then added to a secondary liquid phase, forming the secondary emulsion. The solvent for the

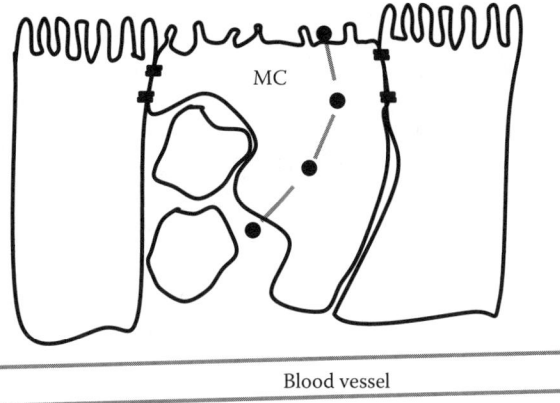

**FIGURE 11.3** Structure of the intestinal wall including an M cell (MC). M cells are part of the immune system and have been shown to be phagocytotic, engulfing particles and imparting an immune response.

polymer, usually methylene chloride, chloroform, or other organic solvent evaporates leaving hardened microparticles filled with water and protein. The evaporation step is one of the most important steps for control of the size and morphology of the microparticles, as heat and agitation have profound effects (Jeyanthi et al. 1997; Capan et al. 1999). Despite numerous attempts, the loading efficiency of the poly(ester) microparticles is quite low. Sufficient loadings are possible for low dose drugs, but large amounts of the peptide are needed to get the high loadings necessary. Due to this, much research is still needed in the formulation of poly(ester) microparticles, and other methods may be shown to be better despite the acceptability of these systems (Kompella and Koushik 2001).

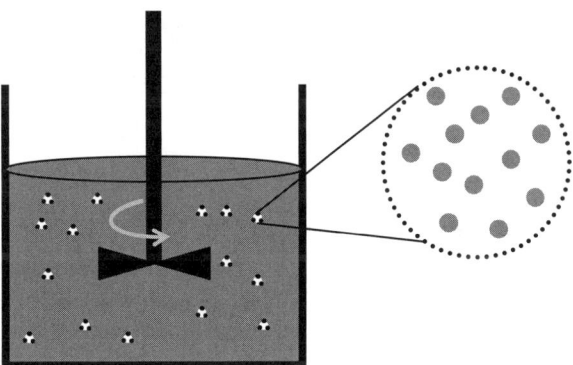

**FIGURE 11.4** Schematic representation of a water-in-oil-in-water emulsion. The *gray* represents the water phase, where the protein or poly(nucleic acid) would be present. The *black dots* surrounding the white oil phase represents a surfactant, typically poly(vinyl alcohol) or serum albumin.

The final obstacle for delivery of biomacromolecules from PLGA microparticles is sterility. Oral delivery of microparticles allows for more bacterial contamination than parenteral or pulmonary dosage forms as the gastrointestinal system is well equipped to control bacterial outgrowth. However, if the particles traverse the intestinal lining and begin circulating, bacterial incorporation will be problematic. To combat bacterial incorporation in the microparticles, one must work under sterile conditions, but the PLGAs may bring bacteria that cannot be removed. Particles sterilized by γ-irradiation after production contain protein with decreased activity and polymers of lower molecular weight than in unirradiated particles. This has caused an increase in delivery rate based upon polymer degradation over the initial days following irradiation with a combined decrease in active protein (Shameem et al. 1999).

*Other Poly(esters)*

Poly(ε-caprolactone) (PCL) is another of the degradable poly(esters) that has been utilized to deliver biomacromolecules orally. The polymer is more hydrophobic than any of the PLGAs, and can be used by itself or as a copolymer with lactic acid, glycolic acid, or any hydroxy acid (Pitt 1990). Due to its higher hydrophobicity, the degradation rate of the polymer is somewhat slower than PLGA; PLGAs are typically the fastest degrading of the poly(esters) as hydrophobicity is directly related to degradation rate. The uptake of particles that are formed with PCLs was also proposed to be higher than that of PLGAs (Eldridge et al. 1990). Other poly(esters) (Table 11.2) have been used for peptide incorporation in the form of microparticles. The delivery from these materials is long term so they can be used to decrease dosing upon injection or implantation (Engelberg and Kohn 1991); however, few have been successfully used in oral delivery of biomacromolecules. All of the poly(esters) are excellent candidates for the delivery of biomacromolecules in the form of microparticles based on the ability of the polymeric particles to protect the proteins and peptide from degradation in the digestive tract. The uptake of the particles by the gut-associated lymphoid tissue can increase the amount of protein that reaches the bloodstream. Unfortunately, a sufficient amount of microparticles is not retained for substantial blood levels of most proteins for this method to be used on a regular basis. Regardless of the type of polyester, much research is still needed to improve microparticle preparation for protein incorporation.

## Poly(cyanoacrylate)

An alternative hydrophobic microparticulate dosage form can be produced using poly(alkyl cyanoacrylates) also referred to as simply poly(cyanoacrylates) (PCAs) (Table 11.3). Poly(cyanoacrylates) are a class of addition polymers that undergo polymerization under mild conditions, and even upon the addition of water or ethanol. Poly(cyanoacrylates) have been widely investigated for delivery of biomacromolecules. Due to their properties, cyanoacrylates can easily be formed into two types of particles: spheres (Couvreur et al. 1982) or capsules (Al-Khouri Fallouh et al. 1986), both of which can be used to deliver biomacromolecules. The most used of the poly(cyanoacrylates) is poly(isobutyl cyanoacrylate) (PBCA). The reason

**TABLE 11.3**
**Structures of Some Common Poly(cyanoacrylates)**

| Polymer | Structure |
|---------|-----------|
| Poly(isohexyl cyanoacrylate) | |
| Poly(isobutyl cyanoacrylate) | |
| Poly(isopropylcyanoacrylate) | |
| Poly(ethyl cyanoacrylate) | |
| Poly(methyl cyanoacrylate) | |

PBCA has been extensively investigated is that it can easily be formed into microparticles with the aid of ethanol as a solubilizer. Other PCAs tend to polymerize in the presence of ethanol, including poly(isohexyl cyanoacrylate) (PIHCA) (Chouinard et al. 1991).

Since the polymerization procedure is mild, there is less activity loss during the incorporation of peptides and proteins into PCA microparticles as compared to poly(ester) microparticles. Also, there is a greater amount of incorporation in the PCA microparticles than the poly(esters) microparticles. One of the major disadvantages of the PCA over the poly(hydroxy acids) is the toxicity of the monomers (Muller et al. 1988). The degradation products of PCAs are alkyl cyanoacetate and formaldehyde, both of which are more toxic than the lactic acid, glycolic acid, or other hydroxy acids formed from the degradation of poly(esters). The relative toxicity may not be as much of a factor as anticipated since the PCA degradation rate can be controlled to limit the amount of degradation product that is present.

The rate of degradation of PCAs is much faster than PLGAs which do not necessitate the overloading of the body with excess polymer that is still present long after initial treatment (Lherm et al. 1992). The alkyl cyanoacetate that is formed and the rate of formation are dependent on the length of the alkyl chain of the monomer used to produce the microparticles. The degree of toxicity is also dependent on the length of the alkyl chain. Poly(cyanoacrylates) have been approved by the FDA in several systems, so it is one of the more acceptable of the polymeric systems.

## Poly(ortho esters)

Poly(ortho esters) (POEs) were one of the first polymers produced that exhibited delivery predominantly via surface degradation (Figure 11.2B). Four distinct types of POEs have been developed since the early 1970s (Heller et al. 2000) (Table 11.4). The design of each poly(ortho esters) is inherently different, and has specific properties. Type I poly(ortho esters) were developed at Alza Corporation (Mountain View, CA) and are described by a series of patents (Choi and Heller 1976, 1978a,b, 1979a,b). Upon degradation, type I poly(ortho esters) form the appropriate alkane diol and γ-butyrolactone. The lactone easily hydrolyzes to form γ–hydroxybutyric acid. The acid accelerates the degradation of the polymer unless neutralized with a basic excipient. Polymers of this family are not well described in the literature, so little information can be given for the specific structures of the polymer. No oral delivery devices have been produced with this type of poly(ortho ester). Insulin-like growth factor has been delivered using this type of poly(ortho ester) indicating that loading and release from this class of polymer is possible (Busch et al. 1996).

Type II poly(ortho esters) are very stable due to their hydrophobic nature (Heller 1990). The degradation rate can be controlled using acidic and basic excipients; acidic excipients increase the degradation rates and facilitate a zero-order release rate over a 2-week period (Sparer et al. 1984). Basic additives increase the degradation time of the polymers and create a polymer that degrades specifically at the surface (Heller 1985). By careful choice of the excipient added, the degradation rate can be closely controlled. No experiments have shown the use of these polymers with proteins or peptides. This is not, however, indicative of the fact that these polymers are not compatible with proteins or peptides, but they are probably not the most appropriate polymeric carrier for oral delivery of biomacromolecules.

Type III poly(ortho esters) are very similar to the type I poly(ortho esters) in that they are based on a five member ortho ester ring (Heller et al. 1990). The linkages are quite different with the degradation process not forming a lactone, but an acid and a triol. Type III poly(ortho esters) developed to date are more hydrophilic

**TABLE 11.4**
**Structures of Poly(ortho esters) Used for Drug Delivery**

| Polymer | Structure |
|---|---|
| Poly(ortho ester) type I | |
| Poly(ortho ester) type II | |
| Poly(ortho ester) type III | |
| Poly(ortho ester) type IV | |

Adapted from Heller et al. 2000.

than other poly(ortho esters), and for this reason, they erode quite quickly (Merkli et al. 1996). Some polymers of this type are semisolid at room temperature (Einmahl et al. 1999), which can facilitate their mixing with biomacromolecules. The semisolid nature has been exploited for specific applications; however, no significant effort has been identified for advancing these materials for oral biomacromolecule delivery.

Type IV poly(ortho esters) are very similar in structure to type II poly(ortho esters), but they do not need to have excipients in the formulation due to the incorporation of no acidic moieties in the polymer backbone (Ng et al. 1997). Rods of poly(ortho ester) loaded with recombinant human-growth hormone and bovine serum albumin have been created. The rods are the products of polymer-protein mixture extrusion at a temperature between 50° and 70°C. Particles have also been produced from these rods (Heller et al. 2000). The size of these particles, >106 μm, was much larger than would be expected to be absorbed by the gastrointestinal lining (Florence 1997). If the particle size can be reduced, this type of polymer system may be made to be acceptable for oral administration.

Despite the fact that these polymers show delivery patterns comparable to gastric transit, there have not been any substantial reports of oral delivery of peptides or proteins based on this type of polymer. Further investigations into this polymer may be desirable since the degradation properties of the polymers can be very tightly controlled.

## Poly(phosphazenes)

Poly(phosphazenes) are another class of biodegradable polymers that have varying degradation rates (Scopelianos 1994). The rate of degradation of poly(phosphazenes), however, is not dependent on the changes to the backbone of the polymer, but rather to the changes of the pendant group properties. A wide variety of polymers can be formed by the molecular substitution on poly(dichloro phosphazene). In fact, poly(phosphazenes) can be produced as either hydrophobic or hydrophilic polymers depending on the exact nature of the pendant groups (Table 11.5). Almost unlimited possibilities exist concerning the structure and properties. The production of the various poly(phosphazenes) is very simple, an alcohol or 1° or 2° amino terminated molecule can be incorporated onto poly(dichlorophosphazene). The hydrophobic polymers will be examined in this section, and the hydrophilic polymers formed with poly(phosphazenes) will be discussed later.

The degradation of the poly(phosphazenes) is very well understood, with the production of ethanol, phosphate, ammonium salts, and the pendant groups (Andrianov and Payne 1998). When the pendant group is an amino acid, all of the degradation

## TABLE 11.5
## Structures of Some Example Poly(phosphazenes) Used in Drug Delivery

| Hydrophobic | Hydrophilic |
|---|---|
| NHCH$_2$COOEt <br> $-[-P=N-]_n-$ <br> NHCH$_2$COOEt | OC$_6$H$_4$COOH <br> $-[-P=N-]_n-$ <br> OC$_6$H$_4$COOH |
| NHCH(CH$_2$C$_6$H$_5$)COOEt <br> $-[-P=N-]_n-$ <br> NHCH$_2$COOEt | NHC$_6$H$_4$COOH <br> $-[-P=N-]_n-$ <br> NHC$_6$H$_4$COOH |
| NHCH$_2$(CH$_3$)COOCH$_2$C$_6$H$_5$ <br> $-[-P=N-]_n-$ <br> NHCH$_2$(CH$_3$)COOCH$_2$C$_6$H$_5$ | OCH$_2$CH$_2$OCH$_2$CH$_2$OCH$_3$ <br> $-[-P=N-]_n-$ <br> OCH$_2$CH$_2$OCH$_2$CH$_2$OCH$_3$ |

products are natural products normally found in the body. The exact properties of these types of polymers are controlled by the selection of the amino acid substituents added (Crommen et al. 1992). Other labile groups have been used to create degradable polymeric systems including ethylamino (Tanigami et al. 1995), imidazolyl, oligopeptide, amino acid esters, and depsipeptide groups. As with other degradable polymers, the more hydrophobic, or bulky, side groups are more hydrolitically stable (Allcock et al. 1994). The degradation and release related directly to the relative mixture of the multiple components on the backbone. Combinations of the various poly(phosphazenes) also show a direct relation between the addition of faster degrading species and the degradation and delivery pattern of the polymer.

Poly(phosphazene) microparticles have shown the potential for oral protein or peptide delivery (Vandorpe et al. 1996; Veronese et al. 1998; Passi et al. 2000). The processing of these materials is very similar to poly(esters) and the biocompatibility of the polymers is exceptional. After the implantation of matrices of poly(phosphazene), no gross areas of inflammation were observed at explanation (Laurencin et al. 1987). The only negative aspect of these materials is that FDA approval of this polymer class has little precedent.

## HYDROPHILIC POLYMERS

Hydrophilic polymers are currently undergoing investigation for improving the transport of biomacromolecules across the intestinal walls. Hydrophilic polymers have been shown to protect proteins and peptides from proteolysis. Multiple methods utilize the properties of polymers to protect biomacromolecules without removing them from the aqueous environment of the intestines.

### Polymer-Protease Inhibitor Conjugates

Polymers may also inhibit proteolytic enzymes or be used to augment the activity of proteolytic inhibitors (Table 11.6). The administration of small molecule protease inhibitors in conjunction with the peptide, however, has been examined with some success (Fujii et al. 1985); however, the ability of these molecules to protect the protein has been hampered by the fact that they tend to cause systemic side effects (Yagi et al. 1980; McCaffrey and Jamieson 1993; Plumpton et al. 1994). Many protease inhibitors have been specific for the proteases in the stomach and intestine, but some of these factors are not specific and some control over absorption of the

## TABLE 11.6
## Polymers that Act as Protease Inhibitors

| Polymer Backbone | Protease |
|---|---|
| Polyacrylate | Trypsin, chymotrypsin, carboxypepsidase A, elastase |
| Polymethacrylate | Trypsin, chymotrypsin, carboxypepsidase A, elastase |
| Chitosan | Trypsin, carboxypepsidase A, aminopepsidase N |
| Carboxymethylcellulose | Elastase, pepsin |

inhibitors was desired. The activity of the inhibitors varied leading to more problems in determining methods for controlling the inhibitory effects.

To alleviate problems with administration of small molecule protease inhibitors, it was hypothesized that a polymer-based inhibitor could prevent peptide degradation. The fact that the inhibitor was attached to a polymer would localize the inhibitory affect, minimizing the effect in the intestine and allowing normal digestion (Melmed et al. 1976; Otsuki et al. 1987). This would also prevent the systemic effects of the inhibitors by decreasing the possibility for absorption if poorly absorbed polymers are used. The only problems with conjugating the inhibitors with proteins were possible degradation, loss of activity, and changes in activity. By carefully choosing the polymers for conjugation, it was thought that inhibition could actually be augmented with little loss or change in the activity of the inhibitors.

Various groups have proposed using this mechanism for protection of peptides with protease inhibitor-polymer conjugates within the digestive tract (Bernkop-Schnürch 1998). Some synthetic polymers have been proposed to be advantageous since they possess mucoadhesive properties (Lueßen et al. 1996), but some natural polymers also have mucoadhesive properties and have also been used (Bernkop-Schnürch and Apprich 1997). Poly(acrylates) have been well described as mucoadhesive polymers because they have hydrogen bonding and chain entanglement with the mucin of the stomach and intestine (Park and Robinson 1987). The mucoadhesive properties of the polymer would be beneficial in that the dosage form would localize at the surface of the intestinal walls. This would have multiple benefits, including minimizing the diffusional distance that the protein or peptide prior to absorption and increase the residence time of the dosage form in the gastrointestinal tract.

The carboxylic acid moieties of the poly(acrylate) are also involved in the inhibition of proteolytic enzymes. Divalent cations are necessary for many proteases to act, specifically $Ca^{2+}$ and $Zn^{2+}$. Whether these polymers can be effective *in vivo* at protecting peptides has been of some concern (Bernkop-Schnürch and Göckel 1997), but the poly(acrylates) do exert an inhibitory effect on many of the proteases present in the gastrointestinal tract (Lueßen et al. 1996; Madsen and Peppas 1999). Polycarbophil, Carbopol® (Noveon, Cleveland, OH), and synthetic poly(methacrylic acid-g-ethylene glycol) hydrogels have been shown to decrease the activity of trypsin by simply adding the polymer to a solution of the enzyme and a substrate (Figure 11.5). Poly(acrylates) have also been shown to increase the permeability of the intestine (Borchard et al. 1996). Since these polymers, e.g., carbophil, have received GRAS (generally regarded as safe) status, the use of these polymers as inhibitor conjugates should be examined closely.

Bowman-Birk inhibitor (BBI), a peptide analog from soybean, was conjugated to poly(acrylic acid) using carbodiimide chemistry as were chymostatin, bacitracin, and elastinal (Bernkop-Schnürch and Göckel 1997). Of these, bacitracin has been the only conjugate that did not possess any bioadhesive properties, while all of the conjugates showed decreased protease activity. The use of the polymer-inhibitor conjugates is not the only mechanism that has been used for oral biomacromolecule delivery. A selection of hydrophilic polymers that have potential for oral biomacromolecule delivery are described below while occasionally revisiting the idea of protease inhibition.

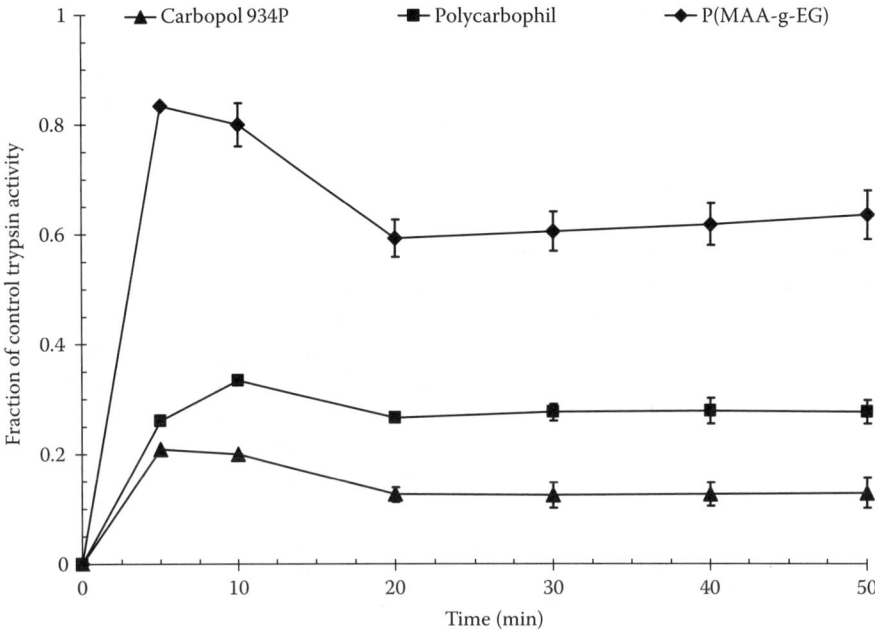

**FIGURE 11.5** Degradation of N-a-benzoyl-l-arginine ethyl ester by trypsin. All three carboxyl containing polymers have antitrypsin activity, but the activity of trypsin in the presence poly(ethylene glycol) modified poly(methacrylic acid) is not greatly reduced. (Adapted from Madsen and Peppas 1999.)

## Poly(alkyl methacrylates)

Eudragit® (Röhm/Pharma Polymers, Darmstadt, Germany) is a class of polymer that will dissolve at specific pH values. Eudragit polymers are members of the poly(alkyl methacrylate) family of polymers. This family of polymers has a wide degree of properties due to the many modifications used in the formation of the polymers. The dissolution properties are based on the modification of the carboxylic moieties of methacrylic acid. The degree of modification and the specific modification determine at what pH and how quickly the polymers dissolve.

Microparticles prepared using Eudragit polymers were able to deliver protein in the intestine based upon the solubility properties of the Eudragit used (Morishita et al. 1991). Entrapment efficiency in the microparticles has been as high as 78% for the emulsion method used. These particles delivered their contents in the intestine due to the low solubility of both Eudragit polymers used in acidic conditions; Eudragit S100 and L100 are soluble at pH values above 7.0 and 6.0, respectively. The *in vivo* activity of insulin delivered by oral force feeding rats, showed decreased levels of glucose for several hours following the administration. These levels were reported as a percent decrease, but it is still clear that active insulin was being absorbed into the bloodstream of the animals.

The delivery of proteins from non-crosslinked microparticles has been hampered by the fact that the amount of protein that crosses the gastrointestinal lining is not augmented. Many of these systems only allow the material to be delivered in the vicinity of the intestinal wall. If particular proteins can easily traverse the GI lining, it will have increased *in vivo* activity.

## Poly(methacrylates) and Poly(acrylates)

Chemical crosslinks have been added to materials similar in nature to the Eudragit polymers to further control the delivery of biomacromolecules. Most of these systems would fall under the category of hydrogel particles. Hydrogels are polymers that are physically or chemically crosslinked and are unable to dissolve in water. Hydrogels retain their shape upon swelling and usually swell in an isotropic manner (Peppas 1986 ). The rate of delivery of the protein or peptide has been shown to be dependent upon the swollen state of the hydrogel, the size of the diffusing peptide, and the mesh size of the hydrogel (Lustig and Peppas 1987). A general equation describing the diffusion coefficient of a solute as a factor of the swollen state of the hydrogel indicates that the more swollen the hydrogel, the faster the molecule will be delivered. The diffusion coefficient of the solute, D, can be related to the mesh size of the hydrogel, $\xi$, the molecular radius of the protein, r, a constant that typically can be assumed to be 1 for most systems, Y, and the swollen ratio of the hydrogel, Q. For insulin in a poly(diethylaminoethyl methacrylate-g-ethylene glycol) hydrogel microparticle, the swollen diffusion coefficient was found to be 21 times that of the collapsed diffusion coefficient. This would correlate with a significant increase in diffusional drug delivery for insulin in the swollen hydrogel particle (Podual et al. 2000).

$$D \cong \left(1 - \frac{r}{\xi}\right) e^{[-Y/(Q-1)]} \qquad (1)$$

These hydrogels contain a large number of ionic pendant groups leading to increased electrostatic repulsion on the chains when they are in the ionized form. Since the pKa of poly(acrylic acid) is approximately 4.5, the hydrogels swell the most when they are placed in media that has a pH greater than 5.5. Poly(ethylene oxide) chains have been grafted to the hydrogel to increase the sensitivity of the hydrogels to their ionic environment. Whereas poly(diethylaminoethyl methacrylate) has a pKa of approximately 7, hydrogels of this polymer are highly swollen in media with a pH lower than 6.0. Since there is such variability between the pH of the stomach and intestine in humans, these polymers can be used as a delivery device dependent on the delivery site. Many other pH sensitive polymers have been examined to be used in drug delivery devices, but the majority that are currently used are in the form of either poly(acrylic acid) or poly(methacrylic acid).

The main reason that poly(acrylic acid) and poly(methacrylic acid) hydrogels and hydrogel particles have been utilized in oral drug delivery is the fact that they also possess bioadhesive properties (Park and Robinson 1987). Bioadhesive polymers adhere to biologic matter, specifically the walls of the gastrointestinal tract. Bioadhesion may be specific or nonspecific in nature depending upon the mechanism of adhesion to the gastrointestinal tract (Ponchel and Irache 1998). Nonspecific interactions that may play a role in bioadhesion are van der Waals, hydrogen bonding, and ionic interactions. Mucin in the gastrointestinal tract is comprised of high molecular weight glycoproteins that vary in thickness throughout the gastrointestinal tract (Allen et al. 1982). Nonspecific bioadhesion has been shown to increase the gastrointestinal transit time; however, the increase is not great due to turnover of the mucin layer of the GI tract every 2 hours (Lehr et al. 1991).

Another type of hydrogel based on methacrylic acid that has been used in oral delivery of proteins is a hydrogel system that is degradable at a specific point in the intestinal tract. Since there are flora in the colon that are significantly different than that of the small intestine, polymers have been designed to be degraded by the enzymes only produced in this region of the intestine. One type of enzyme specific to the colon is azoreductase (Saffran et al. 1986). The azo-bonds of a specially designed crosslinker are degraded specifically in the colon (Brondsted and Kopecek 1992). Proteins can be delivered specifically to the colon using these polymers and this has been shown to be effective when penetration enhancers are added to hydrogel disks (Yeh et al. 1995). Penetration enhancers are needed because the absorption capacity of the colon is significantly lower than that of the small intestine. The fact that the protein and polymer would be delivered in the colon decreases the chances for proteolytic degradation. Colon specific delivery of biomacromolecules may be the preferred route due to this decreased degradation, but the decreased absorptive capacity of the colon may limit the delivery by this route.

## Alginates

Physically crosslinked systems are somewhat similar to chemically crosslinked systems with the major difference being the type of crosslinking that actually takes place. The major polymer that is crosslinked using a physical crosslinking mechanism is alginate. Alginate is a naturally occurring polymer derived from seaweed. The benefit of the physical crosslinking systems as opposed to chemical crosslinking is the mild conditions needed to form a microparticle. Alginate is present as single polymer chains when in the presence of monovalent cations, such as $Na^+$ or $K^+$. However, when in the presence of divalent cations, i.e., $Ca^{2+}$, $Zn^{2+}$, and $Mg^{2+}$, the polymers associate into an ordered structure that is solid (Rees and Welsh 1977). Particles of alginate can be formed by simply adding a solution of alginate polymer containing the biomacromolecule to a solution of divalent cations. The exact form of the particles produced depends on the conditions used.

Large particles can be produced by simply injecting viscous solutions of sodium alginate from a large diameter needle into a solution containing calcium or other divalent cations (Badwan et al. 1985). It was found that anything above 5% (w/v) alginate was too viscous to prepare. Poly(L-lysine) (Dupuy et al. 1994) and chitosan (Takahashi et al. 1990) have been shown to increase the aggregation of alginates by forming a complex with the alginate, thereby strengthening the beads. Microparticles have been produced by several methods, but three methods have predominated: atomization (Matsumoto et al. 1986), emulsification (Wan et al. 1992), and coacervation (Arneodo et al. 1987). Each of these methods for the production of alginate microparticles has its advantages and disadvantages according to the specific proteins or peptides being delivered. Since the emulsification method used harsher chemicals, it is thought to be the least useful for most biomacromolecules.

Alginates are well accepted and are generally accepted as safe by the Food and Drug Administration. Because they are safe and produce particles with high encapsulation efficiency, alginates have been well studied for drug delivery. Unfortunately, alginates do not possess the mucoadhesive properties of the acrylic polymers and hydrogels. It should be possible to modify the alginate polymers to impart an adhesive ability onto the chain so that the high encapsulation efficiency and mild encapsulation conditions can be combined with a mucoadhesive system.

## Chitosan

Chitosan is a polymer produced from hydrolysis of natural chitin. Chitosan is not readily soluble in aqueous solutions, but can be solubilized and is thus considered with other water soluble polymers. In the hydrophobic form, chitosan has been treated in a similar manner to other hydrophobic polymers with microparticles produced by emulsion and phase separation techniques. Microparticles can be taken up by the gastrointestinal lining in a manner similar to that discussed for other hydrophobic microparticles.

The biggest difference between chitosan and other polymers is that it has both chelating and bioadhesive properties. Chitosan has been conjugated to antipain, chymostatin, elastatinal, ethylene diamine triacetate (EDTA), and combinations thereof (Bernkop-Schnürch and Scerbe-Saiko 1998). Cation binding inhibits proteins without the addition of proteolysis inhibitors in a similar manner to poly(acrylates), but the inhibitory effect of chitosan-EDTA toward $Ca^{2+}$-dependent proteases was not always found to be significant (Bernkop-Schnürch and Krajicek, 1998). These polymers can be incorporated into conventional tablets that will slowly dissolve and deliver the protein. When chitosan/EDTA/BBI conjugate was included at only 10% (w/w) of a formulation, more than half of the insulin activity can be maintained when administered orally compared to less than 10% insulin activity when no conjugate is added to a tablet. The inner portion of the tablet contained somewhat more active insulin, but even on the outer surface of the tablet, most of the activity still remained.

Polymers that are protease inhibitors and polymer-inhibitor conjugates are now widely investigated for their ability to protect proteins and peptides from proteolytic degradation. These molecules are effective in the immediate area surrounding the delivery device, so the effects on proteins that have diffused far from the delivery device are limited. Due to the fact that bioadhesives were used as the conjugating polymer, the delivery device may adhere to the intestinal lining. If this does happen, the diffusional distance of the protein from the device to the intestinal wall will be quite short. One barrier that the protease inhibitors do not affect is the cellular barrier. Biomacromolecules must still find a method to enter the cells or be taken up by phagocytosis.

## Polyphosphazene Hydrogels

Polyphosphazene hydrogels, as indicated earlier, are hydrogels prepared using the polyphosphazene backbone but that contain hydrophilic moieties pendant from the backbone. A matrix and cross-linking can be produced by incorporating multifunctional groups into the mixture. Mild conditions are needed for production of the hydrogels, so biomacromolecules are not exposed to harsh conditions. Release from polyphosphazene hydrogels is by diffusion as has been described for other hydrogel microparticulate systems, but can be controlled by appropriate control of the polymer and secondary components. As with any of the microparticulate systems, a coating can be applied to the particles, thus preventing initial release of encapsulated molecules. By making the surface of the hydrogel semipermeable, the release of bovine serum albumin can be greatly reduced, but not completely stopped. Further investigations using semipermeable or nonpermeable coatings should be investigated for various hydrophilic systems to decrease the release of biomacromolecules prior to reaching the lower intestines or the area of the intestine desired (Andrianov and Payne 1998).

## Poly(ethylene glycol) or Poly(ethylene oxide)

The use of conjugates of polymers to proteins has been investigated extensively for parenteral administration of proteins (Veronese et al. 1985). A polymer conjugated to a protein prevents proteolytic enzymes from contacting the protein by steric hindrance. Since proteolysis is minimized, circulation of the peptide can be increased greatly. Some degradation is still possible either by alternate proteolytic mechanism, degradation of the conjugated polymer, or hydrolytic degradation. The protein or peptide would still be in contact with the acidic environment of the gastrointestinal tract and this may be the reason behind the limited investigation of this type of polymer for oral delivery.

Another reason that oral polymer-protein conjugates have not been investigated widely is that there may be some difficulty in the ability of the conjugate to traverse the intestinal wall. The hydrophilic protecting chain of the conjugate would decrease the solubility of the protein in the lipid bilayer of the cells of the intestine. The decrease in solubility could cause a decrease in the bioavailability of the protein. The fact that particles are well absorbed by the intestine could indicate that further research is necessary in this area. Conjugates of proteins and polymers are very effective when administered parenterally, and the increase in half-life of circulating protein conjugates could also indicate an increase in the half-life of absorbed proteins delivered orally. This area of oral drug delivery should be investigated more carefully in the future as it could greatly increase the absolute amount of active protein and peptide that can be delivered.

## CONCLUSIONS

The use of polymers to deliver protein and peptides has had some success. Most of these systems have been based on encapsulating the protein or peptide within a polymer. The protection that the polymer gives the peptide from proteolysis was the major benefit that was originally sought. Studies have shown, however, that polymers also play a role in uptake of biomacromolecules, as the polymeric carrier particles are themselves taken up by the intestinal cells. The majority of the particles do not reach the systemic circulation, and in fact, only a rare few make it to circulate in the blood. Blood levels of active biomacromolecules have been shown to increase to significant levels following oral administration, encouraging further investigation into methods that would increase the fraction of particles that are absorbed by the intestinal cells, or at least increasing the amount of time that the particles spend in the digestive tract prior to elimination from the body. Bioadhesive particles are currently thought to be a major breakthrough for oral peptide delivery and novel bioadhesive particles are being developed (Lehr 2000). The increase in absorption and residence time in the gastrointestinal tract could greatly increase the delivery of proteins.

Inhibitors of proteases have also been developed from polymers. These molecules are as simple as a polymer chain, or much more complex. The mechanisms for protease inhibition are variable depending upon the type of molecule used; specific protease inhibitors are available for conjugation to a polymer while other polymeric inhibitors inhibit all divalent cation dependent proteases. The development of these inhibitory polymers and polymer conjugates greatly increase the possibility to protect proteins from degradation in the gastrointestinal tract.

With all of the advances in polymer science and conjugation technology, many methods have been developed to increase the feasibility of oral peptide and protein delivery. There is still no single mechanism that can be used to protect a protein or peptide from degradation and increasing oral availability, but with the multitude of new methods for allowing a protein to negotiate natural barriers, oral delivery of any systemically active protein is a definite possibility at some point in the future.

## REFERENCES AND FURTHER READING

Agrawal, C.M. and Athanasiou, K.A. (1997). Technique to control pH in vicinity of biodegrading PLA-PGA implants. *J. Biomed. Materials Res.,* 38,105–114.

Al-Khouri Fallouh, N., Roblot-Treupel, L., Fessi, H., Devissaguet, J.P., and Puisieux, F. (1986). Development of a new process for the manufacture of polyisobutylcyanoacrylate nanocapsules. *Int. J. Pharmaceut.,* 28, 125–132.

Allcock, H.R., Pucher, S.R., and Scopelianos, A.G. (1994). poly[(amino acid ester)phosphazenes] as substrates for the controlled-release of small molecules. *Biomaterials,* 15, 563–569.

Allen, A., Bell, A., Mantle, M., and Pearson, J.P. (1982). The structure and physiology of gastrointestinal mucus. In: Chantler, E.N., Elder, J.B., Elstein, M., eds. *Mucin in Health and Disease.* Plenum Press, New York, 15–133.

Andrianov, A.K. and Payne, L.G. (1998). Protein release from polyphosphazene matrices. *Adv. Drug Delivery Rev.,* 31, 185–196.

Arneodo, C., Benoit, J.P., and Thies, C. (1987). Characterization of complex coacervates used to form microcapsules. *Polym. Mater. Sci. Eng.,* 57, 255–259.

Badwan, A.A., Abumalooh, A., Sallam, E., Abukalaf, A., and Jawan, O. (1985). A sustained release drug delivery system using calcium beads. *Drug Dev. Indust. Pharm.,* 11, 239–256.

Bernkop-Schnürch, A. (1998). The use of inhibitory agents to overcome the enzymatic barrier to perorally administered therapeutic peptides and proteins. *J. Control. Release,* 52, 1–16.

Bernkop-Schnürch, A. and Arich, I. (1997). Synthesis and evaluation of a modified mucoadhesive polymer protecting from a-chymotrypsin degradation. *Int. J. Pharmaceut.,* 146, 247–254.

Bernkop-Schnürch, A. and Göckel, N.C. (1997). Development and analysis of a polymer protecting from luminal enzymatic degradation of α–chymotrypsin. *Drug Dev. Ind. Pharm.,* 23, 733–740.

Bernkop-Schnürch, A. and Krajicek, M.E. (1998). Mucoadhesive polymers for peroral peptide delivery: Synthesis and evaluation of chitosan-EDTA conjugates. *J. Control. Release,* 50, 215–223.

Bernkop-Schnürch, A. and Scerbe-Saiko, A. (1998). Synthesis and *in vitro* evaluation of chitosan/EDTA/protease inhibitor conjugates which might be useful in oral delivery of peptides and proteins. *Pharmaceut. Res.,* 15, 263–369.

Borchard, G., Lueßen, H.L., Verhoef, J.C., Lehr, C.-M., De Boer, A.G., and Junginger, H.E. (1996). The potential of mucoadhesive polymers in enhancing intestinal peptide drug absorption III: Effects of chitosan-glutamate and carbomer on epithelial tight junctions *in vitro. J. Control. Release,* 39, 131–138.

Brondsted, H. and Kopecek, J. (1992). Hydrogels for site-specific drug delivery to the colon— *in vitro* and *in vivo* degradation. *Pharmaceut. Res.,* 9, 1540–1545.

Busch, O., Solheim, E., Bang, G., and Tornes, K. (1996). Guided tissue regeneration and local delivery of insulinlike growth factor I by bioerodible polyorthoester membranes in rat calvarial defects. *Int. J. Oral Maxillofac. Implants,* 11, 498–505.

Capan, Y., Woo, B.H., Gebrekidan, S., Ahmed, S., and Deluca, P.P. (1999). Influence of formulation parameters on the characteristics of poly(D,L-lactide-co-glycolide) microspheres containing poly(L- lysine) complexed plasmid DNA. *J. Control. Release,* 60, 279–286.

Choi, N.S. and Heller, J. (1976). Poly(Carbonates). Alza Corporation, Mountain View, CA.

Choi, N.S. and Heller, J. (1978a). Drug delivery devices manufactured from poly(orthoesters) and poly(orthocarbonates). Alza Corporation, Mountain View, CA.

Choi, N.S. and Heller, J. (1978b). Structured orthoester and orthocarbonate drug delivery devices. Alza Corporation, Mountain View, CA.

Choi, N.S. and Heller, J. (1979a). Novel orthoester polymers and orthocarbonate polymers. Alza Corporation, Mountain View, CA.

Choi, N.S. and Heller, J. (1979b). Erodible agent releasing device comprising poly(orthoesters) and poly(orthocarbonates). Alza Corporation, US.

Chouinard, F., Kan, F.W.K., Leroux, J.C., Foucher, C., and Lenaerts, V. (1991). Preparation and purification of polyisohexylcyanoacrylate nanocapsules. *Int. J. Pharmaceut.,* 72, 211–217.

Couvreur, P., Roland, M., and Speiser, P. (1982). Biodegradable submicroscopic particles containing a biologically active substance and compositions containing them, USA Patent 4,329,332, 1982.

Crommen, J.H.L., Schacht, E.H., and Mense, E.H.G. (1992). Biodegradable Polymers. 2. Degradation Characteristics of Hydrolysis-Sensitive Poly[(Organo)Phosphazenes]. *Biomaterials,* 13, 601–611.

Cutright, D.E., J.D. Beasley, I., and Perez, B. (1971). Histological comparison of polylactic and polyglycolic acid sutures. *Oral Surg.,* 32, 165–173.

Dupuy, B., Arien, A., and Minnot, A.P. (1994). FT-IR of membranes made with alginate/poly-lysine complexes—variations with the mannuronic or guluronic content of the polysaccharides. *Artif. Cells. Blood Substit. Immobil. Biotechnol.,* 22, 143–151.

Einmahl, S., Zignani, M., Varesio, E., Heller, J., Veuthey, J.L., Tabatabay, C., and Gurny, R. (1999). Concomitant and controlled release of dexamethasone and 5-fluorouracil from poly(ortho ester). *Int. J. Pharmaceut.,* 185, 189–198.

Eldridge, J.H., Hammond, C.J., Meulbroek, J.H., Staas, J.K., Gilley, R.M., and Tice, T.R. (1990). Controlled vaccine release in the gut-associated lymphoid tissues. I. Orally administrated biodegradable microspheres target the Peyer's patches. *J. Control. Release,* 11, 205–214.

Engelberg, I. and Kohn, J. (1991). Physicomechanical properties of degradable polymers used in medical alications—a comparative study. *Biomaterials,* 12, 292–304.

Florence, A.T. (1997). The oral absorption of micro- and nanoparticulates: neither exceptional nor unusual. *Pharmaceut. Res.,* 14, 259–266.

Flory, P.J. (1953). *Principles of Polymer Chemistry.* Cornell University Press, Ithaca, NY.

Frazza, E.J. and Schmitt, E.E. (1971). A new absorbable suture. *J. Biomed. Res. Symp.,* 1, 43–58.

Fu, K., Pack, D.W., Klibanov, A.M., and Langer, R. (2000). Visual evidence of acidic environment within degrading poly(lactic-co-glycolic acid) (PLGA) microspheres. *Pharmaceut. Res.,* 17, 100–106.

Fujii, S., Yokoyama, T., Ikegaya, K., Sato, F., and Yokoo, N. (1985). Promoting effect of the new chymotrypsin inhibitor Fk-448 on the intestinal absorption of insulin in rats and dogs. *J. Pharm. Pharmacol.,* 37, 545–549.

Gabor, F. and Wirth, M. (2003). Lectin-mediated drug delivery: fundamentals and perspectives. *Stp Pharma Sci.,* 13, 3–16.

Grizzi, I., Garreuau, H., Li, S., and Vert, M. (1995). Hydrolytic degradation of devices based on poly(dl-lactic acid) size dependence. *Biomaterials,* 16, 305–311.

Hauser, H. and Wagner, R. (1997). *Mammalian Cell Biotechnology in Protein Production.* Walter de Gruyter, New York.

Heller, J. (1985). Controlled drug release form poly(ortho esters)—A surface eroding polymer. *J. Control. Release,* 2, 167–177.

Heller, J. (1990). Development of poly(ortho esters)—a historical overview. *Biomaterials,* 11, 659–665.

Heller, J., Ng, S.Y., Fritzinger, B.K., and Roskos, K.V. (1990) Controlled drug release from bioerodible hydrophobic ointments. *Biomaterials,* 11, 235–237.

Heller, J., Barr, J., Ng, S.Y., et al. (2000). Poly(ortho esters)—their development and some recent alications. *Eur. J. Pharmaceut. Biopharmaceut.,* 50, 121–128.

Jeyanthi, R., Mehta, R.C., Thanoo, B.C., and Deluca, P.P. (1997). Effect of processing parameters on the properties of peptide-containing PLGA microspheres. *J. Microencapsulation,* 14, 163–174.

Kompella, U.B. and Koushik, K. (2001). Preparation of drug delivery systems using supercritical fluid technology. *Crit. Rev. Ther. Drug Carrier Syst.,* 18, 173–199.

Laurencin, C.T., Koh, H.J., Neenan, T.X., Allcock, H.R., and Langer, R. (1987). controlled release using a new bioerodible polyphosphazene matrix system. *J. Biomed. Materials Res.,* 21, 1231–1246.

Lehr, C.M. (2000). Lectin-mediated drug delivery: The second generation of bioadhesives. *J. Control. Release,* 65, 19–29.

Lehr, C.M., Poelma, F.G.J., Junginger, H.E., and Tukker, J.J. (1991). An estimate of turnover time of intestinal mucus gel layer in the rat *in situ* loop. *Int. J. Pharmaceut.,* 70, 235–240.

Lherm, C., Muller, R.H., Puisieux, F., and Couvreur, P. (1992). Alkylcyanoacrylate drug carriers 2. Cytotoxicity of cyanoacrylate nanoparticles with different alkyl chain-length. *Int. J. Pharmaceut.,* 84, 13–22.

Lueßen, H.L., De Leeuw, B.J., Pérard, D., et al. (1996). Mucoadhesive polymers in peroral peptide drug delivery. I. Influence of mucoadhesive excipients on the proteolytic activity of intestinal enzymes. *Eur. J. Pharmaceut. Sci.,* 4, 117–128.

Lustig, S.R. and Peas, N.A. (1987). Solute and penetrant diffusion in swellable polymers. 7. A free volume based model with mechanical relaxation. *J. Alied Polymers Sci.,* 43, 533–549.

Madsen, F. and Peas, N.A. (1999). Complexation graft copolymer networks: swelling properties, calcium binding and proteolytic enzyme inhibition. *Biomaterials,* 20, 1701–1708.

Matsumoto, S., Kobayashi, H., and Takashima, Y. (1986). Production of monodispersed capsules. *Microencapsulation,* 3, 25–31.

McCaffrey, G. and Jamieson, J.C. (1993). Evidence for the role of a cathepsin D-like activity in the release of gal-beta-1-4glcnac-alpha-2-6-sialyltransferase from rat and mouse-liver in whole-cell systems. *Comp. Biochem. Physiol. B-Biochem. Mol. Biol.,* 104, 91–94.

Melmed, R.N., Elaaser, A.A.A., and Holt, S.J. (1976). Hypertrophy and hyperplasia of neonatal rat exocrine pancreas induced by orally administered soybean trypsin-inhibitor. *Biochimica Et Biophysica Acta,* 421, 280–288.

Merkli, A., Heller, J., Tabatabay, C., and Gurny, R. (1996). Purity and stability assessment of a semi-solid poly(ortho ester) used in drug delivery systems. *Biomaterials,* 17, 897–902.

Morishita, I., Morishita, M., Takayama, K., Machida, Y., and Nagai, T. (1991). Controlled release microspheres based on Eudragit L100 for the oral administration of erythromycin. *Drug Design Delivery,* 7, 309–319.

Muller, R.H., Lherm, C., Jaffray, P., and Couvreur, P. (1988). Toxicity of cyanoacrylate particles in L929 fibroblast cell-cultures–relation between toxicity and *in vitro* characterization parameters. *Archiv Der Pharmazie,* 321, 681–681.

Ng, S.Y., Vandamme, T., Taylor, M.S., and Heller, J. (1997). Controlled drug release from self-catalyzed poly(ortho esters). *Bioartificial Organs,* 168–178.

Odonnell, P.B. and Mcginity, J.W. (1997). Preparation of microspheres by the solvent evaporation technique. *Adv. Drug Delivery Rev.,* 28, 25–42.

Otsuki, M., Ohki, A., Okabayashi, Y., Suehiro, I., and Baba, S. (1987). Effect of synthetic protease inhibitor camostate on pancreatic exocrine function in rats. *Pancreas,* 2, 164–169.

Park, H. and Robinson, J.R. (1987). Mechanisms of mucoadhesion of poly(acrylic acid) hydrogels. *Pharmaceut. Res.,* 4, 457–64.

Passi, P., Zadro, A., Marsilio, F., Lora, S., Caliceti, P., and Veronese, F.M. (2000). Plain and drug loaded polyphosphazene membranes and microspheres in the treatment of rabbit bone defects. *J. Materials Science-Materials Med.,* 11, 643–654.

Peas, N.A. (1986). *Hydrogels in Medicine and Pharmacy.* CRC Press, Inc., Boca Raton, FL.

Pitt, C.G. (1990). Poly(ε-caprolactone) and its copolymers. In: Chasin, M., Langer, R., eds. *Biodegradable Polymers as Drug Delivery Systems.* Marcel Dekker, New York.

Plumpton, C., Kalinka, S., Martin, R.C., Horton, J.K., and Davenport, A.P. (1994). Effects of phosphoramidon and pepstatin-a on the secretion of endothelin-1 and big endothelin-1 by human umbilical vein endothelial-cells—measurement by 2-site enzyme-linked immunosorbent assays. *Clin. Sci.,* 87, 245–251.

Podual, K., Doyle, F.J., and Peas, N.A. (2000). Preparation and dynamic response of cationic copolymer hydrogels containing glucose oxidase. *Polymer,* 41, 3975–3983.

Ponchel, G. and Irache, J.M. (1998). Specific and non-specific bioadhesive particulate systems for oral delivery to the gastrointestinal tract. *Adv. Drug Delivery Rev.,* 34, 191–219.

Rees, D.A. and Welsh, E.J. (1977). Secondary and tertiary structure of polysaccharides in solution and gels. *Angewandte Chemie—International Edition in English,* 16, 214–224.

Saffran, M., Kumar, G.S., Savariar, C., Burnham, J.C., Williams, F., and Neckers, D.C. (1986). A new aroach to the oral administration of insulin and other peptide drugs. *Science,* 233, 1081–1084.

Scopelianos, A.G. (1994). Polyphosphazenes as New Biomaterials. In: Shalaby, S.W. (ed.) *Biomedical Polymers: Designed-to-Degrade Systems.* Hanser/Gardner Publishers, Inc., Cincinnati, OH, 153–171.

Seetharam, R. and Sharma, S.K. (1991). *Purification and Analysis of Recombinant Proteins.* Marcel Dekker, New York, 324.

Shameem, M., Lee, H., Burton, K., Thanoo, B.C., and Deluca, P.P. (1999). Effect of gamma-irradiation on peptide-containing hydrophilic poly (d,l-lactide-co-glycolide) micro-spheres. *PDA J. Pharmaceut. Sci. Tech.,* 53, 309–313.

Sparer, R.V., Shi, C., Ringeisen, C.D., and Himmelstein, K.J. (1984). Controlled release from erodible poly(ortho ester) drug delivery systems. *J. Control. Release,* 1, 23–32.

Spenlehauer, G., Vert, M., Benoit, J.P., and Boddaert, A. (1989). In vitro and in vivo degra-dation of poly(DL-lactide/glycolide) type microspheres made by solvent evaporation method. *Biomaterials,* 10, 557–563.

Takahashi, T., Takayama, K., Machida, Y., and Nagai, T. (1990). Characteristics of polyion complexes of chitosan with sodium alginate and sodium polyacrylate. *Int. J. Phar-maceut.,* 61, 35–41.

Tanigami, T., Ohta, H., Orii, R., Yamaura, K., and Matsuzawa, S. (1995). Degradation of poly[bis(ethylamino)phosphazene] in aqueous- solution. *J. Inorganic Organometallic Polymers,* 5, 135–153.

Vandorpe, J., Schacht, E., Stolnik, S., et al. (1996). Poly(organo phosphazene) nanoparticles surface modified with poly(ethylene oxide). *Biotechnol. Bioengineer,* 52, 89–95.

Veronese, F.M., Largajolli, R., Boccu, E., Benassi, C.A., and Schiavon, O. (1985). Surface modification of proteins: activation of monomethoxy-polyethylene glycols by phe-nylchloroformates and modification of ribonuclease and superoxide dismutase. *Alied Biochem. Biotechnol.,* 11, 141–152.

Veronese, F.M., Marsilio, F., Caliceti, P., De Filiis, P., Giunchedi, P., and Lora, S. (1998). Polyorganophosphazene microspheres for drug release: polymer synthesis, micro-sphere preparation, in vitro and in vivo naproxen release. *J. Control. Release,* 52, 227–237.

Wan, L.S., Heng, P.W., and W., C.L. (1992). Drug encapsulation in alginate microspheres by emulsification. *J. Microencapsulation,* 9, 309–316.

Wang, H.T., Palmer, H., Linhardt, R.J., Flanagan, D.R., and Schmitt, E. (1990). Degradation of poly(ester) microspheres. *Biomaterials,* 11, 679–685.

Yagi, T., Ishizaki, K., and Takebe, H. (1980). Cytotoxic effects of protease inhibitors on human-cells 2. Effect of elastatinal. *Cancer Lett.,* 10, 301–307.

Yeh, P.Y., Berenson, M.M., Samowitz, W.S., Kopeckova, P., and Kopecek, J. (1995). Site-specific drug-delivery and penetration enhancement in the gastrointestinal-tract. *J. Control. Release,* 36, 109–124.

# 12 Vaccines: Ancient Medicines to Modern Therapeutics

*H. O. Alpar, Ph.D. and M. J. Groves, Ph.D., D.Sc.*

## CONTENTS

# INTRODUCTION

## A Brief History of Vaccine Development

The concept that individuals who survive an infectious disease do not get infected a second time is as old as humankind. Thucydides recorded that in the Peloponnesian War (431 BC) survivors of the plague took care of the sick believing they would not get the disease again. We now know that this is due to our immune systems which recognize invading materials as foreign and organizes a defense against them. This will be discussed in the following chapter although it has to be admitted that not all of the details are clearly defined even today. Pliny the Elder reported in his encyclopedia of natural science that the Romans explored the use of the livers from dogs who had died from rabies in order to prevent the disease in man but this effort remains an antique curiosity.

Although two centuries have gone by since the investigations by Jenner in the 1790s into *Vaccinia* that began the modern phase of vaccine development, vaccination as such can be traced back much earlier.

Around 1716 Lady Mary Pierrepont, wife of the then British Ambassador to the Turkish Porte or Court in Istanbul, Sir Edward Wortley Montagu, and herself scarred from an earlier attack of smallpox, reported that Turkish village women exposed healthy individuals to scabs and pustules obtained from patients who manifested mild cases of the disease. Two points need to be made from this observation, namely that the disease can vary in intensity from patient to patient and that immunity can be transferred. Since she had also lost a brother to the disease she reported this observation in letters home and, when she eventually arrived back in London, she introduced the practice to society and had her own children successfully vaccinated.

It might be worth pointing out here that the Turkish practice probably originated in China centuries prior to this observation. Effectively what these ancients had discovered was the variability of the intensity of the infection (*virulence*), a key issue in the development of a safe vaccine.

Toward the end of the 18th century Edward Jenner, a country doctor, and coincidentally already vaccinated against smallpox by the then fashionable process introduced by Lady Pierrepont, had noted that there was a legend amongst country people that milk maids suffered a mild form of smallpox called cowpox that was nonfatal and did not leave the debilitating scars associated with smallpox itself. To his eternal credit Jenner decided to investigate this phenomenon and discovered that the legend was true. Cowpox protected against the more virulent smallpox and could be used as a vaccine. At the time the British medical establishment mocked these findings, but doctors across Europe followed through and confirmed that cowpox made an effective and safe vaccine to protect patients at risk from smallpox. Jenner is now regarded as the "Father of Vaccination." Later Louis Pasteur developed and tested clinically an attenuated form of rabies that was less virulent but cured the disease as opposed to only protecting against subsequent infection. In 1901 von Behring received the first Nobel Prize in Medicine for his discovery of what would come to be known as *antibodies.*

Although *Vaccinia* was eliminated from the international scene as recently as two decades ago, terrorism fears have generated renewed interest in the large-scale protection of an unprotected population against the disease, which has occasionally reappeared as a result of laboratory accidents and, in some cases, of deliberate dissemination of the virus. This is one example of the primary need for protection against the disease and for suitable vaccine delivery systems.

## BCG—THE ONLY TUBERCULOSIS VACCINE

The vaccine known as BCG, or Bacille Calmette-Guérin, is officially defined as a living culture of an attenuated form of *Mycobacterium bovis,* an organism similar to *Mycobacterium tuberculosis* which is the leading cause of tuberculosis in humans. The vaccine was developed at the turn of the 20th century as part of an attempt to halt the spread of tuberculosis in European industrial cities. This was, and remains, a significant disease that caused a great deal of human misery but, at that time, it was also becoming a significant factor in slowing the economic revolution then being experienced by the Europeans.

Little was known about the cause of the disease up to the last quarter of the 19th century but tuberculosis had become known as the "White Plague" and was a major cause of premature death in the working classes. However, it was by no means confined to the poor and there was little that could be done to stop the spread of the disease except to isolate the patients. Treatment consisted of providing good nutrition and sometimes heroic surgical procedures like collapsing or removing lungs. The clinics or sanitaria were often located in isolated areas such as the tops of mountains. To get a good idea of what conditions were like at that time, the novel *The Magic Mountain* (1924) by Thomas Mann provides an excellent description of the contemporary treatment.

The causative organism was isolated by Robert Koch in 1882 and named *M. tuberculosis* although, at the time, this observation proved controversial. Koch went on to claim that a sterile filtrate of a growing virulent strain of the tuberculosis organism could act as a vaccine against the disease. Unfortunately this claim proved to be invalid, although the filtrate ("Old Tuberculin") became a valuable diagnostic agent, producing a marked dermal reaction in a patient who had tuberculosis antibodies even if clinical signs of the disease were not present.

Spread mainly as an airborne droplet infection, the disease remains prevalent in overcrowded communities, such as those found in industrial cities of the time. Today the disease has often become associated with AIDS. In France the disease was endemic among the working classes of the city of Lille. In 1894 the physician Albert Calmette was sent from the Institut Pasteur in Paris to set up a branch laboratory devoted to studying the disease in the community and, if possible develop a vaccine against it. Calmette was mainly interested in the social impact of the disease and spent much of his time organizing clinics and obtaining relief for the factory workers affected by the disease.

He was joined in 1897 by Camille Guérin, a microbiologist. In 1908 they discovered accidentally that the addition of a bile extract to a growing culture of tuberculosis isolates allowed the hydrophobic cells to be dispersed more evenly, thus allowing reproducible samples to be taken more readily from the growing culture. Using an extremely virulent bovine strain of tuberculosis (the original culture has been lost so the designation may not be correct) they started what turned out to be a long process of progressively affecting the virulence of the organism by passage. The organism is very slow growing so this process involved growing a culture for 3 weeks and placing a sample in a fresh sterile broth which was then grown for another 3 weeks and so on. It took a total of 13 years and 231 transplants for the organism to lose its virulence, initially to a series of different animals such as mice, rats, and guinea pigs. Eventually the investigators became convinced that it was safe to test in humans. It is interesting to note that virulence of this organism has never been restored. The immunological protection, although it has never been lost, has been affected sometimes by cultural conditions. Another interesting observation about this work is that the vaccine was initially administered orally and was proven effective by this route. The dermal scarification procedure now used was only introduced into medicine much later.

Like all other crowded industrial manufacturing centers, Chicago was badly affected by tuberculosis, especially among the poorest members of the community. There was a small laboratory in Cook County Hospital (the local hospital in Chicago set up in 1835 for the poor community that still serves this function) under the direction of Dr. Frederick Tice who was studying the disease in the 1930s. The main efforts were devoted to controlling the spread of the disease by setting up and maintaining a sanitarium south of Chicago. However, Dr. Tice sent a young doctor, Sol Roy Rosenthal, to study at the Institut Pasteur from 1933 to 1934; in particular, to study the new and controversial BCG vaccine as part of his doctoral studies. Dr. Rosenthal returned and brought back with him samples of the Pasteur BCG vaccine given to him by Guérin for further study. This was part of an enlightened policy adopted by the Institut Pasteur to allow the attenuated organism to be both freely and readily available to all countries.

Scientifically this may have had some unforeseen effects in the fact that each national laboratory began to grow the vaccine under different cultural conditions from those laid down by the originators. The net result was that a number of different BCG vaccines, each identified by its country of origin, began to appear and, with hindsight, these different cultural procedures resulted in vaccines with different potencies.

Dr. Rosenthal was aware of this and, in fact, by 1950 had developed his own BCG vaccine which he claimed was more potent than most of the other brands then available. To show his gratitude he named the improved vaccine after his mentor and it is still known as the Tice strain BCG. In all fairness, in many tests undertaken since that time the Tice vaccine has usually been demonstrated to be superior to most of the other sources of vaccine, with the possible exception of the original Pasteur strain. In the United States the use of BCG vaccination in the community proved to be different from almost every other country. Rosenthal encountered an entrenched medical opinion, especially with members of the U.S. Public Health Service who assumed a defeatist attitude toward BCG. The medical establishment had decided by the late 1930s, based on some incorrect or otherwise unconvincing evidence, that BCG was unreliable and too dangerous to be used on the American public. Rosenthal spent most of his working life trying to convince them otherwise. He organized clinical trials of the vaccine on a large scale on the south side of Chicago, then a major industrial hub associated with meat processing. These trials are interesting because he managed to organize follow-up examination of the patients, in some cases for as long as 25 years, and, in the process, made some interesting parallel discoveries. He may have been among the first to report in 1936 that BCG was a potent stimulant of the reticuloendothelial system, finding that neoplasms could be suppressed by BCG. For example, an unexpected observation from the tuberculosis trials in Chicago involving over 1500 patients was that diseases such as leukemia and soft tissue tumors were significantly suppressed among the patients treated with BCG, quite apart from a reduction in the development of active tuberculosis.

The significance of this observation has only been appreciated more recently. Rosenthal wrote a book in 1957 called *BCG Vaccination Against Tuberculosis*. The second edition of 1980, substantially rewritten from the first, was called *BCG Vaccine: Tuberculosis–Cancer,* and this simple change graphically illustrates the change in emphasis in the use of BCG. By 1960 the menace of tuberculosis was perceived as being overcome by the use of antibacterials such as streptomycin and isoniazide, and the need for a tuberculosis vaccine was generally considered to be less relevant to the needs of the time.

However, since then there has been a major resurgence of drug resistant-tuberculosis, associated with the AIDS pandemic, and this has caused increasing concern over the past decade. BCG vaccination of medical and nursing personnel involved in these areas has now become an accepted precaution since there is no other approved vaccine available and any acellular vaccine, for example, will probably not receive regulatory approval for at least another decade, if at all.

The Tice substrain of BCG is still the only licensed source of the vaccine in the United States, although manufacturing is now carried out in North Carolina. The fresh liquid form of the vaccine, with a shelf life of only 10 days, was later improved by freeze drying the product, extending the shelf life to 18 months.

## REQUIREMENTS OF AN IDEAL VACCINE FOR TODAY

Before discussing the way in which the immune system functions and how vaccines can work, it might be a useful exercise for the student to consider what an ideal vaccine should consist of. For example, most vaccines today are administered parenterally and this can sometimes be a painful or distressing experience, especially for young children.

At the 1990 International Task Force for Vaccine Development meeting in New York, a children's vaccine initiative (CVI) was set up with the aim of developing an ideal vaccine. The following criteria were established although the list may not be totally inclusive.

1. A vaccine should only require a single dose.
2. A vaccine should be given early in life.
3. The route of administration should be nonparenteral.
4. Vaccines should be combined in order to reduce the number of visits to a doctor or medical center.
5. Vaccines should be heat stable and retain activity during transport and storage, especially in tropical climates.
6. Vaccines need to be developed against diseases with high mortality rates, such as AIDS, pneumonic plague, acute respiratory infections, diarrhea, and parasitic diseases such as malaria.
7. And, above all, the cost must be low throughout the world.

The ultimate vaccines in the future will be required to fit these criteria and it is not difficult to see that the cost may be the biggest problem.

In the 15 years since these criteria were promulgated by declaration, it will be evident that most vaccines are still administered parenterally with the exception of polio and typhoid vaccines. In many ways this can be attributed to the physicochemical characteristics of vaccine antigens themselves, which are large molecules susceptible to proteolytic degradation, denaturation, and rapid clearance from plasma. Some combination vaccines are available which reduce the number of injections. However, the MMR (measles, mumps, and rubella) combination vaccine has gained an unsafe image in the popular press, mainly due to a reputed link with autism in some children that as yet remains unproven scientifically. In some quarters the autism was associated with the use of thiomersalate as a mercurial preservative in multidose injections but, again, this supposition remains unproven.

The controversy has reappeared with the introduction of a five component children's vaccine containing diphtheria, polio, measles, mumps, and rubella; although this should be much more convenient and contains a killed polio instead of an attenuated virus which is known to occasionally revert to the active form, albeit in single numbers per million injections. In this case the children's vaccine should be safer and more convenient.

Throughout this chapter the reader should bear in mind the possibility that modern biotechnology, the subject of the book as a whole, is making progress toward the ideal criteria outlined above.

## TYPES OF MODERN VACCINES

Modern vaccines can be classified into groups according to whether the organism is alive or dead or whether the vaccine is prepared from naturally occurring fragments or synthetically derived components (Figure 12.1).

**Natural or modified microorganisms**

|  |  |
|---|---|
| | killed whole cells |
| attenuated strains | purified (detoxified) toxins |
| rabies, M. tuberculosis | purified antigens |
| (BCG) | purified adjuvants |

**Live** ─────────────────────────────────────── **Inert**

|  |  |
|---|---|
| recombinant attenuated | recombinant antigen |
| recombinant virus | peptide subunit (epitopes) |
| recombinant commensal organisms | recombinant detoxified (mutant) antigen/adjuvant |
| recombinant phage | synthetic adjuvants |
| | non-particulate antigen and adjuvant mixtures |
| | particulate antigen in or on synthetic carrier |
| | micro-encapsulated killed virus |

**Synthetic or bioengineered products**

**FIGURE 12.1** Division of current vaccines.

## ATTENUATED LIVE VACCINES

Attenuated live vaccines are less common than they were a century ago, examples being the original Pasteur rabies vaccine and BCG, but the recent approval of a nasal influenza vaccine, FluMist® (MedImmune Vaccines, Inc., Gaithersburg, MD) shows that the approach remains valid.

Other examples include polio, mumps, and rubella vaccines. The attenuated polio vaccine can be given orally and is known to occasionally revert to an active form which does represent a slight but measurable issue with this vaccine.

The attenuated mumps, measles, and rubella viruses are usually administered in one combined vaccine, MMR. The three viruses are grown separately, lyophilized with various cryoprotectants such as sorbitol and amino acids or hydrolyzed gelatin, and combined in a final pack, usually with neomycin as a preservative. The combined vaccine has to be kept refrigerated prior to use.

Some viruses and bacteria have been explored as genetically engineered organisms to express appropriate peptides or proteins or even specific epitopes to raise the protective immunity of the entire organism against aggressive pathogens.

## KILLED INACTIVATED VACCINES

To some extent the simplest vaccines are those in which the bacterial or viral pathogens has been killed by chemicals or heat so that they themselves cannot cause disease but can confer protection against invasion. This has been used, for example, for the combined diphtheria-tetanus-pertussis vaccine. In this case there were concerns about the presence of entire cells that could cause complications other than the needed protection. Recently acellular systems have been introduced which may be more effective and safer. This concept of using only the fragments of an invading organism that cause the problems is attractive because very often side effects due to other components of the intact organism are avoided. Such components may be proteins or fragments of protein, carbohydrates, lipids, or even DNA fragments, but while they are often effective, their immunological potential may well be weak. This emphasizes the need for adjuvants to increase their effectiveness and this will be dealt with in a later section.

## CONJUGATE VACCINES

Conjugate vaccines combine toxoids with polysaccharides and attempt to train the immune system to recognize the polysaccharides as being foreign. Young children cannot react to pathogens covered with polysaccharide and Hib conjugates have lowered the infection rate in children from 1:60 to 1:100,000, a worthwhile achievement.

Combination vaccines are more convenient in use and are exploited in the familiar DPT and MMR vaccines. However, there is always the possibility of autoimmune reactions and an increased risk of side effects. As noted, these issues have been obscured legally by claims that autism in children is caused by such combinations or the use of thiomersalate as a mercurial preservative. Scientifically it is probably safe to say that these side effects have not been demonstrated convincingly but these issues have caused difficulties for the manufacturers.

DNA vaccines are being explored but it has proved to be difficult to deliver the naked DNA to cellular sites where the appropriate antigen will be produced. Gold particles coated with DNA plasmids have been evaluated. Naked DNA is wasteful in some senses because relatively large quantities are required for even small quantities of translation and adjuvants are required. In addition there is no protection against environmental DNAase enzymes. However, DNA and the appropriate protein antigen may have more potential.

## SUBUNIT VACCINES

In the early 20th century it was recognized that some components of a microbacterial cell were more important than others for protection and thus came the concept of subunit vaccines. When combined with the discovery that bacterial toxins could be inactivated with formaldehyde, the result was the introduction of a diphtheria subunit vaccine in 1923 and a tetanus subunit vaccine in 1927.

Modern subunit vaccines contain one or more selected antigenic subunits that have been found to provide protection against a particular pathogen. They are better defined from a physicochemical aspect and have fewer side effects than vaccines, which contain intact cells, whether inactivated or attenuated. Current subunit antigens include viral and bacterial proteins as well as bacterial capsular polysaccharides.

Toxins from *Corynebacterium diphtheriae* or *Clostridium tetani* are water soluble proteins, which effectively constitute the respective vaccine antigens. However, they are treated with formaldehyde to eliminate or reduce the associated toxicity to form *toxoids*. It is also necessary to preserve the epitopes responsible for antibody formation and the process is usually a balance between the two requirements.

Toxoids generally have a poor immunogenicity and they are usually adsorbed onto aluminum salt suspensions which act as adjuvants.

Genetic manipulation of *B. pertussis* has resulted in acellular toxoids which are claimed to be devoid of toxicity, and acellular vaccines have been introduced in the United States and Japan for the treatment of older children. Trials have demonstrated that the efficacy of acellular vaccines is comparable to whole-cell vaccines but have virtually no side effects.

The genetic engineering of hepatitis B subunit vaccine in yeast cells has resulted in a subunit vaccine replacing the conventional whole cell vaccines obtained from the plasma of infected humans. The main advantages of recombinant DNA (rDNA) vaccines when compared with human plasma products is that they offer higher yields, are of more consistent quality, and are safer, thereby being easier to produce and cheaper.

## EMERGING VACCINE TYPES

### PROTEIN VACCINES

The majority of pathogenic antigens in nature are proteins with specific biological functions. For example, the human immunodeficiency virus (HIV) glycoprotein (gp) 120 binds onto the surface receptor cluster designation 4, CD4, on leukocytes to facilitate entry of the virus into the cell. The antigen 85 protein complex of *M. tuberculosis* is needed for the synthesis of factors that mediate bacterial cell wall integrity and immunomodulation. In vaccine manufacture pathogen proteins are either purified from the pathogen itself or are synthesized by recombinant methods.

However, in practice, it is generally found that protective immunity is rarely provided by just a single component from the pathogen, multiple components being needed as in the DPT vaccine. For regulatory reasons it is usually better to explore single proteins or protein complexes. Unfortunately, with a few exceptions such as the *Bacillus anthracis* protective antigen (PA) and the urease enzyme from *Helicobacter pylori,* most single protein antigens tend to have weak immunoprotective activity and require the use of *adjuvants* or immunostimulators (see later). In addition, without these additional materials protein antigens tend to induce a type 2 immune response, which may not always provide adequate protection, especially in the case of intracellular pathogens.

### DNA VACCINES

Outside the remit of this present chapter, it might be noted that a significant number of DNA sequences have had no identifiable function and were initially labeled as "junk." However, gradually some functions for this so-called junk DNA have been identified, especially those of a regulatory nature. This has resulted in a greater interest in the possibility of providing DNA sequences that synthesize antigenic proteins.

The issue was one of getting the DNA into a cell and this gave rise to the use of viral delivery systems and bacterial plasmid DNA (pDNA) vectors. pDNA vectors are useful since they are much safer to manufacture on a large scale, inexpensive, and easily customized—apart from being safer for the patient. However, the down side is that these systems are poorly immunogenic although, as noted above, this might be improved by the addition of an appropriate adjuvant.

In the body, expressed proteins are processed through the major histocompatibility class 1 systems which generally results in a type 1 immune response and this is usually effective against intracellular pathogens. In the case of the transgenic product eliciting a type II response the expressed protein is taken up by phagocytes and antigen presenting cells.

One of the main safety aspects of using a pDNA vaccine is that, unlike the viral delivery systems, these cannot replicate and infect the host. In addition, it does not combine with the host genome which could otherwise lead to a potential cancerous reaction. Trials of so-called naked DNA suggested that a reduced cellular residence time may have been responsible for a poor protective response but this may also have been due to other physical and biological barriers.

Loss of vaccine by any first-pass effects could be minimized by avoiding the intravenous route and careful selection of mucosal routes would be of benefit. Indeed, unlike small molecular weight drugs, vaccines are not administered through the IV route and alternatives are required.

Retention times within the cell may be increased through the use of nuclear retention signal peptides and this with other approaches are certain to be tried in the immediate future.

## LIPID AND CARBOHYDRATE ANTIGEN VACCINES

As will be discussed in a later section, adaptive immune responses are mediated through the major histocompatibility complex (MHC) class I or II antigen presentation pathways. However, some MHC-independent antigen presentation pathways have been identified for pathogenic lipids and carbohydrates. Antigens presenting leukocytes express the CD1 molecules in a structure similar to that identified in the MHC class I, except that there are two deep cavities that create a hydrophobic environment suitable for lipids, glycolipids, or lipoproteins. Two groups of CD1 molecules elicit the production of type I cytokines such as interferon (through the group 1 CDI pathway) that can recognize mycobacterial lipid/glycolipid antigens which lyse infected dendritic cells and secrete bactericidal cytokines. Since group 2 CDI molecules have been detected in the human gastrointestinal epithelial cells, this has some implications for the viability of oral vaccines.

## RECOMBINANT LIVE CARRIERS

The main invasive pathogens are bacteria and viruses and recombinant live carriers using bacterial and viruses have been described. Viral carriers rely on the established and efficient methods for invading and infecting eukaryotic cells and their *in vivo* replicative process improves the induction of type I and type II immune responses.

These are discussed elsewhere but include a recombinant vaccinia virus that expresses rabies virus glycoproteins used for oral vaccination.

## THE IMMUNE SYSTEM AND MECHANISMS OF ACTION

Although some aspects of the human immunological system remain obscure and beyond direct explanation, much of this very subtle protective mechanism has been elucidated and used.

Many diseases in humans and the animal kingdom are caused by invading pathogenic microorganisms that proliferate and disturb the normal physiological functions of the intact body. There has been a tendency for the medical community to label microorganisms as pathogenic or nonpathogenic, but the sad truth of the matter is that almost all bacteria and viruses in the wrong place at the wrong time will manifest some type of disease. It is also true that some organisms are much more aggressive than others so that only a small number of anthrax cells, as an example, introduced into an animal or human will often result in various forms of the disease. This concept of aggressiveness or pathogenicity is therefore easily understood. Some microorganisms vary in their effect, as noted above, but here there may be two causes for this: (a) the organism itself losing the ability to affect the body, and (b) the body being able to resist the invasion by some mechanism.

The story is told of Mithradates, the 1st century BC king of Pontus, who was so afraid of being poisoned that he immunized himself by taking increasing doses of known poisons in order to develop a natural resistance to them. It is this concept that underlies vaccines and their mode of action. The body is "vaccinated" against a disease causing microorganism so that, after a first infection it is able to resist any subsequent invasions. Here the first invasion occurs in the form of a highly controlled infection in the form of a vaccine which may consist of an attenuated bacterium or one that is modified in some way to produce some symptoms of the disease but, in the process, stimulates the body to defend itself.

This is the concept. The questions remain as to how and why this happens, a subject that has puzzled scientists since Jenner's time. The subject has become known as immunology and in recent years following the elucidation of the gene coding and the way proteins are synthesized in the body has taken up a significant degree of current scientific effort.

Considering a theoretical pathogen consisting of a bacterial cell attached to some surface such as a blood cell, on its own it is unlikely that anything will happen. However, the pathogens may grow and start to release material from their cells (*toxins*) which interfere with the normal physiological functioning of the body—the "disease." These secretions, the bacterial toxins, are often countered by the body producing *antitoxins* which adsorb or neutralize the toxic components. If there are few pathogenic cells this will often be sufficient to protect the body but the system also has "memory" so that if the same type of invasion occurred again a successful defense could be readily mounted. The pathogen may then make some slight changes in the design of the toxins so that the body now has to reinvent itself and come up with new antitoxin or *antibody* molecules in order to renew the defense. The process is ongoing and we are all subject to the vagaries of bacterial combat throughout our lives.

## ANTIGENS

The materials capable of stimulating the lymphoid tissues to produce antibodies are termed *antigens* and comprise bacterial and viruses as well as some smaller molecular entities. However, the response is not to the intact organisms but rather to some specific parts which have characteristic three-dimensional structures, the *epitopes*, and this sensitivity to structure is a characteristic feature of the immune response. Once an animal is in contact with an epitope the response can be in the circulatory or *humoral* system or directly as a cell-mediated response, but it is exquisitely sensitive to the specific antigen and rarely to any other.

Antigens themselves consist of two classes, the first being substances which are capable of generating an immune response on their own—*immunogens*—and smaller molecules that can react weakly with antibodies, so-called *haptens*. However, if haptens are attached to larger molecular moieties they begin to function as immunogens.

## ANTIBODIES

Antibodies remove antigens by binding directly to the three-dimensional epitope. Cell-mediated responses are against sites within antigens. The main effectors cells are the T-lymphocytes which have T-cell receptors capable of binding antigens when presented to them. The T-cell receptor is able to recognize both the antigen fragment and the structure to which it is bound, effectively being able to recognize small pieces of foreign molecules bound to the host-cell surface molecules. There appears to be a molecular weight cut-off for immunogenic foreign entities of greater than approximately 5 kDa. Haptens with molecular weights below this limit usually need to be attached to larger molecular weight entities or form complexes with tissue proteins acting as carriers, which would then have the overall required properties.

Epitopes on invading organisms cause the formation of corresponding antibodies, often more than one epitope being present on the organism and each responsible for its own antibody. The overall and characteristic epitope shape can be formed from a single segment of the antigenic molecular moiety or it can be formed in three dimensions by folding of the molecule in its native environment. The interacting section of the host antibody or T-cell receptor is called the *paratope* and forms part of the terminal groups of the corresponding *immunoglobulin*.

## IMMUNOGLOBULINS

Antibodies or immunoglobulins (these terms are interchangeable) are *glycoproteins* which bind specifically to the antigens that induced their formation and each one is formed as a unique response to that particular antigen. In the body they are present in body fluids and certain types of cell. Serum, the fluid that separates when blood is allowed to clot, contains no cells but it does contain the immunoglobulins which can be separated electrophoretically. They have different molecular weights and charge, and there are five basic classes or isotypes: IgG, IgA, IgM, IgD, and IgE. Of these, IgG is the most common by weight and has a molecular weight of around 160 kDa. There is a pentameric variant, IgM, which has a molecular weight of around 970 kDa; and the rest have molecular weights of 160-188 kDa and are present in varying amounts in normal human serum.

Structurally antibody molecules are similar in that they have the same basic structure consisting of two *light* chains and two *heavy* chains lying together in the shape of a Y. The chains are help together by disulfide bridges and some noncovalent interactions. Along the molecules some areas are conserved and others are totally unique. Different carbohydrates are attached to the main molecule which determines some of the subsequent properties of the antibody. The most common antibody in serum is IgG but it has at least four subclasses in humans. It is the major antibody of the secondary immune system and is found in both serum and tissue fluids. With all immunogens the antigen-recognizing paratope is contained in the so-called Fab end of the molecule and, as noted, the remainder of the molecule is the effector portion. Because the reaction between the epitope and its corresponding paratope involves for the most part short-range noncovalent forces (in three-dimensions) the two must have close proximity and therefore close fitting structures in order to interact. This heightens the specificity of the reaction. (For a more detailed description, see Figure 4.1 and Chapter 4.)

## T-Cell Receptors

The T-lymphocytes carry a number of glycoproteins on their surfaces that are involved in antigen recognition. These molecules or receptors are responsible for the recognition of the specific *major histocompatibility complex* (MHC) and antigen complexes and will be different for every T-cell so there are mechanisms in place for diversity.

## Major Histocompatibility Complex (MHC)

The major histocompatibility complex (MHC) is part of the system that codes for molecules important in immune recognition, including graft rejection. MHC class I and II molecules present antigen fragments to T-lymphocytes. For example, class I molecules bind viral proteins and present them to the CD8+ T cells. Exogenous antigens such as proteins taken into the cell by endocytosis are processed within the cell and presented to CD4+ cells.

There are a number of MHC class I and class II polymorphic molecules, all with substantially the same basic structure. The MHC class I molecule is a dimer consisting of a glycoslylated transmembrane peptide of molecular weight 45 kDa covalently linked to a 12 kDa peptide; it is found on the surface of most nucleated cells within the body. It is believed that the polypeptide backbones fold in such as way as to form a platform of $\beta$-pleated sheet structures to support a peptide binding cleft in which the antigen fragments are held and presented to the T cells.

The class II molecules are similar but are held together by noncovalent links. These are less widely distributed, being mainly found on the surface of some cells

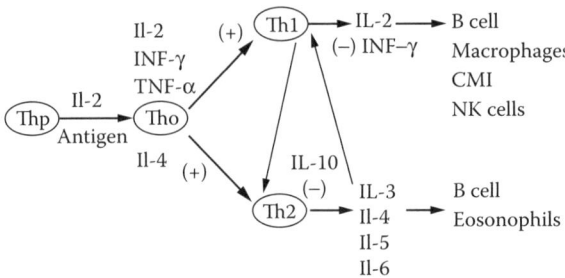

**FIGURE 12.2** Cytokines and cells activating Th1 and Th2 cells.

of the immune system such as β-lymphocytes, macrophages, monocytes, and activated T-lymphocytes. MHC antigens are essential for recognition by T-lymphocytes since they will only recognize epitopes when presented in the cleft of the MHC.

Naive CD4 T-cells need to be activated by MHC class II antigen and then start to secrete a wide range of cytokines including interleukins (IL) 2, 3, 4, 5, and 10 as well as interferon (IFN)( (Figure 12.2). These cells are then converted to Th1 (inflammatory T-cells) which secrete IL-2, IFN-γ, and tumor necrosis factor (TNF). Alternatively they can be converted to the Th2 cells which secrete IL-4, 5, 6, and 10. The Th1 pathway results in a powerful stimulation of macrophages to kill phagocytosed microorganisms and encourages other macrophages, lymphocytes, and neutrophils to come to the site of activation or invasion. The Th2 cytokines, on the other hand, activate B-cells that can differentiate into antibody-secreting cells. In some disease states the two Th1 and Th2 subsets can get out of balance. It has been suggested that autoimmune diseases, including rheumatoid arthritis, are due to an excess of Th1 activity with the associated cytokines inducing an inflammatory response.

Antigen MHC class I complexes have a different function and present their antigen to CD8+ cytolytic T cells, which produce an appropriate group of effector molecules leading to apoptosis of target cells.

## TYPES OF IMMUNE DEFENSE MECHANISMS

It will be evident from the above historical background that empirically it was appreciated that immunity to a disease could be both active and passive. Immunity was shown to be specific and nonspecific with two main responses, humoral (body fluids) and cellular.

**Passive immunity** only lasts a short while and occurs when the mother passes protective agents such as immunoglobulins to the child in, for example, breast milk. Tetanus immunization is another example since it only lasts 10 years. The Rhesus response where the first child immunizes the mother against other children is also well known. Passive immunity also has the potential to produce undesirable immune responses such as allergic reactions or anaphylactic shock.

**Active immunity** includes an efficient set of mechanisms to deal with infection and these are highly adaptable to situations as they arise.

**Systemic immunity** protects the blood and organs and interior tissues of the body. Antibodies and specialized cells circulate throughout the body looking to destroy foreign cells or tissues. A recognition pattern is remembered for subsequent invasions by the same organisms.

**Innate or nonspecific immunity** has probably developed during evolution and includes anatomical barriers such as the skin, cilia in the lungs, and bronchial tubes, as well as the presence of specific components in specialized tissues that combat invasion.

**Adaptive or specific immunity** is generally slower to respond but recognizes nonself molecules or invaders and is capable of destroying them if they have particular molecules on their surface.

**Mucosal immunity** represents the first line of defense against invasion of viruses and bacteria by utilizing immunoglobulins and other materials as antibodies. This will be discussed later.

Other materials that assist in resisting invasion include:

**Fibronectin (Fn)** is ubiquitous throughout the body and serves a number of functions, including coating bacterial and foreign particles with *apoprotein* which promotes recognition and destruction by circulating cells of the immune system.

**Lysozyme** is an enzyme that occurs in tears and serum and breaks down bacterial cell walls.

**Interferon** is capable of inhibiting viral replication processes and activates cells that kill pathogens.

**Tumor necrosis factor (TNF)**$\alpha$ is capable of suppressing viral replication and also activates phagocytes.

**Transferrin** and **lactoferrin** deprive organisms of the trace quantities of iron needed for metabolism.

## THE MUCOSAL SYSTEM

In recent years it has become evident that many pathogens invade the body through the mucosal system, which in itself appears to provide a first line of defense against invasion. All mucosal surfaces are accessible to pathogens, which are effectively living particles, but if these can access the body through this route then it follows that particulate vaccine carriers could also enter by the same means. This site is an attractive alternative to parenteral administration of vaccines because it is relatively large in surface area (some 400 m$^2$ as opposed to 2 m$^2$ for the external skin), although not all of it is directly accessible. It includes the nasal passages, eye lids and surrounding tissues, mouth and lungs, the whole of the gastrointestinal tract, and the vagina. These surfaces are of interest for direct drug delivery since particulate systems can be delivered to specific targets and in many cases made to adhere to the mucosa.

The net effect is that mucosal administration of a vaccine is an attractive and more effective alternative to parenteral administration which, in any case, tends to only provide systemic protection.

## ASSOCIATED LYMPHOID TISSUES

The common mucosal immune system (CMIS) is now well established as a separate component of the host's immune apparatus, quite distinct from and independent of the systemic immune system described above. Moreover, if an immune response is induced at one site in the mucosal system this generally leads to responses at distal mucosal sites of the CMIS, presenting a potentially large advantage. It should be noted that there are approximately $6 \times 10^{10}$ antibody producing cells in mucosal tissues and $2.5 \times 10^{10}$ lymphocytes in the entire lymphatic system.

Needless to say, there are issues associated with mucosal vaccine delivery, including in some places a harsh environment. One example would be the low pH of the stomach and upper part of the gastrointestinal (GI) tract and the widespread availability of proteolytic enzymes, especially along the whole of the GI tract. There is the potential to develop tolerance to materials delivered orally although the oral route is considered to be the safest of all. The nasal route, on the other hand, is currently under investigation in a number of laboratories and appears to be safe, although there is some evidence of antigen transfer to neuronal tissues through the olfactory bulb in mice.

## MUCOSAL-ASSOCIATED LYMPHOID TISSUES

Mucosal-associated lymphoid tissues (MALT) differ in various sites as follows.

Nasal-associated lymphoid tissues (NALT) consist of lymphoid follicles with overlying ciliated epithelium to sweep the mucus along the site, mucus goblet cells producing the mucus and numerous membranous or microfolded cells, the M cells (Figure 12.3 and Figure 12.4). These organized cellular structures are found at the entrance to the nasopharyngeal duct in the mouse but have also been described in the human proximal nasal passage. Nasal administration demonstrates higher permeability

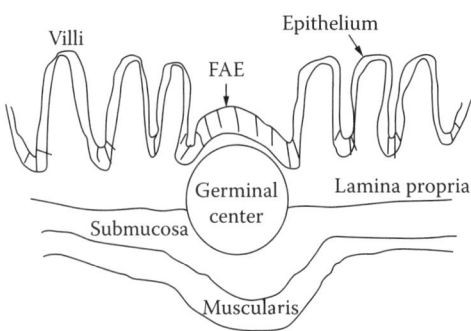

**FIGURE 12.3** Follicle-associated epithelium (FAE).

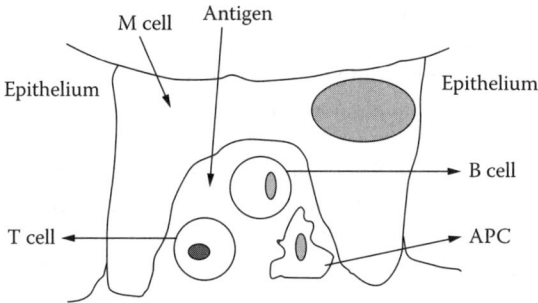

**FIGURE 12.4** Movement of antigen through the M cell.

than other mucosal sites and offers an alternative site for vaccine delivery. The nasal cavity has a large and readily accessible mucosal surface. There are fewer challenges to absorption and a milder environment than that experienced while traveling down the GI tract. In addition, any material passing through the nasal wall is distributed systemically without first passing through the liver, avoiding the first pass effects associated with most drug absorption pathways. Nevertheless, the exact mechanisms for drug or vaccine absorption through the nasal mucosal wall remain at present obscure, especially in humans.

Gut associated lymphoid tissues (GALT) contain the so-called Peyer's patches which are organized lymphoid follicles with overlying M cells and are considered to be the main entry point for particulate matter during passage down the GI tract (Figure 12.3 and Figure 12.4).

Bronchus associated lymphoid tissues (BALT).

Rectal associated lymphoid tissue (RALT).

The mucosal system has been shown to form direct protective mechanisms for the body as a whole against pathogenic invasion. These include the secretion and movement of mucus along the tissue concerned ("trafficking"), the secretion of stomach acids, enzymatic degradation, peristaltic movements, and the presence of tight junctions. However, it might also be argued that the main function of these is in the digestion of food, especially from the gut, but food is often absorbed as small particles or, in the case of lipids, droplets. If this process could be mimicked by vaccine delivery systems, this would be a clear advantage. However, the mucosal system also has its own immunoprotective action irrelevant to food processing.

Mucosal immunization prevents pathogens from infiltrating or infecting the body whereas systemic immunization resolves an infection after the invasion has occurred, thereby suppressing the disease. There is also clear evidence that both systemic and mucosal immunity is induced by mucosal immunization. As noted earlier, there is also evidence that immunization at a single site in the body can result in protection of the entire mucosal system so there must be some form of communication between sites; and this justifies the term common mucosal immune system.

Mucosal immunity is divided into two main components, the *inductive phase* and the *effector phase*. In the inductive phase antigen is presented which results in lymphocytes being primed and moved from their inductive site to the lymph nodes and the circulating blood. This may represent the means by which the distant mucosal components communicate and allows the lymphocytes to reach the effector phase sites. Here the secretory immunity, sigA, is induced by the production of appropriate antibodies. In the intraepithelial CD8+ T-cell mediated immunity and the CD4+ T-cytokine production occurs in the *lamina propria* cells of the intestine.

M cells are potentially important targets for vaccine delivery and are found in all inductive mucosal sites. Some pathogens are capable of entering the body through these cells whose primary function is the sampling of luminal antigens and moving antigenic material to underlying lymphoid tissues, *transcytosis*. They are characterized by their lack of brush border microvilli and there is no mucus secreted although, in the GI tract, for example, there is slow movement of mucus across their surface. They are also capable of efficiently endocytosing adherent micromolecules and particles, dead or alive.

## VACCINE ADJUVANTS

Although there are a number of advantages associated with the use of subunit vaccines (e.g., highly purified peptides, proteins or DNA) as vaccines (e.g., specificity), one feature they all have in common is that they are generally poorly immunogenic. The more traditional vaccines contain many other components, some of which elicit additional T-cell assistance or function as adjuvants. An adjuvant is a substance that acts as an immunostimulator, one example being the bacterial DNA in a whole cell vaccine. The overall result is a more robust immune response than that provided by the antigen alone.

Adjuvants improve the antigenic response by a number of mechanisms, such as

- Increasing the immunogenicity of weak antigens
- Enhancing the speed and duration of the immune response
- Modulating the antibody activity
- Stimulating the cellular mediated immunity
- Promoting the induction of mucosal immunity
- Enhancing the immune response in immunologically immature individuals
- Reducing the dose of an antigen required for a response
- Increasing safety and reducing production costs

A wide variety of materials have been explored for their adjuvant activity, although not all are equally effective or nontoxic, especially in humans. Alum and other aluminum salts were first recognized in 1926 and remain the most effective agents licensed for human use by the FDA, although some French products also use calcium phosphate. However, in recent years it has become evident that new and improved vaccine adjuvants are needed.

Although widely approved and effective as adjuvants, alum and other aluminum salts do have some issues since they require relatively large quantities of antigen, which requires repeated dosing, are nonbioadhesive, cannot elicit cell-mediated immunity, and require constant refrigeration. In addition, alums are not effective by the mucosal route as adjuvants and there are concerns relating to the production of IgF when alum is used. These are becoming more relevant as vaccines for use in tropical underdeveloped nations become of increasing concern. There is now a movement toward heat-stable, single-dose vaccines and this may be achieved by using microparticle formulated vaccines.

Until very recently the development of adjuvants has remained substantially a trial-and-error process. This probably accounts for the wide variety of materials described as being suitable for the purpose.

Bacterial DNA, long a component of the earlier whole cell vaccines, has been shown to have an immunostimulatory effect on immune cells and is a potent inducer of cytokines such as IL-1, IL-6, and IL-12. Monophosphoryl A, a component of mycobacterial cell walls, reacts with receptors on antigen producing cells and generates a Th1 response due to the production of IL-2 and IFN-$\gamma$.

Perhaps one of the most potent immunostimulants is the Freunds Complete Adjuvant (FCA). This consists of a water-in-oil emulsion of killed mycobacteria in mineral oil. The exact mode of action is uncertain although it appears to be connected with the way in which the mycobacterial cells are presented to the surrounding tissues when injected. The reaction is rapid and devastating to the point where it cannot be used in humans and there is a move to stop using it in animals. There is an Incomplete Freunds Adjuvant (IFA), which is an emulsion without the mycobacteria and the reaction is less severe. In addition, MF59 is a squalene oil-in-water emulsion without additional immunostimulatory materials and has proved to be a potent adjuvant currently under testing. Some components of mycobacterial cells may also have immunostimulatory action. Synthetic and semisynthetic derivatives have been tested, including muramyl di- and tripeptides, MDP and MTP. MDP, –acetyl-muramyl-L-alanyl-D-isoglutamine, is a small glycopeptide which appears to represent the smallest structure essential for mycobacterial adjuvanticity. However, synthetic MDP and some other analogues have the ability to enhance nonspecific resistance against diverse microbial infections and are capable of conferring resistance against a wide variety of pathogens, including influenza, herpes simplex, vaccinia, and Sendai virus. A purified monophosphoryl A in an emulsion has been evaluated clinically although it does not appear to have progressed to the market place.

Some plant materials have been evaluated, including the soap-like saponins from *Quilaja saponaria* and the highly purified Quil A extract obtained from saponins which induce the production of cytokines. In common with all saponins these materials have hemolytic activity that limits their direct use in humans although it is used in veterinary medicine. However, combination of Quil A with cholesterol, phospholipids, and antigens forms a human-compatible adjuvant called immuno-stimulatory complexes (ISCOMs). ISCOMs have an interesting range of adjuvant activities resulting in an increased Th1 response but have the advantage that they are active orally. More recently other purified Quil A extracts have been prepared and shown to be less toxic.

The most potent mucosal adjuvants have been shown to be the toxins derived from *Vibrio cholerae* or *Escherichia coli,* which should not be surprising since these organisms invade the body through the GI tract. Obviously too toxic for human use because they are the source of cholera or diarrhoea, heat labile enterotoxins have been tested in mice and shown to be potent adjuvants for orally or nasally administered influenza vaccine. The potency of heat-labile enterotoxin mutants may also be enhanced by formulation into bioadhesive particulate delivery systems, and this is an area under current exploration.

## MODERN MICROPARTICULATE VACCINE VEHICLES

A wide variety of particulate systems have been explored as drug delivery systems and are now being evaluated for the delivery of vaccines or vaccine components. Examples of particles are latex emulsions, carbon particles, liposomes, and polystyrene and poly(lactide-co-glucoside) particles. As noted earlier, antigens or antigenic epitopes attached to a particulate carrier are more likely to bring about a successful immunological reaction and some, such as chitosan particles, can act as adjuvants in their own right. Liposomes have been discussed in an earlier chapter and will not be reviewed here.

Natural polymers such as gelatin or albumin have been used as particulate drug delivery systems, although they are of uncertain purity and certainly have the potential for immunogenicity.

### BIODEGRADABLE POLYMERS

Synthetic polymers, especially polyesters of lactic and glycolic acids, have been used since the 1960s as resorbable sutures and microparticulate drug delivery systems. Extensive studies of these materials and other similar biologically acceptable polymers have demonstrated that they can degrade by random cleavage in the particle matrix or simply hydrolyse at the particle surface, producing organic acids at the interface. This hydrolysis means that the surrounding environment will be acidic and this can sometimes have implications for the stability of any incorporated drug substance. Other than this caveat, polylactide (PL) and polylactide-co-glucoside (PLGA) polymers have become accepted as safe and effective as controlled-release vehicles.

Low molecular weight PLA and PLGA can be produced by direct catalysis of mixtures of lactic or glycolic acid using antimony trioxide as catalyst; higher molecular weight entities being produced using antimony, tin, or lead catalysts. The composition of the polymer and the molecular weight are controlled by selection of appropriate molar ratios of the two primary acids and polymerization conditions. Factors such as crystallinity, polydispersity geometry, and polymer structure are all controlled during the manufacturing process. Breakdown in the body occurs by hydrolysis, with high molecular weight material reverting to polymers of lower molecular weight. It is generally accepted that enzymic degradation is hardly involved in the breakdown, the main process being by simple hydrolytic breakdown. However, polymers with a high lactic acid content are more stable to hydrolytic attack than those with intermediate ratios of lactic:glycolic acids. Hydrolysis is also likely to be slower through regions of crystallinity in the matrix. A 50:50 lactide:glycolide co-polymer degrades the most rapidly because it is unlikely to contain crystalline blocks of either of the constituent polymers. The glucoside component is thought to encourage the penetration of water into the solid matrix, thereby forming water channels and causing the surface hydrolysis over a wider area. As the molecular weight falls the relative numbers of the end groups increase so that what was initially a hydrophobic block becomes progressively more hydrophillic. This is seen with the higher molecular weight polymers which exhibit two distinct release phases of water uptake, separated by a short lag period. Moreover, in an alkaline medium or solutions of high ionic strength the hydrolysis rate is accelerated. Effectively the hydrolytic process is autocatalytic and the interior of microparticles has been observed to degrade faster than the interface.

The release of incorporated proteins from polymer matrices has been described as a three-stage process, a lag phase followed by a burst and then a steady state. The lag phase is due to water penetrating into the matrix but the burst effect is likely to be due to material at or near the matrix interface which is readily released as soon as it is wetted. After that the release profile settles down into a steady state until most of the drug has been released. These three phases are not always clearly separated, depending on the composition of the matrix and the amount of drug incorporated, but it should be possible to design a system that will manifest a zero order–release pattern over a reasonable period of time ideal for vaccine administration. If antigens are delivered to the mucosal system incorporated into small particles the main enzymatic barrier to absorption in the form of the exo- and endopeptidases in the lumen and the mucosal cellular membranes can be minimized, allowing both a mucosal and systemic immune response to be obtained. Antigen uptake is promoted or antigen can be directly delivered to the lymph nodes, although this will not necessarily result in the induction of an immune response. A full immunological response requires the antigen-presenting cell (APC) in the presence of other immunostimulants and cytokines. Coincidentally, perhaps, microparticles have the same approximate dimensions as invading pathogenic microorganisms with which the immune system has evolved to combat. Typically a vaccination requires one initial dose followed by two or more booster doses and some authors have considered the possibility of developing controlled release formulations which will mimic this effect but work by using a single dose (Alpar et al. 2000).

Vaccine antigens have been encapsulated inside polymeric particles and shown to stimulate production of antigen-specific serum antibody responses as well as mucosal IgA. Since bioadhesive microparticles can be produced this is proving to be an attractive avenue for the exploration of intranasal and other mucosal vaccines.

## LIPID PARTICLES AS ADJUVANTS AND DELIVERY SYSTEMS

Liposomes (see Chapter 9) have been used to deliver vaccines and have been observed to have immunostimulant activity. When administered orally liposomes with encapsulated antigens have been claimed to provide protection from the gastric proteolytic enzymes. Liposomes also have potential as mucosal delivery systems since not only are they immunogenic in their own right but physical association of the antigen with the liposomal structure is not a requirement for intranasal immunostimulation. Simple mixtures of liposomes with other agents such as chitosan have potential for nasal delivery and have facilitated enhanced responses to vaccines administered orally.

Cochleate systems consisting of calcium-precipitated protein–phospholipid complexes are stable solid sheets that roll up into a spiral with no internal aqueous space, and the calcium ions bridge adjacent sheets. Oral administration of vaccines in cochlear delivery systems has been shown to induce strong long-lasting circulation and mucosal antibody responses with long-term immunological memory to influenza glycoproteins.

Virosomes are viral glycoproteins encapsulated in lipid vesicles, which have been shown to be effective as experimental vaccines delivered by both mucosal and systemic routes. Viruses and their surface glycoproteins have a high affinity for receptors on mucosal surfaces, especially along the respiratory tract.

## CHITOSANS

Chitin is almost as common in nature as cellulose and is a main structural element of Crustacea, molluscs, and insects. Because it has limited solubility in industrial solvents, it has limited use; but when deacetylated under alkaline conditions it is converted to chitosan. Chitosan has terminal free amino groups distributed along its molecular chain (Figure 12.5), giving it a higher chemical and biochemical reactivity and thereby allowing it to be applied in a number of areas, including cosmetics and drug delivery. While inexpensive and readily available commercially, this material is also claimed to be nontoxic, biodegradable, and biocompatible. One other advantage, less widely recognized, is that it is a mucoadhesive and it also appears to act as an immunoadjuvant which, in the context of vaccine delivery, might provide a considerable advantage. In addition, the adjuvanticity of chitosan can be enhanced by the addition of secondary adjuvants so, overall, this material is seen to be very promising.

Commercially chitosans can have molecular weights varying from 4 to 2000 kDa and vary in the degree of deacetylation from 66% to 95%. Because of the free amino groups chitosan behaves as a weak base, with a pKa of 6.2–7.0 and is insoluble in water or organic solvents. It is a polyamine and therefore dissolves in hydrochloric

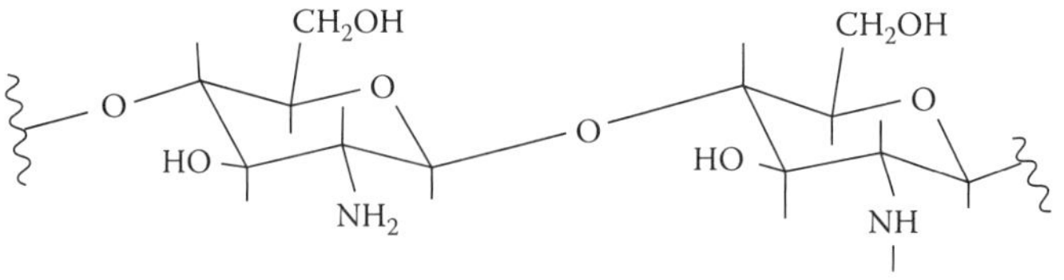

**FIGURE 12.5** Chitosan structure. (From Leone et al. 2004.)

acid and various organic acids including acetic, oxalic, and lactic acids, thereby forming salts. Chitosan salts are soluble in water, the solubility depending on the type of acid involved. For example, sebasic, phosphoric, and sulfuric salts are all less soluble and this provides a means of making dispersed insoluble particles of chitosan. In one method the chitosan powder is dissolved in a dilute acetic acid solution and poured into a solution of sodium sulfate, forming a fine dispersion of chitosan microparticles that can be collected and dried.

The stability of unmodified chitosan particles in an aqueous environment may be questionable and some authors have made covalent cross-linked chitosan micro-particles by taking advantage of the formation of Schiff bases with the free amino groups using a reactive aldehyde such as glutaraldehyde. These cross-linked particles may be less soluble in water and they are more stable physically but need to be loaded with any drug only after the remaining glutaraldehyde is thoroughly washed out and neutralized with sodium metabisulfite.

## THE FUTURE OF VACCINES AND VACCINATION

It is evident that currently vaccine research is a vigorous and developing topic but a number of goals remain elusive. For example, an ideal vaccine should elicit the required immunological response against specified pathogens, whether it requires a specialized delivery system or adjuvants. This entails comparison of the requirements for an ideal vaccine, with progress to date.

In addition it is becoming clear that vaccines should be heat stable since many are required in tropical countries where the cold supply chain used for many current vaccines is not available. Any newly developed vaccine must be completely safe since it should not cause disease or manifest side effects. This consideration is difficult to achieve in many cases because the side effects may not always be evident at first and may only show up after thousands of patients have been treated, often as an idiosyncratic reaction in just a few individuals. This issue is also experienced when developing small molecule drugs for the market place.

An advantage would be obtained if protection could be achieved using only a single dose that would be effective for the rest of the patients life. In some instances a controlled-release formulation has been tested and pulsatile systems that could mimic the administration of a booster dose may also have the desired effect.

Finally, the vaccine must be inexpensive and this is by no means an easy requirement. To illustrate the point, if a vaccine costs $25 to produce, pack, and deliver to the patient, how can this be acceptable in a country where the amount of money available per patient is only $250 per annum? Of course, money is saved on the subsequent savings in healthcare costs throughout the remainder of the patient's life. The rest of the debate is limited to a discussion of the value of a life, but in human terms this cannot be measured.

An interesting issue has surfaced at the time of writing (October 2004) when the supply of influenza vaccine in both the United States and United Kingdom became severely limited owing to a failure in good manufacturing practices within one organization. This resulted in the closure of the plant involved but, more to the point, a loss of about 50% of available doses for the winter season. Since there are only two plants worldwide making this vaccine, this has affected affluent members of society who would otherwise not consider themselves affected by the economic constraints that affect developing nations. The media has probably contributed to the hysterical discussion of which members of society should or should not receive the vaccine, also promoted by some politicians as the vaccine that could make the difference between life or death. The more responsible media has suggested that there may be insufficient economic return and too high a commercial risk for many pharmaceutical manufacturers to wish to get involved in vaccine production in general so that failures of this magnitude are almost inevitable. This entire issue will not go away and it is certain the discussion will continue in the foreseeable future.

On the basis of current research it may be possible to speculate about future directions for vaccine development. For example, controlled-release drug systems are well recognized and have appeared in clinical practice. It seems reasonable to ask if the same technology could be applied to vaccines. What would be ideal is a vaccine that only required a single dose which incorporated that booster dose so often necessary for complete effectiveness. If this in turn was combined with heat stability to overcome the problem of maintaining an effective cold chain for distribution in tropical countries we would be well on the way to providing an ideal product.

The probability is that acellular and subcellular vaccines will represent the future because they are generally much safer although, without appropriate adjuvants, they may be less effective. Viral shells, without their DNA should not be able to grow *in vivo* but should be capable of triggering an immune response in the form of the production of antibodies and memory cells. Isolated bacterial flagellae may also be capable of the same response. Acellular vaccines against *Haemophilus influenzae B* (Hib) bacterial flagellae have been cultivated and tested. An anthrax vaccine uses the protective antigen of anthrax, and there have been a number of attempts to develop a vaccine against the GP 120 surface protein of the HIV virus.

Recombinant DNA vaccines offer alternatives as subunit vaccines and organisms can be engineered to produce antigens or even epitopes. A hepatitis B vaccine has been engineered using yeast as the host cell. Adverse reactions are rare to subunit vaccines, making them safer for use in immunocompromised patients.

A recent development has been the use of genetically modified plants to produce vaccine components. Examples include common foodstuffs such as tomatoes, bananas, potatoes, and corn. The prospects for oral vaccines in bananas especially would appear to be promising since this is a staple food in many tropical countries. Vaccine epitopes have also been produced in the milk of goats, sheep, and cows although these may be difficult to purify, process, and formulate.

Bacterial toxins such as diphtheria and tetanus can damage host cells but the isolated toxins can also be immunogenic. However, the induced response may not always be very strong and booster shots are required every 10 years. Adjuvants could improve the response and both diphtheria and tetanus toxoids are more effective when combined with pertussis subunit vaccines, the DPT combination at present used clinically.

## SOME FUTURE VACCINE OPPORTUNITIES

There appear to be a myriad of opportunities for vaccines against, for example, cancers, allergies, hepatitis, tuberculosis, and HIV. Vaccines to treat drug addicts, including tobacco products, nicotine, and cocaine are all within the realm of possibility.

There is a perceived need to develop effective vaccines against bovine spongiform encephalopathy (BSE) and the human form, new variant Creutzfeldt-Jacob (nvCJ) disease. Organ transplant vaccines would obviously be beneficial and tetanus toxoid has been shown to lower cholesterol in animals. Tumor-specific antigens have also been explored and heat stress proteins have been evaluated as adjuvants in this case.

Attention has recently become focused on epidermal administration of vaccines, either as polymer rods inserted subdermally or by particulate systems fired into the skin using the high pressure jets developed by companies such as PowderJect Pharmaceuticals (Oxford, UK). The epidermis contains immune cells and less vaccine may be required to achieve a response which also reduces costs.

## RISK-BENEFIT RATIOS

The increased used of vaccines has drawn attention to ethical issues associated with safety and the risk–benefit ratio in some cases has come under scrutiny. Vaccines are usually given to patients who are otherwise healthy and have a lower tolerance for risk. Adverse reactions are either quite common ($>>10\%$) or very rare ($<<0.0001\%$). The question of justifiable risk in a healthy population then becomes problematic and difficult to identify.

To identify a justified risk in a healthy population requires clinical trials and if the risk is very low inordinate numbers of patients are required to identify issues. These become extremely expensive in practice and this is one reason why vaccines are subject to postlicencing and postmarketing surveillance for safety.

Animal models are rarely valuable in predicting human responses. Animals are required to be immunized twice monthly and monitored for at least 6 months. Some reassurance might be provided if there were no deaths or obvious and unusual effects. However, postmortem examination is required to ensure that no macroscopic or microscopic changes have been produced in any of the internal organs.

Once the animal studies are completed human studies can commence, moving through phase I (up to 50 naive patients to establish a sufficient immune response); phase II (several hundred patients in several locations) to phase III in which expanded studies on as many as thousands of patients. The trials require to be randomized and closely controlled by being blinded to avoid bias in interpretation. The patients all have to provide informed consent and strict adherence to good clinical practices in accordance with ethical principles is required to ensure risk-benefit considerations justify the risks associated with all trials of this type. The safety of the patients is an overall requirement and trials must be supervised by the appropriate independent ethics committee or the local Institutional Review Board.

It should be noted that, despite all these precautions, some individuals have become impatient with the system and have tried to take shortcuts, with inevitable severe repercussions when things go wrong. The system is there to protect the patients and the complexity of these issues certainly accounts for the high cost of commercially available vaccines (and drugs in general).

## REFERENCES AND FURTHER READING

Alpar, H.O., Ward, K.R., and Williamson, E.D. (2000). New strategies in vaccine delivery, *S.T.P. Pharma Sci.,* 10(4), 269–278.

Gregoriadis, G. (1995). *New Generation Vaccines: The Role of Basic Immunology.* NATO ASI Series A, Life Sciences, vol. 261. Plenum Press, New York and London.

Kersten, G. and Hirschberg, H. (2004). Antigen delivery systems. *Expert Rev. Vaccines,* 3(4), 453–462.

Leone, M.M., Nankervis, R., Smith, A., and Illum, L. (2004). Use of the ninhydrin assay to measure the release of chitosan from oral delivery systems. *Int. J. Pharmaceut.,* 271, 241–243.

Nijkamp, F.P. and Parnham, M.J., eds. (1999). *Principles of Immunopharmacology.* Birkhäuser Verlag, Basel, Switzerland.

Nugent, J., Li Wan Po, A., and Scott, E.M. (1998). Design and delivery of non-parenteral vaccines. *J. Clin. Pharm. Ther.,* 23, 257–285.

O'Hagan, D.T. (1994). *Novel Delivery Systems for Oral Vaccines.* CRC Press, Boca Raton, FL.

O'Hagan, D.T. (2000). *Vaccine Adjuvants: Preparation, Methods and Research Protocols,* Methods in Molecular Medicine, vol. 35. Humana Press, New York.

Powell, M.F. (1995). *Vaccine Design: The Subunit and Adjuvant Approach.* Plenum Press, New York and London.

Rosenthal, S.R. (1980). *BCG Vaccine: Tuberculosis-Cancer,* 2nd ed. PSG Publishing Company Inc., Littleton, MA.

Scott, C. (2004). Special issue, Vaccines, *Bioprocess Int.,* 2(suppl.1), April.

Shen, W.-C. (1999). *Immunology for Pharmacy Students.* Harwood Academic Press, Amsterdam.

# 13 Gene Therapy: An Overview of the Current Viral and Nonviral Vectors

*Kadriye Ciftci, Ph.D. and Anshul Gupte, Ph.D.*

## CONTENTS

## INTRODUCTION

Gene therapy holds great promise for the treatment of many diseases (e.g., cancer, AIDS, cystic fibrosis, adenosine deaminase deficiency, cardiovascular diseases, Gaucher disease, α1-antitrypsin deficiency, rheumatoid arthritis, and several others) (1,2). Advances in genomics and molecular biology have revealed that almost all diseases have a genetic component. In some cases, such as cystic fibrosis or hemophilia, mutations in a single gene can result in the disease (3–5). In other cases, such as hypertension or high cholesterol, certain genetic variations may interact with environmental stimuli to cause the disease (6–8) or pathological conditions associated with aging frequently result in the loss of gene activity in specific types of cells.

The main targets of gene therapy are to repair or replace mutated genes, regulate gene expression and signal transduction, manipulate the immune system or target malignant and other cells for destruction (1). There are several factors involved in effective gene transfer to somatic cells in patients: (a) the type of vehicle used for gene delivery that will determine efficacy of delivery; (b) interaction of gene vehicle; (c) targeting to the specific area; (d) entrance to the target cell; (e) release from the cytoplasmic compartment, transport to the nucleus; (f) type and potency of regulatory elements; (g) expression (transcription) of the transgene and translation into protein.

Compared to conventional small molecule drug therapies with a transient effect on their molecular targets, gene therapy usually requires an efficient transfer by delivery system to target cells resulting in a permanent change to the genetic constitution. The application of gene delivery technology to a growing roster of clinical indications is predicated on significant advances in both genomics and gene delivery systems.

There is a wide variety of vectors used to deliver DNA or oligonucleotides into mammalian cells, either *in vitro* or *in vivo*. The most common vector systems are based on viral [retroviruses (9, 10), adeno-associated virus (AAV) (11), adenovirus (12, 13), herpes simplex virus (HSV) (14)] and nonviral [cationic liposomes (15, 16), polymers and receptor-mediated polylysine-DNA] complexes (17). Other viral vectors that are currently under development are based on lentiviruses (18), human cytomegalovirus (CMV) (19), Epstein-Barr virus (EBV) (20), poxviruses (21), negative-strand RNA viruses (influenza virus), alphaviruses and herpesvirus saimiri (22). Also a hybrid adenoviral/retroviral vector has successfully been used for *in vivo* gene transduction (23). A simplified schematic representation of basic human gene therapy methods is described in Figure 13.1.

The choice of the appropriate delivery system for successful gene therapy requires understanding of the drawbacks and advantages of each delivery system (Table 13.1 for comparison of viral vectors and Table 13.2 for comparison of nonviral methods for gene therapy), such as limitations in the total length of the DNA that can be introduced, including the plasmid size and control elements. Understanding of the pathophysiology of the disease and the cell targets (IV, IP, intratumoral, SC injection) is required. The type of control elements required for the tissue-specific expression of the construct, the presence of viral or other origins of replication as well as of the cDNA encoding the viral replication initiator protein for an episomal replication of the transgene and sequences that prompt integration is also important for successful gene transfer. While no single vector developed to date is optimal for all clinical indications, the growing number of viral and synthetic vectors will enable gene therapy to be used in treating a wide variety of significant diseases.

Current gene therapy programs apply gene delivery technology across a broader spectrum of disease conditions (2). Since 1989, when the first human gene therapy study was performed, enormous research efforts have followed (3). Although much effort has been directed in the last decade toward improvement of protocols in human gene therapy, the therapeutic applications of gene transfer technology still remain

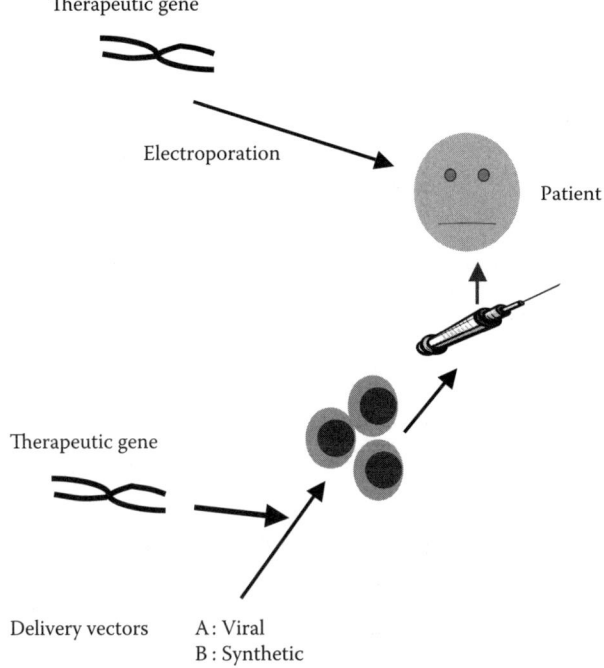

**FIGURE 13.1** Schematic representation of human gene therapy.

mostly theoretical. The weakest point of gene therapy development programs is vector design, followed by gene regulation and avoidance of immune responses. Basic research is cautiously progressing to address these pressing issues. The characteristics of the most developed gene delivery systems are discussed in the following section.

**TABLE 13.1**
**Comparison of Different Viral Vectors for Gene Therapy**

| Vector | Advantages | Disadvantages |
|---|---|---|
| Retrovirus | Integration into host DNA<br>All viral genes removed<br>Relatively safe | Insertional mutagenesis<br>Requires cell division<br>Relatively low titer |
| Adenovirus | Higher titer<br>Efficient in nondividing cells | Toxicity<br>Immunological response |
| Adeno-associated virus | All viral genes removed | Limited size of foreign DNA<br>Labor-intensive production<br>Status of genome not fully elucidated |
| Lentivirus | Provide long-term and stable gene<br>  expression<br>Infect nondividing cells | Similar retrovirus |

**TABLE 13.2**
**Methods of Nonviral Gene Transfer**

| Method | Size of DNA | Target Cells | Transfection Efficiency | Transfection | Cellular Toxicity | Gene Expression | Preparation | Application |
|---|---|---|---|---|---|---|---|---|
| Naked DNA | No limit | Especially myocytes | 10–30% of cells at injection site | Extra chromosomal | Lymphocytic infiltration | Until death of cell | Easy and cheap | In vivo |
| Microinjection | No limit | Mitosis/resting | Stable <0.1–1% | Integration possible | 30% survival | | 200–400 injections/hr | In vitro |
| Electroporation | 150 Kb | Mitosis/resting | Stable <0.1–1% | 1–2 copies | 20–60% survival | | Easy | In vitro |
| Particle bombardment | 10,000 copies | Mitosis/resting | Stable <0.1–1%, transient <20% | Persistent and integration? | 85–95% survival | 2–12 month | Easy | In vitro and in vivo |
| Lipofection | No limit | Mitosis/resting | Stable <0.1–1%, transient 80% | Integration possible | Membrane toxic | | Easy | In vitro and in vivo |
| Ligand mediated | >48 Kb | Mitosis/resting | Up to 50% | Extra chromosomal | High | High, Transient | Labor intensive | In vitro and in vivo |
| Calcium phosphate precipitation | No limit | Mitosis/resting | Stable <0.1% | Often multiple copies | High | High, Transient | Labor intensive and time consuming | In vitro and in vivo |

# GENE DELIVERY SYSTEMS

## Viral Vectors

Viral vectors are the first used vectors for gene therapy research. It has been known that many viruses have the capability of efficiently transferring their nucleic acid genomes to mammalian cells in order to initiate their first step in life cycle. Viral vectors take advantage of the ability of the virus to enter cells and deliver genetic material to the nucleus. Most viral vectors are engineered in such a way that they can enter the cells but they do not have the ability to replicate in the cell. To successfully develop viral vectors, the important consideration includes introducing therapeutic genes into their genomes while concurrently removing the native viral genes that code for harmful viral proteins. To develop viral vectors first viral DNA is removed and is replaced with a therapeutic gene and the recombinant virus is thus produced and functions purely as a delivery system for the therapeutic genes to the nucleus of the target cell without causing cellular damage or subsequent virus propagation (24). Depending on the therapeutic aim of a particular gene therapy, transient or permanent expression may be desirable. There are several different classes of viral vectors including retrovirus, adenovirus, adeno-associated virus, lentiviruses, herpes simplex, and alpha($\alpha$)-viruses used for gene therapy. The characteristics and applications of these vectors are discussed below.

## Retroviruses

The retroviruses are enveloped viruses, roughly spherical, about 120 nm in diameter. They contain a diploid RNA genome of 7 to 11 kb that is converted into a DNA intermediate by the reverse transcriptase upon entry into the cytoplasm of a cell (25, 26). The DNA is then transported to the nucleus, where it integrates randomly into the genome (25). Retroviruses can only accommodate less than 9 kb of foreign genetic information. The use of retroviruses for gene transfer requires a two-component approach as described in Figure 13.2. The first involves the replacement of the genetic material encoding the gag, pol, and env proteins with the DNA to be transferred. This DNA is expressed under the control of the promoter elements in the 5LTR. The second component involves the introduction of this DNA into a retroviral packaging cell line to produce virus able to infect the appropriate host species. This cell line contains a replication-defective helper retrovirus that will provide the gag, pol, and env proteins and an encapsidation signal for efficient viral packaging (27, 28).

Before the *in vivo* gene therapy with retroviruses becomes a successful reality a number of problems must be overcome. The major limitation of retroviruses has been poor gene expression *in vivo*, which has been overcome through the use of tissue-specific promoters. Use of internal ribosome entry sites from picornaviruses in retroviral vectors has provided stable expression of multiple gene enhancers. Another drawback of retroviruses for their exploitation in gene therapy has been the low viral titers obtained, too low to achieve therapeutic levels of gene expression; methods for the efficient concentration from large volumes of supernatant and purification of amphotropic retrovirus particles have been developed in several

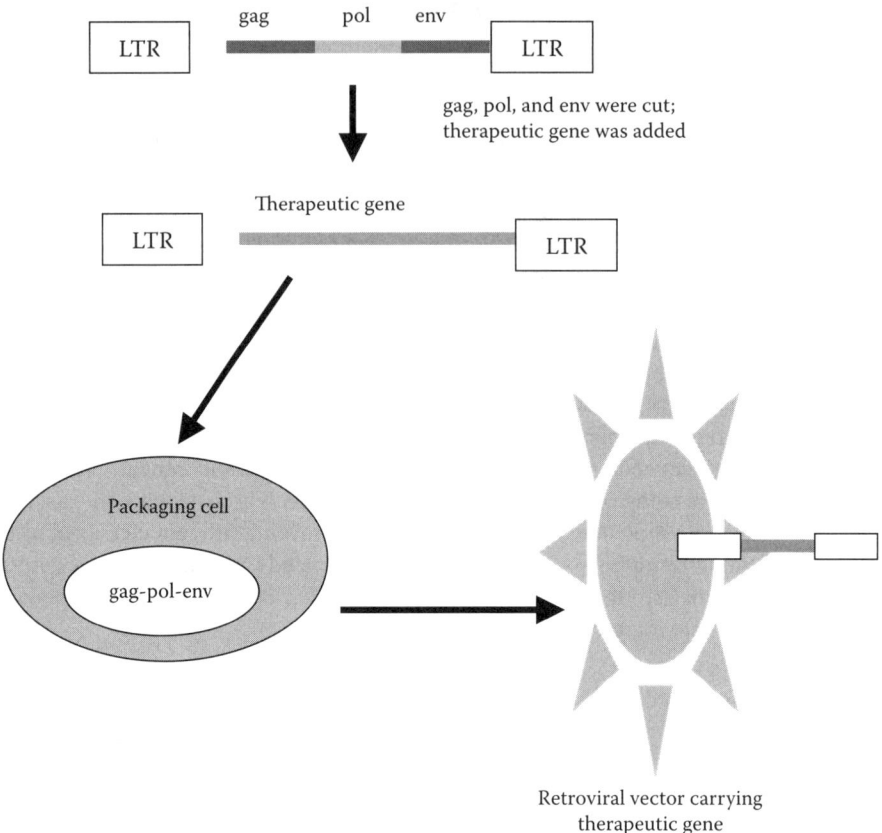

**FIGURE 13.2** Formation of retroviral vectors.

laboratories. For example, Bowles et al. (1996) have used concentration and further purification of virus particles by sucrose banding ultracentrifugation; animal studies have shown that viral transduction increased proportionally with titer of the retrovirus. In addition, retroviruses transfer the gene of interest permanently into the genome of the target cell, which could result in chronic overexpression of the inserted gene or can lead to insertional mutagenesis. Moreover, retroviruses can infect proliferating cells only. This may decrease their usefulness for gene transfer into stem cells, which are largely noncycling. To overcome limitations of host cell tropism, retrovirus vectors have been pseudotyped with envelope proteins from other viruses such as the G glycoprotein from vesicular stomatitis virus (VSV) (30). VSV G pseudotyped retroviruses are less labile and can be concentrated to high titers and also show a much broader host range than the wild-type retrovirus. Another important feature of retroviruses is that although they do not elicit immune responses in the host, they are susceptible to rapid degradation by the complement. This is also a major limitation for *in vivo* retroviral-mediated gene transfer.

The most common retroviral vector is based on the amphotropic Moloney murine leukemia virus (MLV) (31). This system is particularly suitable for efficient *in vitro* cell transduction: the amphotropic MLV has a broad cell tropism, it can be produced at relatively high titers ($10^6$–$10^7$ iu/mL), and allows for long-term transgene expression because of the viral integration in the host chromosomal DNA.

Retrovirus vectors were subjected to the first clinical trial on human gene therapy to correct adenosine deaminase (ADA) deficiency (32). White-blood cells isolated from patients were infected *ex vivo* with an MLV-based vector expressing ADA and a neomycin marker gene. After selection with G418, neomycin-resistant cells were isolated and reintroduced into patients. The treatment improved the physical condition of the patients and the ADA-containing provirus was stable in the blood for several years.

Investigators have been considering the engineering of chimeric retroviruses with specific cell tropism. This would greatly facilitate the in vivo application of retroviral vectors in clinical trials. In this respect, there have been many attempts to alter the cell tropism of ecotropic retroviruses, which do not infect human cells. This approach consists of placing foreign genes (CD4), single-chain antibodies, the polypeptide erythropoietin, short peptides binding to several integrins, and human heregulin (33–35). The retroviral systems used in these studies were: avian leukosis virus, ecotropic MLV and spleen necrosis virus. The foreign genes used in the early studies to generate hybrid envelopes were: In some cases, there has been a partial success in redirecting the cell tropism of ecotropic retroviruses (36), but the transduction efficiency is far from being optimal for *in vivo* applications.

Retrovirus vectors have demonstrated some promising results in cancer therapy and bone marrow transplantation. The introduction of retrovirus particles expressing HSV-TK and administration of GCV suggested that the treatment of graft-versus-host disease was efficient (37). The demonstration of the full correction of the SCID-X1 phenotype in infants is a further indication of the efficacy of retrovirus vectors (38).

## Adenoviruses

Adenoviruses are nonenveloped DNA viruses with 80 to 110 nm diameter icosahedral protein shell containing double-strand DNA genome of 36 kb that encodes four early proteins (E1 to E4) and five late proteins (L1 to L5) (39). In order to enter the host cell the adenovirus first attaches with a high affinity to a cell surface receptor, whose nature still remains elusive, using the head domains of the protruding viral fibers; the fibronectin-binding integrin on the cell surface then associates with the penton base protein on the adenovirus triggering endocytosis of the virus particle via coated pits and coated vesicles (40, 41). The third step in adenovirus entry into the host cell includes penetration of the adenoviral particles by acid-catalyzed rupture of the endosomal membrane involving the penton protein and the integrins and allowing escape to the cytoplasmic compartment; a decrease in endosome pH during internalization expose hydrophobic domains of these adenoviral capsid proteins, which permits these proteins to insert into the vesicle membrane in a fashion that ultimately disrupts its integrity (42). At the final step the adenoviral particle is

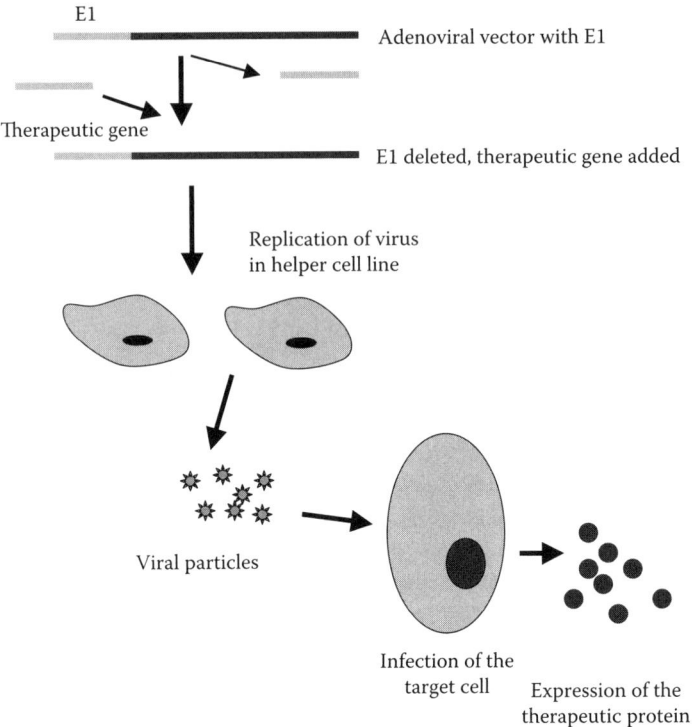

E1

Adenoviral vector with E1

Therapeutic gene

E1 deleted, therapeutic gene added

Replication of virus
in helper cell line

Viral particles

Infection of the
target cell

Expression of the
therapeutic protein

**FIGURE 13.3** Gene delivery with adenoviral vectors.

attached to the cytoplasmic side of pore complexes and the DNA is released to the interior of pore annuli entering the nucleoplasm.

Because replication is controlled by E1, it is usually deleted in adenoviral vectors used for gene therapy and replaced by the gene to be transferred, as shown in Figure 13.3. The resultant recombinant adenovirus is replication incompetent. This recombinant adenoviral DNA is then transferred into a complementing cell line containing E1 sequences in its genome (but lacking other sequences required for replication) to generate viral particles that are infectious but replication defective (43).

Adenoviruses have certain advantages over retroviruses for gene therapy. They can be produced in high titer ($>10^{13}$ viral particles per milliliter) and can transfer genes efficiently into both replicating and nonreplicating cells (44). Adenoviruses possess a linear double-stranded genome which can be manipulated to accommodate up to 7.5 kb of DNA. Although early versions of adenoviruses showed toxic side effects and strong immune responses, newer second- and third-generation vectors with many of the viral genes deleted, have demonstrated significant improvements (45). As the transferred genetic material is located episomally, the risks of permanently altering the genetic material of the cell and of insertional mutagenesis are avoided (46). For safety, replication-deficient, infectious adenoviruses are being used in somatic gene transfer; for example, deletion in a portion of the E3 region of the virus permits encapsidation whereas deletion of a portion of the E1A coding sequence impairs viral replication (47, 48).

A disadvantage of adenoviral vectors is that the viral proteins are immunogenic and can induce nonspecific inflammation and specific cellular responses (43). Also, episomes tend to be lost from infected cells within 2 to 4 weeks, so repeated administration may be necessary (46). The efficacy of adenovirus delivery might be severely hampered because most people have been exposed to natural adenoviruses infections, even when using replication-deficient vectors. In a novel approach, the surfaces of viral particles were coated with a multivalent copolymer based on poly-[N-(2-hydroxypropyl) methacrylamide] (pHPMA) (49). To improve targeting, fibroblast growth factor (FGF) and vascular endothelial growth factor (VEGF) were incorporated in the polymer, which resulted in targeting of bFGF receptor-positive A549 cells. Targeting of endothelial human umbilical endothelial cells with polymer-VEGF coated adenovirus was also highly efficient (49). The *in vivo* targeting of polymer-bFGF ADLacZ virus in nude mice bearing intraperitoneal xenografts of human SUIT2 cells was highly efficient.

In addition, the polymer-coated adenovirus particles were able to shield against antibody recognition. In another approach, the PEGylation of the adenovirus capsid protein prolonged transgene expression after systemic delivery of E1-deleted adenovirus, and allowed partial readministration with native virus (50). Adenovirus has been explored as vector for the treatment of cystic fibrosis (51), for Duchenne muscular dystrophy (52), to deliver tumor suppressor genes for cancer treatment (53), for gene transfer to the brain (54) and for melanoma specific vector (55).

Recombinant adenovirus vectors have been used for variety of applications including, the transfer of factor IX gene in hemophilia B to dogs via vein injection (56) and in mice for the transfer of genes into neurons and glia in the brain for the transfer of the gene of ornithine transcarmylase in deficient mouse and human hepatocytes (57) for the transfer of the very low density lipoprotein receptor gene for treatment of familial hypercholesterolaemia in the mouse model (58) for the transfer of low density lipoprotein receptor gene in normal mice and for the *ex vivo* transduction of T cells from ADA-deficient patients (59). The adenovirus major late promoter was linked to a human α1-antitrypsin gene for its transfer to lung epithelia of cotton rat respiratory pathway as a model for the treatment of α1-antitrypsin deficiency; *in vitro* and *in vivo* infections have shown production and secretion of α1-antitrypsin by the lung cells (60).

To overcome one of the major limitations of the clinical utility of adenoviruses which is the low efficiency of gene transfer achieved in vivo, Arcasoy et al. (1997) found that the presence of the polycations polybrene, protamine, DEAE-dextran, and poly-L-lysine significantly increased the transfection efficiency in cell culture using the lacZ gene; because the polyanion heparin did not significantly alter gene transfer efficiency, but completely abrogated the effects of polycations it supports the idea that the negative charges presented by membrane glycoproteins reduce the efficiency of adenovirus-mediated gene transfer, an obstacle that can be overcome by polycations.

## Adeno-Associated Viruses

Adeno-associated viruses (AAV) are parvoviruses that are not pathogenic in humans. They are extremely small, nonenveloped icosahedral virus of 18 to 26 nm in diameter carrying a single-stranded ~5 kb DNA genome with short, inverted terminal repeats that are required for genome replication and packaging. Unlike adenoviruses, AAV may integrate into the host genome and do so at preferred locations, in particular, at one site on chromosome 19 (43). AAV has established its position as one of the most popular gene delivery systems. This is mainly because of the long-term and efficient transgene expression in various cell types in many tissues such as liver, muscle, retina and the central nervous system (62). Recombinant AAV vectors used for gene transfer contain 145 bp terminal repeat sequence and a polyadenylation site. They have had most of the viral genome deleted and replaced with DNA encoding the therapeutic gene. Since a few viral proteins are expressed, these viruses induce less of an immune response than for adenoviruses. Like adenoviruses, AAV vectors do not require cell replication for integration but high AAV titers are often difficult to obtain because the production of infectious AAV requires the use of an adenovirus, in which case contamination of the AAV with adenovirus is a concern (43).

There are some disadvantages associated with the application of AAV. Gene transfer with AAV vectors has been shown to be low. Difficulties in generating recombinant virus on a large scale sufficient for preclinical and clinical trials and in obtaining high-titer virus stocks after the initial transfection into producer cells is a limiting factor for the widespread usage of AAV vectors. AAV vector particles in cell lysates could be concentrated by sulfonated cellulose column chromatography to a titer higher than $10^8$ cfu/mL or $5 \times 10^{10}$ particles/mL (63). A method for transfecting cells at extremely high efficiency with a rAAV vector and complementation plasmid while simultaneously infecting those cells with replication competent adenovirus using adenovirus-polylysine-DNA complexes has been developed (64).

After infection of cell cultures with recombinant AAV there is a decline in the percentage of cells expressing the transferred gene with time in culture. This decline was associated with ongoing losses of vector genomes (65). The packaging capacity is relatively restricted and the large-scale production inefficient. In addition, the pre-existing immunity to human AAV vectors is comparable to adenovirus and the integration into the host genome is random, which can lead to unexpected activation or inhibition of endogenous gene expression.

Different AAV serotypes have shown remarkably different expression patterns because of differences in cell entry and intracellular activities (66, 67). Application of the dimerizer-inducible transcriptional regulatory system for AAV has allowed pharmacological regulation of heterologous gene expression *in vivo* (68).

AAV normally contains a single-stranded copy of its genome. Transduction with AAV can be enhanced in the presence of adenovirus gene products through the formation of double-stranded, nonintegrated AAV genomes. AAV has been reported to have advantages over other viruses for gene transfer to hematopoietic stem cells due to their high titers and relative lack of dependence on cell cycle for target cell integration. A robust CMV/LacZ reporter gene expression in primary human CD34+CD2- progenitor cells induced to undergo T-cell differentiation was obtained without toxicity or alteration in the pattern of T-cell differentiation. Seventy percent to 80% of the cells isolated from either adult bone marrow or umbilical cord blood were efficiently transduced with AAV, however, the expression was transient without integration; this limits the potential use of AAV in gene therapy strategies for diseases such as AIDS (69).

Gene transduction by AAV vectors in cell culture can be stimulated over 100-fold by treatment of the target cells with agents that affect DNA metabolism, such as irradiation or topoisomerase inhibitors (70), great improvements in transduction efficiency can also be achieved *in vivo*: previous g-irradiation increased the transduction rate in mouse liver by up to 900-fold, and the topoisomerase inhibitor etoposide increased transduction by about 20-fold after direct liver injection or after systemic delivery via tail vein injection; up to 3% of hepatocytes could be transduced after a single systemic vector injection (71). This is a significant advantage compared to stealth liposomes which, although concentrating in the liver, spleen and tumors, can transduce Kupffer cells but not hepatocytes after systemic delivery.

In another study, AAV-mediated delivery of the lacZ gene by direct injection to brain tumors which were induced from human glioma cells in nude mice showed that 30% to 40% of the cells along the needle track expressed b-galactosidase; subsequent delivery of the HSV-tk/IL-2 genes to these tumors with AAV and administration of GCV to the animals for 6 days resulted in a 35-fold reduction in the mean volume of tumors compared with controls by a significant contribution from the bystander effect (72).

## Other Viruses

### Lentiviruses

Although lentiviruses belong to retroviral class, gene therapy vectors derived from lentiviruses offer many potentially unique advantages over more conventional retroviral gene delivery systems. Many of the lentivirus vectors used in gene therapy are based on the human immunodeficiency virus (HIV) (73). An advantage of HIV vectors has been the broad range of tissues and cell types they can transduce, a property granted because lentiviral vectors are pseudotyped with vesicular stomatitis virus G glycoprotein. Human lentiviral (HIV)-based vectors can transduce nondividing cells *in vitro* and deliver genes *in vivo*; expression of transgenes in the brain has been detected for more than 6 months. HIV vectors have been also used to introduce genes directly into liver and muscle; 3% to 4% of the total liver tissue was transduced by a single injection of $1\text{-}3 \times 10^7$ infectious units (IU) of recombinant HIV with no inflammation or recruitment of lymphocytes at the site of injection. Whereas expression of green fluorescent protein (GFP), used as a surrogate for therapeutic protein, was observed for more than 22 weeks in the liver and for more than 8 weeks in the muscle using lentiviral vectors, little or no GFP could be detected in liver or muscle transduced with the Moloney murine leukemia virus (Mo-MLV), a prototypic retroviral vector (74).

The development of a stable noninfectious HIV-1 packaging cell line capable of generating high-titer HIV-1 vectors is another important step towards use of HIV vectors in gene therapy (75). HIV-mediated gene transfer was used to transfer the

GFP gene under control of CMV to retinal cells by injection into the subretinal space of eyes in rats; the GFP gene was efficiently expressed in both photoreceptor cells and retinal pigment epithelium; predominant expression in photoreceptor cells was achieved using the rhodopsin promoter. The transduction efficiency was high and photoreceptor cells in >80% of the area of whole retina were expressing GFP (76). Intron-containing constructs have been successfully introduced into recent versions of lentivirus vectors (77).

Recently, a series of lentivirus vectors were developed for transduction of hepatocytes *in vivo* (78). Various promoters, such as the human CMV, the human phosphoglycerate kinase (PGK) and the mouse albumin promoter, were introduced into the HIV-1–based vector. These vectors showed enhanced nuclear translocation in hepatocytes and improved transgene expression. Interestingly, targeted expression to the liver could be accomplished by the use of the albumin promoter. Therapeutic levels of human factor IX were achieved after a single injection.

However, the use of lentiviral-based vectors in the clinic raises specific safety and ethical issues. Concerns include the possible generation of replication competent lentiviruses during vector production, mobilization of the vector by endogenous retroviruses in the genomes of patients, insertional mutagenesis leading to cancer, germline alteration resulting in transgenerational effects and dissemination of new viruses from gene therapy patients (79). One approach to address safety issues has been to develop lentivirus vectors incapable of replication in human cells. Gene transfer into hematopoietic stem cells using lentiviral vector (80) have been developed that are able to deliver and express genes in nondividing cells *in vitro* and *in vivo*.

*Herpes Simplex Viruses*
Herpes simplex virus (HSV-1) has a capacity of inserting up to 30 kb of exogenous DNA, which is a clear advantage over the adenovirus (up to 7.5 kb of exogenous DNA). High-titer viral stocks can be prepared from HSV-1. HSV-1 also displays a wide range of host cells and can infect nonreplicating cells such as neuron cells in which the vectors can be maintained indefinitely in a latent state. However, infection with HSV-1 is cytotoxic to cells because of residual viral proteins produced by the virus. Strategies to circumvent this drawback led to the development of viral vectors with a very large capacity for insertion (almost as large as the size of the virus), which depend on defective helper virus for replication and packaging into infectious virions (see below). A mini viral vector can combine the advantage of cloning the gene in bacterial plasmids, the high efficiency of virus-mediated gene transfer, and the possibility to transfer large genomic DNA fragments including far upstream, downstream and intronic regulatory elements.

Two types of viral vectors have been used for gene transfer to cancer cells: replication-incompetent vectors expressing a gene product that leads to the destruction of the tumor or replication-competent vectors that are inherently cytotoxic to the tumor cells. In order to combine the two modes of action Miyatake et al. used a defective HSV vector that consisted of a defective particle, containing tandem repeats of the HSV-tk gene, and a replication-competent, non-neurovirulent HSV mutant as a helper virus. When glioma GL261 cells were infected with the tk-defective vector/helper virus the HSV-TK activity was significantly higher than that in helper virus-infected cells which contained a single copy of HSV-tk; subcutaneous injection of these cells to C57BL/6 mice inducing gliomas led to a significant decrease in tumor size after GCV treatment.

*Epstein Barr Viruses*

Epstein Barr virus (EBV) is an episomaly-replicating virus in synchrony with the cell cycle. EBV infects human cells causing mononucleosis; the presence of the unique latent origin of replication (oriP) in EBV allows for episomal replication of the virus in human cells without entering the lytic cycle. The presence of oriP and of the replication initiator protein EBNA1 cDNA on a vector allows episomal replication in human cells; in addition, plasmids containing only oriP can replicate episomally into cell lines expressing EBNA-1 (82). Infection of tumor-derived fibro-blast and epithelial cell lines in culture and local injection of human liver tumors in nude mice was used to demonstrate 95% to 99% efficiency of infection and transfer of the reporter b-galactosidase gene.

## NONVIRAL VECTORS

An alternative to the use of viral vectors for gene delivery is to deliver genetic material in the form of bacterial plasmid DNA. In the simplest form, naked plasmid DNA can be injected into skeletal muscle leading to transfection of muscle fibers close to the site of delivery (83). Though the transfection efficiency by nonviral vectors is relatively lower than that by viral vectors, synthetic nonviral vectors are designed to overcome many of the problems associated with viral vectors, such as risk of generating the infectious form or inducing tumorigenic mutations, risk of immune reaction, limitation to the size of genes incorporated, and difficulty for the production to scale up (84, 85).

The advantages of nonviral carriers over their viral counterparts are: (1) they are easy to prepare and to scale-up; (2) they are generally safer *in vivo*; (3) they do not elicit a specific immune response and can therefore be administered repeatedly; (4) nonviral vectors allow for the delivery of large DNA fragments and are also particularly suitable to deliver oligonucleotides to mammalian cells, which is an excellent feature for the application of antisense strategies to downregulate the expression of certain genes; and (5) they are better for delivering cytokine genes because they are less immunogenic than viral vectors (84, 86, 87).

Nonviral vector systems are usually either composed of a plasmid based expression cassette alone ("naked" DNA), or are prepared with a synthetic amphipathic DNA-complexing agent (84, 88). Gene delivery systems based on nonviral vectors mainly comprise cationic liposomes, DNA-polymer–protein complexes, and mechanic admin-istration of naked DNA. An idealized/optimized multifunctional nonviral gene delivery system is depicted in Figure 13.4.

Several major barriers need to be overcome for the development of nonviral gene delivery systems into true therapeutic products for use in humans. These barriers fall into three classes: manufacturing, formulation, and stability (extracellular barriers and intracellular barriers) (85). Cationic lipids and cationic polymers self-assemble with DNA to form small particles that are suitable for cellular uptake. At the thera-peutic doses positively charged particles readily aggregate as their concentration increases, and are quickly precipitated above their critical flocculation concentration.

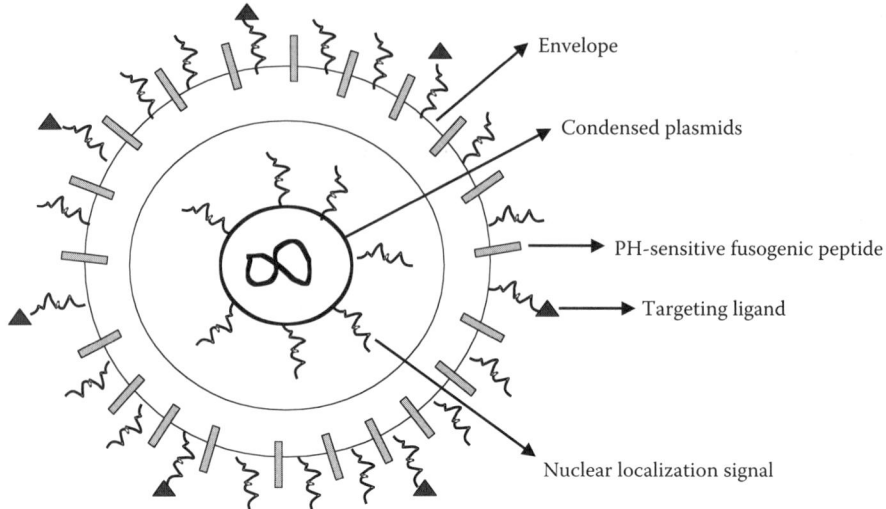

**FIGURE 13.4** An optimized/ideal multifunctional nonviral gene delivery system.

To circumvent this problem, hydrophilic polymers like polyethylene glycol (PEG) have been used to create PEGylated particles to provide steric stabilization. The ability to prepare well-defined particles and uniform morphology at high concentration is essential to the development of a pharmaceutical product. In addition storage stability is important for all gene delivery systems. Lyophilization is a feasible method of preparing nonviral gene delivery systems for storage. Lyophilization of lipoplexes (89) and polyplexes (90, 91) in the presence of lyoprotectants, such as trehalose and sugars, appears to provide for long-term storage. In addition to formulation and stability issues, the ability to scale-up of the nonviral vectors needs to be addressed.

The nonviral gene delivery systems must show low toxicity, escape the immune system, minimize interactions with plasma proteins, extracellular matrices, and nontargeted cell surfaces and not aggregate. Efforts to prepare nonviral gene delivery systems that have ideal characteristics are ongoing. One limitation of nonviral gene delivery systems is their toxicity. Recent studies are involved in preparing carriers that have lower toxicity. For example, recent evidence shows that low molecular weight preparations of polycations such as chitosan (92), polyethyleneimine (PEI) (93) and β-cyclodextrins-containing polymers (94) are significantly less toxic than high molecular weight polycation both in cultured cells and in animals. Additionally, the molecular architecture of the nonviral delivery system can modulate the toxicity, and these data suggest that the toxicity should be controllable.

The stabilization of nonviral gene delivery particles is necessary to extended circulation time that is required to target particular cell. Strong positive charge on the particles facilitates nonspecific interactions to the extracellular matrix, cell surfaces, and plasma proteins (all negatively charged), whereas strong negative charge can cause scavenging by phagocytosis via the macrophage polyanion receptor. Steric stabilization can be achieved by using hydrophilic polymer on the surface of lipoplexes or polyplexes, thereby decreasing interactions. For example, the PEGy-lation of polyplexes (95), covalent grafting of PEG (96) or HPMA (poly [$N$-(2-hydroxy-propyl) methacrylamide]) (97), after particle formation can all increase stability against aggregation and reduce nonself-interactions. Steric stabilization of these systems can be accomplished without alteration of polyplex morphology or, obviously, disruption of the polyplex. For further stabilization, the polymer strands are crosslinked in preformed polyplexes (95). These crosslinked particles do not show sufficient gene expression. Stabilization of lipoplexes and polyplexes should target to specific cell by using surface receptors and ligand-containing nonviral gene delivery systems. These ligands can be small molecules (e.g., folate, galactose, etc.) or peptides and proteins (e.g., transferrin and antibodies). Numerous systems have been investigated. For example, transferrin is a common ligand used to target tumor cells (96) and galactose-containing ligands have been used to target hepatocytes (95).

In most cases nonviral vectors enter cells either by charge-mediated interactions with proteoglycans on cell membranes or by receptor-mediated endocytosis by ligand–receptor binding interactions. Both methods result in uptake into vesicular compartments that ultimately deliver their contents to lysosomes and escape from the lysosomes, trafficking to the nucleus, nuclear entry and vector dissociation that are required for gene delivery. Dissociation of the lipid or polycation from the DNA may occur upon vesicle escape or any time thereafter. Currently, the optimal release rate remains unknown (98). The proposed mechanism for the transfer of lipoplexes/genosomes to the nucleus is schematically shown in Figure 13.5. Other methods of assisting plasmid transport to the nucleus include the use of nuclear localization signal peptides. Nuclear localizing signals (NLS) are present on his-tones, transcription factors, nuclear enzymes, and a number of other nuclear proteins; nascent chains of DNA-binding polypeptides could bind to the supercoiled plasmid in the cytoplasm mediating its translocation to the nucleus. Recent evidence also suggests that particular targeting ligands (e.g., fibroblast growth factor) (99) can

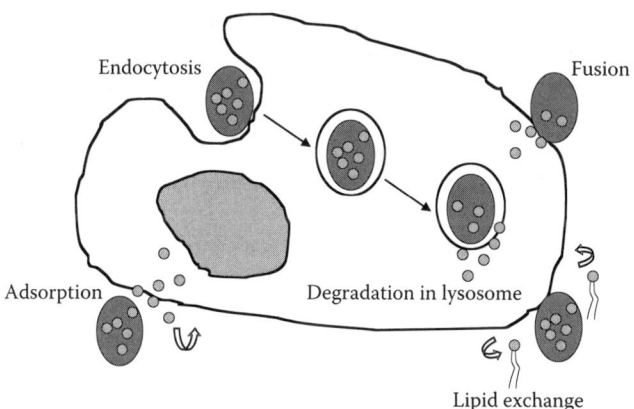

**FIGURE 13.5** Proposed mechanism for transfer of lipoplexes/genosomes to the nucleus.

influence the trafficking of polyplex-mediated delivery to cell nuclei. Much more work on intracellular trafficking and nuclear entry need to be done to obtain efficient nonviral gene delivery *in vivo*.

## Direct Injection of Naked DNA

The simplest nonviral gene transfer system in use for gene therapy is the injection of naked plasmid DNA (pDNA) into local tissues or the systemic circulation (88, 100). Naked DNA systems are composed of a bacterial plasmid that contains the cDNA of a reporter or therapeutic gene under the transcriptional control of various regulatory elements (101, 102). In recent years, work in several laboratories has shown that naked plasmid DNA (pDNA) can be delivered efficiently to cells *in vivo* either via electroporation, or by intravascular delivery, and has great prospects for basic research and gene therapy (101). Efficient transfection levels have also been obtained on direct application of naked DNA to the liver (103, 104), solid tumours (105), the epidermis (106), and hair follicles (106).

One of the obstacles with these systems is, in systemic circulation, that naked DNA is degraded rapidly by nucleases and cleared by the mononuclear-phagocyte system. However, naked DNA injection (into mice) can induce efficient gene transfer in internal organs (101). Since physical pressure (e.g., hydrodynamic or hydrostatic) is probably the major driving force in delivering DNA into cells, it might be possible to treat human cancer patients using simple intramuscular injections of naked pDNA expressing a cytokine gene, delivered on an infrequent basis (107, 108).

The major problem associated with plasmid based gene delivery to skeletal muscle is the relatively low efficiency of transfection. Recent developments have demonstrated improved delivery associated with the application of an electrical field to the muscle after injection of the plasmid DNA (109). It is clear that the application of naked DNA close to the site of pathology and away from degradative elements such as plasma is thus a viable strategy for gene delivery. However, this method is ineffective if DNA dosing to anatomically inaccessible sites (e.g., solid tumors in organs) is desired.

It has been reported that the plasmid vector is unable to translocate to the nucleus unless complexed in the cytoplasm with nuclear proteins possessing NLS. NLS are short karyophilic peptides on proteins that bind to specific transporter molecules in the cytoplasm, mediating their passage through the pore complexes to the nucleus. Examples of these peptides will be given later in this section. DNA can also be presented to cells in culture as a complex with polycations such as polylysine, or basic proteins such as protamine, total histones or specific histone fractions (110), cationized albumin, and others. These molecules increase the transfection efficiency. In addition histone H1 is identified as transfection-enhancing protein in cell culture (111).

Thee future prospects for naked DNA gene therapy include clinical trials for genetic diseases (e.g., Duchenne muscular dystrophy, ischemia, hemophilia), which would be initiated in the next few years, and tail vein injections in rodents, which will become a widely used technique for rapidly testing expression vectors/gene therapy approaches (101).

## Liposomes

Liposomes are well characterized and widely used as gene delivery systems because of their potential advantages such as the transient expression of delivered genes and avoidance of chromosomal integration (112–117). Among others, cationic lipids and liposomes are the most extensively investigated nonviral vectors (118–120). Cationic liposomes react spontaneously with the negatively charged DNA molecules (self-assembling system), forming complexes with DNA molecules participating in the reaction (121). These complexes have become a popular method for delivering therapeutic genes and are being tested in preclinical and clinical trials. Cationic lipoplexes are easy to produce, they are made up of nontoxic and nonimmunogenic precursors, and they have the potential to deliver large polynucleotides into somatic cells. In addition, these reagents are easily manipulated in the laboratory to incorporate novel biological functions, or to produce new formulations that can be screened for *in vivo* gene transfer activity (122–126). The cationic lipid component is amphipathic and can vary in its chemical structure. An example of cationic lipids are dioctadecylamidoglicylspermin (DOGS or "transfectam") (127) or "lipofectin" (128, 129). One of the first cationic lipids used, dioleoxypropyl trimethylammonium chloride (DOTMA) (129) whose polar head group has the quaternary amine moiety. Thus, vesicles formed by DOTMA carry a positive charge on their surface. Each cationic lipid can have single or multiple cationic charges and the overall positive charge must be preserved. In other words cationic lipids may have monocationic head groups, such as DOTMA, dimyristooxypropyl dimethyl hydroxyethyl ammonium bromide (DMRI) (130), dioleoyloxy-3-(trimethylammonio)propane (DOTAP) (130, 131), DC-Chol or polycationic head groups (DOSPA and DOGS) (130, 131). It is shown that two processes are involved in the complex formation. A fast exothermic process is attributed to the electrostatic binding of DNA to the liposome surface (132). A subsequent slower endothermic reaction is likely to be caused by the fusion of the two components and their rearrangement into a new structure. During this process, the homogenous and physically stable suspensions are often formed (133).

Successful gene delivery by use of cationic liposomes requires the following conditions (134): (1) condensation of DNA into the complex and its protection from degradation by intracellular nucleases; (2) adhesion of DNA–lipid complex onto the cellular surface; (3) complex internalization; (4) fusion of an internalized DNA–cationic liposome complex with the endosome membrane; (5) escape of DNA from the endosome; (6) entry of DNA into the nucleus followed by gene expression.

Entry of the DNA liposomes into the cell may occur by endocytosis with subsequent destruction of an endosome within the cell (135, 136) and direct fusion with cellular membrane (129). It was shown (137) that the major parts of the complexes are internalized by endocytosis and only 2% of cells are transfected through direct complex-membrane fusion. The positive charge on the particle surface ensures their binding to the negatively charged cellular membranes. Adhesion of the complex, containing the positively charged cationic liposomes onto the negatively charged outer membrane of the cell occurs through electrostatic interactions. Removal of the negatively charged glycoprotein from the cellular membrane diminishes the transfection efficiency, whereas the treatment of cells with poly-L-lysine (PLL) prior to transfection enhances it. PLL probably helps the formation of a protective layer, consisting of positively charged polypeptide residues on the negatively charged cellular surface, or to the enhanced adhesion of the complex (136). To facilitate endocytosis, the incorporation of proteins such as anti–MNS-antibodies, transferrin and Senday virus into liposomes may be accomplished, which will allow for plasmid DNA penetration from the endosome into the cytoplasm, thereby avoiding degradation. In addition, incorporation of only small amounts of anionic lipid into liposomes leads to DNA association with the inner surface of the liposomal membrane, which protects DNA from enzymatic degradation (135, 136).

It was believed that the main factors affecting transfection efficiency were the structure of the cationic lipid, the type of helper lipid used and their susceptibility to disruption by serum proteins. For gene transfer *in vivo*, apart from DOTMA-based liposomes, other complexes (in equimolar ratios) are also used—such as dioctadecylamidoglicylspermidin (DLS)/DOPE (137), DOPE/DOTMA (1:1), DOPE/DOTAP (1:1) (138, 139), dimethyloctadecylammonium bromide (DDAB), and DOTAP with cholesterol (1:1) (mol/mol) (139).

Cationic liposomes are usually formed from a variety of cationic lipids that are incorporated with a neutral lipid such as DOPE (dioleoylphosphatidyl-ethanolamine) (excluding DOGS-based liposomes which cannot be used this way) to facilitate membrane fusion. In many cases, the equimolar mixture of a cationic lipid and DOPE ensures the optimally efficient transfection (130). However, it has been shown in some studies that inclusion of DOPE into the complex essentially lowers the transfection efficiency (131–139). The commercially available liposomes such as Lipofectamine™ and Lipofectin® (Invitrogen Corporation, Carlsbad, CA) have been developed as DNA delivery vehicles using DOTAP (N-1(-(2,3-dioleoyloxy)propyl)-N,N,N-trimethylammoniumethyl sulphate) with DOPE, and DOTAP with DOTMA (N-(1-(2,3-dioleoyloxy)propyl)-N,N,N-trimethylammonium chloride), respectively (140–142). Usage of neutral lipids such as cholesterol and its derivatives allows one to attain higher transfection levels *in vivo* (131, 133, 143). Reliably higher expression in many organs was observed upon application of cholesterol-containing liposomes (143) as compared to other liposomes. DOPE-containing liposomes, as well as various galactosylated cholesterol derivatives, exhibit low toxicity and high transfection efficiency in human hepatoma cells, Hep G2. The efficiency enhancement is caused apparently by affinity to the asyaloglycoprotein receptor, specific for parenchymal liver cells (144). Taking into account an important role of cholesterol in efficient gene transfer, a series of oligocations containing one, two, or three cholesteryl moieties (143), provide a possibility to vary hydrophobicity/hydrophilicity ratio inside the same group of transfection agents, which could be important to provide targeted delivery of therapeutic genes. They represent a new group of transfection mediators with high transfection efficiency (145). In a recent study liposomes formed by *O,O*-ditetradecanoyl-*N*-(-trimethylammonioacetyl) diethanolamine chloride (DC-6-14) and DOPE or cholesterol as helper lipid, exhibit high transfection efficiency with regard to disseminated peritoneal tumour cells. They are more efficient than commercial cationic liposomes such as "lipofectin," "lipofectACE," and "lipofectamine" (146).

The major disadvantage of cationic liposomes as a gene delivery system is their much lower transfection efficiency compared to viral vector systems. The limitations of cationic liposome usage for transfection purposes (as a gene therapy tool) are closely connected to a short lifetime of the complexes, as well as to their inactivation by serum proteins and toxicity of cationic lipids in high concentrations. Another drawback of cationic lipid-mediated gene transfer is the lipoplexes colloidal insta- bility, which leads to the formation of large aggregates. This creates difficulties for *in vivo* studies and clinical trials, which require homogenous lipoplexes with size compatible with *in vivo* delivery. It has been found that addition of high sucrose and trehalose concentrations in freeze dried products of DOTAP/DNA lipoplex-enhanced transfection efficiency and physical characteristics of the complex. Although many cell culture studies have been documented, systemic delivery of genes with cationic lipids *in vivo* has been very limited (135–140). After systemic application, lipoplexes were rapidly cleared from the bloodstream due to aggregation of the complexes and the highest expression levels are observed in first-pass organs, particularly the lungs. Therefore, all clinical protocols use subcutaneous, intradermal, intratumoral and intracranial injection as well as intranasal, intrapleural, or aerosol administration but not IV delivery because of the toxicity of the cationic lipids and DOPE (147). Liposomes formulated from DOPE and cationic lipids based on diacyltrimethylam- monium propane (dioleoyl-, dimyristoyl-, dipalmitoyl-, disteroyl-trimethylammo- nium propane or DOTAP, DMTAP, DPTAP, DSTAP, respectively) or DDAB were highly toxic when incubated *in vitro* with phagocytic cells (macrophages and U937 cells), but not toward nonphagocytic T lymphocytes; the rank order of toxicity was DOPE/DDAB > DOPE/DOTAP > DOPE/DMTAP > DOPE/DPTAP > DOPE/DSTAP (145–148).

Another factor to be considered before IV injections are undertaken is that nega- tively charged serum proteins can interact and cause inactivation of cationic liposomes. Condensing agents used for plasmid delivery including polylysine, transferrin-polyl- ysine, a fifth-generation poly(amidoamine) (PAMAM) dendrimer, poly(ethylene- imine), and several cationic lipids (DOTAP, DC-Chol/DOPE, DOGS/DOPE, and DOTMA/DOPE) were found to activate the complement system to varying extents. Strong complement activation was seen with long-chain polylysines, the dendrimer, poly(ethyleneimine), and DOGS; complement activation was considerably reduced by modifying the surface of preformed DNA complexes with polyethyleneglycol (148).

The interaction of plasmid DNA with protamine sulfate followed by the addition of DOTAP cationic liposomes offered a better protection of plasmid DNA against enzymatic digestion and gave consistently higher gene expression in mice via tail vein injection compared with DOTAP/DNA complexes, gene expression was detected in the lung as early as 1 hour after injection, peaked at 6 hour and declined thereafter. Intraportal injection of protamine/DOTAP/DNA led to about a 100-fold decrease in gene expression in the lung as compared with IV injection; endothelial cells were the primary locus of lacZ transgene expression (147). Protamine sulfate enhanced plasmid delivery into several different types of cells *in vitro* using the monovalent cationic liposomal formulations (DC-Chol and lipofectin); this effect was less pronounced with the multivalent cationic liposome formulation, lipo- fectamine (149).

Spermine has been found to enhance the transfection efficiency of DNA-cationic liposome complexes in cell culture and in animal studies; this biogenic polyamine at high concentrations caused liposome fusion most likely promoted by the simultaneous interaction of one molecule of spermine (four positively charged amino groups) with the polar head groups of two or more molecules of lipids. At low concentrations (0.03–0.1 mM) it promoted anchorage of the liposome–DNA complex to the surface of cells and enhanced significantly transfection efficiency.

A number of factors for DOTAP-cholesterol/DNA complex preparation including the DNA/liposome ratio, mild sonication, heating, and extrusion were found to be crucial for improved systemic delivery; maximal gene expression was obtained when a homogeneous population of DNA/liposome complexes (200–450 nm) was used. Cryoelectron microscopy showed that the DNA was condensed on the interior of liposomes between two lipid bilayers in these formulations, a factor that was thought to be responsible for the high transfection efficiency *in vivo* and for the broad tissue distribution (150).

Steric stabilization of liposomes generally increases biocompatibility, reduces immune response, increases *in vivo* stability and delays clearance by the reticulo–endothelial system. However, toxicity remains a major barrier to the use of cationic lipids in clinical trials (135). Cationic lipids increase the transfection efficiency by destabilizing the biological membranes including plasma, endosomal, and lysosomal membranes; indeed, incubation of isolated lysosomes with low concentrations of DOTAP caused a striking increase in free activity of b-galactosidase, and even a release of the enzyme into the medium demonstrating that lysosomal membrane is deeply destabilized by the lipid. The mechanism of destabilization was thought to involve an interaction between cationic liposomes and anionic lipids of the lysosomal membrane, allowing a fusion between the lipid bilayers; the process was less pronounced at pH 5.0 than at pH 7.4 and anionic amphipathic lipids were able to prevent partially this membrane destabilization (151). In contrast to DOTAP and DMRIE (100% charged at pH 7.4), DC-Chol was only about 50% charged as monitored by a pH-sensitive fluorophore; this difference decreases the charge on the external surfaces of the liposomes and promotes dissociation of DC-Chol from the plasmid DNA and an increase in release of the DNA-lipid complex into the cytosol from the endosomes (152). Encapsulation of oligonucleotides into liposomes increased their therapeutic index, avoided degradation in human serum and reduced toxicity to cells (153–155). In addition, conjugation to a fusogenic peptide enhanced the biological activity of antisense oligonucleotides (156). It has been found that complexes of oligonucleotides with DOTAP liposomes entered the cell using an endocytic pathway, oligonucleotides redistributed from cytoplasmic regions into the nucleus and the process was independent of acidification of the endosomal vesicles. The nuclear uptake of oligonucleotides depended on charge of the particle, and negative charges on the cell membrane are required for efficient fusion with cationic liposome-oligonucleotide complexes to promote entry to the cell (157). Transfection by the use of anionic or neutral liposomes is not very efficient. Anionic and neutral liposmes are transfection efficient when applied *in vitro* and much less efficient *in vivo*. In recent studies, transfection was conducted based on the use of amphiphilic phospholipid vesicles in the presence of high concentration of bivalent metal ions

(Ca, Mg) (158). Phosphate groups of DNA within the neutral liposome are bridged with phosphoryl groups of phospholipids by divalent metal ions. Such complexes form a network consisting of nucleic acid and liposomes, bridged by metal (II) ions. The structure of the complexes was investigated by electron microscopic and the results showed that DNA is enwrapped into cylindrical phospholipid bilayers (159–160), allowing it to retain stability in transfection conditions *in vivo*. Neutral phospholipids are not subject to the major drawback of cationic liposomes which form complex with the DNA too tightly and prevent DNA release and its nuclear uptake. The addition of anionic liposomes to the DNA–lipid complex is known to lead to the rapid destabilization of a membrane and to DNA release (160).

## Polymers

Polymer-based systems (e.g., lactic or glycolic acid, polyanhydride or polyethylene vinyl coacetate and collagen) provide several potential advantages for the therapeutic delivery of DNA including: (a) DNA can be delivered to target tissues in sustained, controlled, predictable manner; (b) DNA is effectively protected before being released; (c) site-specific delivery is possible with simple implantation or direct injection; and (d) repeated injection is not necessary. For polymer-based gene delivery system, three different mechanisms are reported: (1) cationic polymers execute their effect by both condensing large genes into smaller structures as well as masking the negative DNA charges to necessitate transfecting to most types of cells; (2) neutrally-charged polymers such as PVA (polyvinyl alcohol), which cannot condense DNA, are considered to protect DNA from extracellular nuclease degradation therefore to retain their integrity and biological function at the site of action; (3) polymeric nanoparticles or microparticles based on PLGA (polylactic-coglycolic acid), gelatin, alginate, chitosan, etc., which absorb or encapsulate genes, are regarded as important matrices for sustained release of gene materials.

### Cationic Polymer-Based Gene Delivery Systems

Under a variety of conditions, plasmid DNA undergoes a dramatic compaction in the presence of condensing agents such as multivalent cations and cationic polymers. Naked DNA coils, typically with a hydrodynamic size of hundreds of nanometers, after condensation it may become only tens of nanometer in size. Contrary to proteins which show a unique tertiary structure, DNA coils do not condense into unique compact structure. Cationic polymers execute their gene carrier function by their condensation effect on gene materials and, furthermore, their protection effect on DNA from nuclease digestion. Currently, the most widely used cationic polymers in research include linear or branched PEI (poly (ethyleneimine) (161–165), polypeptides such as PLL (poly-L-lysine) (166–169), PLA (poly-L-arginine) (170).

Poly (ethylenimine) (PEI) has been demonstrated as an efficient gene delivery vehicle both *in vitro* and *in vivo* (161–163). Linear (22 kDa) and branched PEI formulations of varying molecular weights (0.6–800 kDa) have been reported. While polyplexes from higher molecular weight branched PEIs (70–800 kDa) were found to be more efficient *in vitro* but on intravenous administration the smaller and linear PEIs seem in general to be more efficient (171). However, questions as to the mechanism of PEI-mediated transfection remain largely unanswered. Recently, it was discovered that PEI–DNA complexes enter cell nuclei intact during the transfection process. This finding brings to light the question of what effect the polycationic polymer has on cells after nuclear entry. Polycations (such as PEI) act to spontaneously bind with and condense plasmid DNA in the test tube, so it is not unwarranted to predict that PEI in cell nuclei might also interact with host DNA (161–163). Such an alteration of the nuclear environment has the potential to alter host transcriptional processes and thereby affect the well-being of the cell (or organism) as a whole. The charge ratio is usually defined as the molar ratio of positive charges on the polymers to the negative charge of the DNA phosphate group, for PEI (where protonated amine groups depend on pH) the charge ratio could be defined as the molar N/P ratio of PEI nitrogen to DNA phosphates (172). Poly-(ethylenimine) (PEI) is a highly branched cationic polymer with a ratio of 1:2:1 of primary:secondary:tertiary amines. The high-density amine groups give PEI several advantages over other cationic carriers, such as to help the formation of tighter and smaller complexes with negative DNA through charge interactions to act as proton sponge to facilitate the release of DNA from the endosomes and to aid the delivered plasmid to enter the nucleus (173). The polycation has terminal amines that are ionizable at pH 6.9 and internal amines that are ionizable at pH 3.9 and because of this organization can generate a change in vesicle pH which leads to vesicle swelling and, eventually, release from endosome entrapment. PEI polyplexes have been used to achieve gene expression in experimental animals by direct application to various anatomical sites such as rat kidneys by intrarterial injection (174), mouse brains (175,176) and mouse tumors (177, 178) by direct injection and rabbit lungs by intratracheal administration (179, 180). Overall the gene expression seen with linear PEI is superior to that seen with cationic liposomes both on intravenous (181) and intratracheal administration.

PEI with molecular weight 25kDa has been successfully shown to mediate intracellular DNA delivery and transgene expression in terminally differentiated neurons, when transgene expression was investigated in rat spinal cord through repeated intrathecal administration of plasmid DNA complexed with 25 kDa PEI into the lumbar subarachnoid space, with a single injection, DNA/PEI complexes could provide transgene expression in the spinal cord 40-fold higher than naked plasmid DNA (182). It has been shown that linear 22 kDa PEI gives better lung delivery than branched 25 kDa PEI or liposomes but gene expression is mainly restricted to alveolar cells, such as pneumocytes (183).

Poly-L-Iysine (PLL) and poly-L-arginine (PLA) are synthetic poly-cations that consist of repeating lysine and arginine residues, respectively, thus possess a biodegradable nature (166–169). The biodegradable nature is an important factor for *in vivo* use. PLL has been extensively studied for gene and oligonucleotide delivery due to several advantages: (a) easily available and cost effective; (b) presence of side substituents for the ligand or drug incorporation. Encapsulation of DNA–PLL complex in controlled release drug delivery devices such as PLGA microspheres, nanoparticles have been reported earlier as an alternative for improving gene delivery. But due to drawbacks such as low plasmid DNA incorporation and formation of acidic byproducts during the polymer degradation and release of DNA have hampered the development of these encouraging strategies (184). PLL or PLA can be synthesized in various sizes ranging from 15 to over 1000 residues in length. The binding of the DNA to PLL or PLA occurs between the positive charge of the amino groups and the negative charge of the phosphate groups on the DNA. This contributes to complete charge neutralization on the DNA molecule. Although poly-L-lysine polyplexes prevent the degradation of DNA by serum nucleases (185) in a similar manner to liposomes (186, 187); on intravenous injection, these polyplexes, are bound by plasma proteins and rapidly cleared from the plasma (188), again like cationic liposomes (189). Polyplex opsonization by plasma proteins may be suppressed by coating the polyplexes with a hydrophilic polymer such as hydroxypropyl methacrylic acid, and the cellular uptake of the polyplexes may once again be promoted by the conjugation of targeting ligands such as transferrin (190) or fibroblast growth factor (191) to the surface of the coated polyplexes. In fact PLL has very low transfection ability when applied alone without any modification, one of the most common modification is coating with PEG which not only increases the half-life of these vectors but also the transfection efficiency.

In addition the efficiency of synthetic vectors like PLL can be improved using artificial nucleic-acid carriers incorporating functional elements that mimic viruses. For example, the adenovirus hexon protein enhances the nuclear delivery and increases the transgene expression of polyethyleneimine–pDNA vectors (192). Furthermore, artificial viral envelopes that mimic the lipid composition of retroviruses can be used to encapsulate condensed pDNA. Nonviral vectors have also been designed to mimic the receptor-mediated cell entry of adenoviruses; early attempts to deliver genes by nonviral receptor-mediated endocytosis had failed to mediate gene transfer effectively, mainly owing to a lack of cellular endosome escape. The inclusion of viral fusogenic peptides or endosomolytic agents in nonviral vectors has enhanced the gene delivery capacity of these systems. The adenovirus penton protein has also been used in this way *in vitro*. In addition, formulations that combine the merits of viral and nonviral systems have been developed, such as a virus–cationic-liposome–DNA complex (193), "haemagglutinating virus of Japan" liposomes (194), and cationic-lipid–DNA mixed with the G glycoprotein from the "vesicular stomatitis virus" envelope (195).

It has been shown that poly-(2-(dimethylamino) ethyl methacrylate) (pDMAEMA) as well as other copolymers are able to introduce DNA into cells (196). Freeze drying was shown to be an excellent method to preserve the size and transfection potential of pDMAEMA/plasmid complexes (polyplexes), even after aging at $40°C$ (197). In another study, it was demonstrated that the DNA topology affected pDMAEMA-mediated transfection: the circular forms of DNA (supercoiled and open circular) had a higher transfection activity than the linear forms (198). Dextran, a branched polymer consisting of repeating units of glucose, is a commonly used biopolymer in drug delivery systems to enhance the circulation time of drug. Its branched structure might not only provide better shielding effect than those linear polymers, such as PEG, to minimize the charge interactions with serum proteins but also allow the conjugation of multiple ligands on each dextran molecule to increase the valance of the modified vector. The conjugation of dextran onto PEI has been shown to improve the stability of the DNA–polymer complex in the presence of serum (199). Nevertheless it is unclear how dextran molecular weight and the degree of dextran grafting affect the stability of DNA–polymer complexes.

Poly-(3-hydroxybutanoic acid) (PHB), belongs to the large family of poly-(hydroxyalkanoates) (PHAs), high molecular weight natural polymers produced by various microorganisms and stored in cell cytoplasm (200). Low molecular weight PHB, also present in bacteria and are primarily involved in transport of ions and DNA across inner bacterial membrane (201). PHB could be developed as a valuable biocompatible material with possible applications in gene delivery after cytotoxic, safety, and efficacy evaluations.

Another cationic polymer, poly-{-(4-aminobutyl)-L-glycolic) acid} (PLAGA) has been shown to condense DNA efficiently and also to be less cytotoxic than PLL. PLAGA is biodegradable, not toxic to the cells, and enhances transfection in cultured cells (202).

## Neutrally Charged Polymer-Based Gene Delivery System

Intrinsic drawbacks with cationic carriers, such as solubility, cytotoxicity, and low transfection efficiency, have limited their use *in vivo*. These vectors sometimes attract serum proteins and blood cells when entering the circulation, resulting in dynamic changes in their physicochemical properties. Therefore recent studies have focused on the development of neutrally charged gene delivery systems. In this category, PVP (polyvinylpyrolidone) and PVA (polyvinyl alcohol) are the representative reagents. When used for gene delivery, their basic function is considered to be a protective effect for the DNA from degradation by nuclease. Because of their neutrally charged structure, the transfection efficiency of this group of polymers is always relatively low. But they could be very useful and an important composition for gene delivery when combined with other cationic polymers.

These stabilizers are added to the formulation in order to stabilize the emulsion formed during particle preparation. These stabilizers, however, can also influence the properties of the particles formed. The type and concentration of the stabilizer selected may affect the particle size. Being present at the boundary layer between the water phase and the organic phase during particle formation, the stabilizer can also be incorporated on the particle surface, modifying particle properties such as particle zeta potential and mucoadhesion (203). Other polymers have also been evaluated as stabilizers in earlier studies such as cellulosic derivatives methylcellulose (MC), hydroxyethylcellulose (HEC), hydroxypropylcellulose (HPC), and hydroxypropylmethylcellulose (HPMC), as well as gelatin type A and B, carbomer and poloxamer (203).

## Polymeric Nanoparticles or Microparticles-Based Gene Delivery System

Particulate systems (e.g., collagen, lactic or glycolic acids, polyanhydride or poly-ethylene vinyl coacetate) have several potential advantages for the therapeutic delivery of DNA (or of drugs), such as DNA encapsulation within the polymer can protect DNA degradation until release; injection or implantation of the polymer into the body can be used to target a particular cell type or organ; drug release from the polymer and into the tissue can be designed to occur rapidly (a bolus delivery) or over an extended period of time, thus, the delivery system can be tailored to a particular application. The choice of polymer and its physical form determine the DNA release characteristics.

Control over DNA delivery can be achieved by the formation of both synthetic and natural polymers in a variety of dosage forms such as matrix, reservoir, nano/microparticles.

Nanoparticles based on their size can be endocytosed by the cells, which would effectively increase the cellular uptake of the entrapped DNA. The DNA entrapped in the nanoparticles would be released slowly with the hydrolysis of the polymer matrix; thus achieving a sustained gene expression in the target tissue. PLGA (D,L-lactide co-glycolide) is a biocompatible and biodegradable polymer frequently employed to formulate nanoparticles/microspheres for gene therapy systems, although other polymers such as chitosan, gelatin and other biodegradable polymers have also been investigated for nonviral gene therapy. Chitosan is a biodegradable polysaccharide, which has an established toxicity profile. Due to its good biocompatibility and toxicity profile, it has been widely used in pharmaceutical research and in industry as a carrier for drug delivery and as biomedical material for artificial skin and wound healing bandage applications (204). Chitosan has also been shown to effectively bind DNA in saline and acetic acid solution and partially protect it from nuclease degradation (205).

Nanoparticles (NPs) formulated from the biodegradable PLGA is able to cross the endosomal barrier and deliver the encapsulated therapeutic agents into the cytoplasm. NPs are colloidal systems that typically range in diameter from 10 to 1000 nm, with the therapeutic agent either entrapped into or adsorbed or chemically coupled onto the polymer matrix (206). The PLGA NP formulation with a therapeutic agent entrapped into the polymer matrix provides sustained drug release. The degradation products of PLGA are lactic and glycolic acids that are formed at a very slow rate and are easily metabolized in the body via the Krebs cycle and are eliminated.

Nanoparticles formulated with PLGA have been shown to be rapidly uptaken by the endothelial cells, the uptake was shown to depend on the nanoparticle concentration and the particles where mainly shown to localize in the cytoplasm (207). These nanoparticles were also shown to be biocompatible with the cells with no effect on cell viability (207). This is important due to the fact that endothelium is an important target for gene therapy in a number of disorders including angiogenesis, atherosclerosis, tumor growth, myocardial infarction, limb and cardiac ischemia, restenosis (207).

Polymers like polyalkylcyanoacrylate (PACA), polybutylcyanoacrylate (PBCA), polyisohexylcyanoacrylate (PIHCA), and polyhexylcyanoacrylate (PHCA) have also been used in nanoparticle formulations. Because of the negative surface potential on the particles, a cationic polymer or cationic detergent was generally combined with the polymers to facilitate the binding of the DNA (208). Diethylaminoethyl (DEAE)-dextran is a polycationic derivative of dextran containing diethyl amino groups coupled with glucose residues with ether linkage has also been used in conjunction with PHCA and other polymers in the formulation of positively charged nanoparticles (209).

Microspheres can deliver DNA in two ways: DNA can be physically entrapped in the microspheres, or DNA can be bound through electrostatic interaction with the positive charged surface of the particles. Microparticles in particular have a number of advantages over other delivery forms for gene delivery including, depending on their composition they can be delivered through different ways (oral, subcutaneous, intramuscular), they are suitable for scale up to industrial size production. Microspheres can be delivered in a minimally invasive manner (e.g., by direct injection or by oral delivery), and matrices can be implanted at the appropriate site, for example, for applications in tissue repair and wound healing. Targeting of gene transfer has also been achieved by modification of gene carriers using cell targeting ligands, such as asialoglycoproteins for hepatocytes, transferrin for some cancer cells, insulin, or galactose. In addition, a targeted folate-expressing, cationic-liposome–based transfection complex has been shown to specifically transfect folate-receptor–expressing cells and tumors, suggesting that this is a potential therapy for intraperitoneal cancers (210).

Dendrimers are synthetic hyperbranched polymers which are highly soluble in water. Although they are less efficient than the viral vectors they can be used in gene therapy based on their safety and the lack of immunogenicity (211). The branched structure terminating in charged groups, results in the binding of DNA. Dendrimers with positively charged terminal groups can bind DNA forming complexes called as dendriplexes (211). The overall positive charge on the dendriplexes is thought to affect its interaction with the cell surface, thus these groups can be modified to incorporate molecules for targeting (211).

## Receptor-Mediated Gene Delivery

As an extension of lipofection or polyfection, targeting ligands have been incorporated into DNA complexes to serve both aims of (a) targeting specific cell types and (b) enhancing intracellular delivery. The receptor-mediated gene delivery provides an opportunity to achieve cell specific delivery of DNA complexes. The structure–function relationships for many receptor ligands are known, so it is possible to obtain receptor ligands with high binding affinities (1–10 nM). Most of the receptor ligands are proteins and in general the receptor binding ligand domain of the protein has been identified by site specific mutagenesis. The mechanisms and routes of internalization of several receptor–ligand complexes *in vivo* have been partially characterized, therefore, the biodistribution of the DNA complex can be predicted.

The asialoglycoprotein receptor has been widely used because it is a part of the surveillance system in the circulation and it removes proteins as they age by the spontaneous loss of sialic acid (212, 213). This hepatic receptor has been widely used (214) to deliver reporter genes to the rat liver *in vivo*. In a study, a ligand for the asialoglycoprotein receptor (ASOR) was covalently linked to poly(L-lysine). After hepatic uptake of the complex of DNA and the ASOR–poly (L-lysine) conjugate, the reporter gene product is found in the liver (214). Other examples of receptor-mediated gene delivery include the transferrin receptor (215), epidermal growth factor EGF receptor (216), polymeric immunoglobulin receptor (217), CD3-T cell receptor (218), lectins (219, 220), folate receptor (221), malarial circumsporozoite protein receptor (222), integrins (223), the insulin receptor (224), and the thrombomodulin receptor (225).

Use of fusogenic peptides from influenza virus hemagglutinin HA-2 enhanced greatly the efficiency of transferrin-polylysine–DNA complex uptake by cells; in this case the peptide was linked to polylysine and the complex was delivered by the transferrin receptor-mediated endocytosis. Curiel (226) used the transferrin receptor on the surface of mammalian cells to deliver plasmid-polylysine-transferrin complexes to cells. These complexes are taken up by endosomes following receptor binding, a method that suffers in that the endocytosed DNA is trapped in the intracellular vesicle and is later largely destroyed by lysosomes; use of the capacity of the adenoviruses to disrupt endosomes as part of their entry mechanism to the cells have augmented over 1000-fold the efficiency of gene transfer.

The obstacles of receptor-mediated gene delivery are the naturally occurring receptor ligands (either proteins or complex carbohydrates), which are extremely difficult to obtain in high purity and in sufficient quantity. These receptor ligands are usually covalently crosslinked to poly-(L-lysine), thereby creating novel antigenic epitopes (227). Because the conformation of the binding surface is formed by chance, binding of DNA to the conjugates is variable. The differences in polydispersity of the commercially available poly-(L-lysine) is one of the major causes of variability in the formulation of reproducible, stable formulations. The polydispersity of commercial poly-(L-lysine) means that the individual molecular species of the polycation interact with DNA with individually distinct kinetics, for both the electrostatic and the hydrophobic interactions. The extreme heterogeneity greatly complicates both the kinetics of DNA–poly-(L-lysine) interaction and the thermodynamic stability of the final DNA complexes. Further, poly-(L-lysine) exists as a random coil at neutral pH, an -helix at alkaline pH and a mixture of conformations at pH 7.4 (228). In addition to the molecular heterogeneity of poly (L-lysines), they are toxic to living cells in nM concentrations, which limits their general applicability.

## Hybrid Vectors (Viral and Nonviral Components)

As shown in Figure 13.6. hybrid vectors incorporate viral components in cationic-amphiphile based vectors. The efficiency of synthetic vectors can be improved using artificial nucleic-acid carriers incorporating functional elements that mimic viruses. For example, the adenovirus hexon protein enhances the nuclear delivery and increases the transgene expression of polyethyleneimine–pDNA complex (229). Nonviral vectors have also been designed to mimic the cell entry of adenoviruses; early attempts had failed to mediate gene transfer effectively, mainly owing to a lack of cellular endosome escape. The inclusion of viral fusogenic peptides or endosomolytic agents in nonviral vectors has enhanced the gene delivery capacity of these systems. The adenovirus penton protein has also been used in this way *in vitro*.

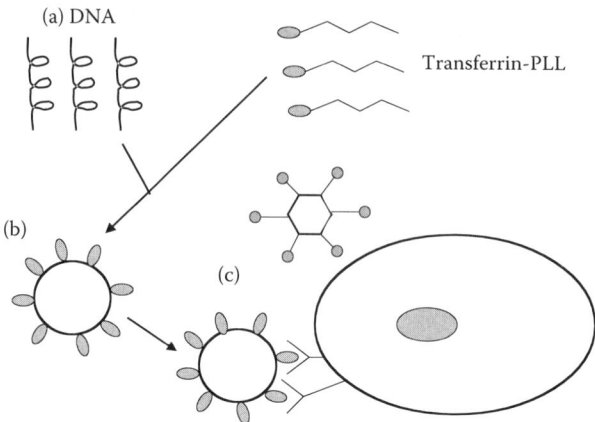

**FIGURE 13.6** Viral–nonviral hybrid vectors. DNA is bound to a poly-L-lysine (PLL)–transferrin conjugate (A) to form a PLL–transferrin–DNA complex (B). Transferrin binds to specific receptors on the surface of some cancer cells, thereby targeting gene delivery to these cells (C).

## Nuclear Localization Signal Peptides and Membrane-Modifying Agents

### NLS Peptides

Translocation of exogenous DNA through the nuclear membrane is a major concern of gene delivery technologies. Translocation of macromolecules across the nuclear membrane is via pores that serve as size-exclusion barriers, with small molecules entering the nucleus by passive diffusion while larger macromolecules only enter via highly regulated active processes. Selective nuclear import of macromolecules larger than 40 kDa is mediated by nuclear localizing signals (NLS). The NLS is recognized by a heterodimeric protein complex of importin-α (also known as karyopherins-α, Kap-α) and importin-β (Kap-β). Importin-α then interacts directly with the cargo NLS, whereas importin-β docks the complex to the nuclear pore complex (NPC) by specifically binding to a subset of hydrophobic phenylalanine–glycine-rich repeats (FG repeats, standard IUPAC single-letter amino acid code), repeat containing NPC proteins (nucleoporins, or nups) (230–231).

The ability of NLS–peptides to mediate delivery of biologically active pDNA has been tested indirectly by monitoring the reporter gene activity following direct introduction into the cell via microinjection and/or transfection of the DNA–NLS conjugate into the cells. Ludtke et al. (1999) showed that microinjection of a construct consisting of a biotinylated 900-bp green fluorescent protein (GFP) expression vector with streptavidin conjugated to a 39 amino acid peptide (H-CKKKSSSD-DEATADSQHSTPPKKKRKVEDPKDFPSELLS) containing a functional SV40 large T antigen NLS resulted in a fourfold increase in GFP expression compared to that seen with constructs coupled to mutant NLS (232). In another study (233) NLS peptide (PKKKRKVEDPYC) was conjugated to closed linear plasmid DNA. This DNA fragment contained the minimal immuno genetically defined gene expression (MIDGE) vector for hepatitis B surface antigen (HbsAg) required to stimulate an antibody response to HbsAg after gene gun immunization of BALB/c mice. Compared to the identical expression vector not conjugated to the NLS peptide, the HbsAg plasmid–NLS conjugate enhanced by 15-fold, both the priming of antibody responses to HbsAg after intramuscular injection and transfection efficiency *in vitro*. The enhanced immune response observed following treatment with import-competent NLS conjugated plasmid was not, however, compared to a control import-deficient conjugated plasmid. There are a number of potential drawbacks in the use of NLS peptides to enhance DNA nuclear import. Firstly, the approach involves the use of chemically synthesized proteinaceous material which triggers the host immune surveillance. Even humanized NLS peptides might not be able to escape this fate as most common conjugation chemistries necessitate the addition of non-native cysteine residues. Secondly, and even though some recent evidence has demonstrated the sequence-specific interaction of the "classic" SV40 TAg NLS with DNA/DNA complexes, much still remains to be learned.

## Membrane-Modifying Agents

Gene transfer was found to be also strongly enhanced by the endosome-destabilizing activity of replication-defective adenovirus particles or rhinovirus particles which were either added to the transfection medium (234) or directly linked to the DNA complex (235, 236). A variety of membrane-modifying agents including pH-specific fusogenic or lytic peptides, bacterial proteins, lipids, glycerol, or inactivated virus particles have been evaluated for the enhancement of DNA–polycation complex-based gene transfer. The enhancement depends on the characteristics of both the cationic carrier for DNA and the membrane-modifying agent. Peptides derived from viral sequences have been used, such as the N-terminus of influenza virus hemagglutinin HA-2, the N-terminus of rhinovirus HRV2 VP-1 protein, and other synthetic or natural sequences such as the amphipathic peptides GALA, KALA, EGLA, JTS1, or gramicidin S.

Investigations with influenza HA2-derived sequences showed that the transfection levels largely correlate with the capacity of peptides to disrupt liposomes of natural lipid composition or erythrocytes in a pH-specific manner (237). Few changes in the influenza peptide sequence (introduction of acidic residues; effect of residues tryptophane-14 and -21 and length of amphipathic sequence; effect of the conserved stretch of glycines) have a strong effect on membrane-disruption and polylysine-mediated gene transfer. The following observations further confirm the hypothesis that the peptides enhance transfection activity by endosomal release: the block of endosomal protonation by the specific inhibitor bafilomycin reduces gene transfer activity; influenza peptides can trigger the pH-specific endosomal release of fluorescent molecules into the cytoplasm of cultured cells.

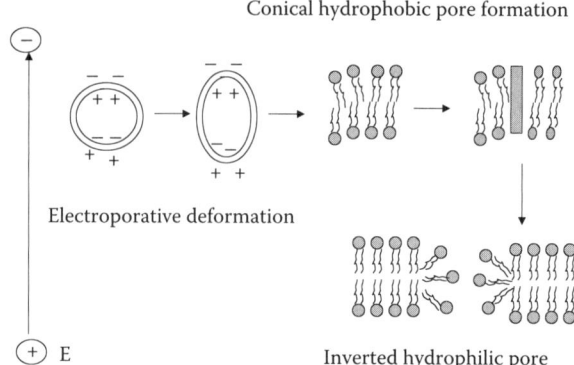

**FIGURE 13.7** Electroporation occurs when an applied external field exceeds the capacity of the cell membrane.

## ELECTROPORATION

In recent years, the physical techniques of gene transfer have gained increasing importance in gene therapy. As shown in Figure 13.7 electroporation designates the use of short high-voltage pulses to overcome the barrier of the cell membrane (238). Compared with other methods, electroporation has numerous advantages. It is simple, rapid, and relatively nontoxic to target cells (239). In general, electroporation can deliver any charged molecule such as chemical compounds, peptides, or even large proteins. Moreover, instruments for electroporation operation are commercially available and are easy to operate. Other advantages include the minimal need for manipulation of the transgene; the ability to transfect nondividing cells and the nonlimitation of transgene size (240). Despite these favorable features, limitations of electroporation include low transfection efficiency compared to virus-mediated gene transfer and the possibility of integration of the transgene. Over the past decades, the electroporation technology has been progressing in regards to equipments and sophistication of electroporation protocols.

Application of strong electric field pulses to cells and tissue is known to cause some type of structural rearrangement of the cell membrane. These rearrangements consist of temporary aqueous pathways with the electric field playing the dual role of causing pore formation and providing a local driving force for ionic and molecular transport through the pores.

Electroporation is a universal bilayer membrane phenomenon (241,242). Short (μs to ms) electric field pulses cause electroporation. For isolated cells, the necessary single electric field pulse amplitude is the range of 103–104 V/cm, with the value depending on cell size (243). A very general consideration is that the smaller the cell radius, the larger the electric field needed to achieve permeabilization. The extent of permeabilization can be controlled by pulse amplitude. For instance, the higher the pulse amplitude, the greater the area through which diffusion can take place (244). The degree of permeabilization can be controlled by the pulse duration, the longer the pulse the greater the perturbation of the membrane in a given area (245). Reversible electrical breakdown (REB) then occurs and is accompanied by greatly enhanced transport of molecules across the membrane.

Because of the nature of electroporation, virtually any molecule can be introduced into cells. For transfer of DNA, the electroporation forces are important. An electrophoretic effect of the field causes the polyanion DNA to travel toward the positive electrode. Fluorescence studies have shown that DNA enters the cell through the pole facing the negative electrode, where the membrane is more destabilized and where the field will drive the DNA towards the center of the cell (245). Membrane resealing occurs after pore formation. Whereas pore formation happens in the microsecond time frame, membrane resealing happens over a range of minutes with variations depending on electrical parameters and temperature (246).

Commercial instruments for electro cell manipulation (ECM) have been available for more than 20 years. For *in vitro* application as shown in Figure 13.8, ECM systems consist of a generator providing the electric signals and a chamber in which the cells are subject to the electric fields created by the voltage pulse from the generator. A third optional component is a monitoring system, which measures the electrical parameters as the pulse passes through the system. The instruments for *in vivo* electroporation include a voltage generator, applicators that transform the voltage into an efficacious electric field and electrodes. The generator provides a voltage output to the electrodes. This voltage between electrodes results in the generation of an electric field in the volume between the electrodes. The voltage needs to be selected so that in the volume between the electrodes the efficacious field strength is achieved. New ECM systems are under development. For example, a square wave electroporation system is designed for all mammalian *in vitro* and *in vivo* electroporation applications. The generator uses the new power platform technology and an all-new digital user interface. It has wide applications such as mammalian cell transfections/gene therapy, mammalian cell protein/drug electroincorporation, *in vivo* applications, nuclear transfer, plant bacterial and yeast applications, and bacterial and yeast electroporation.

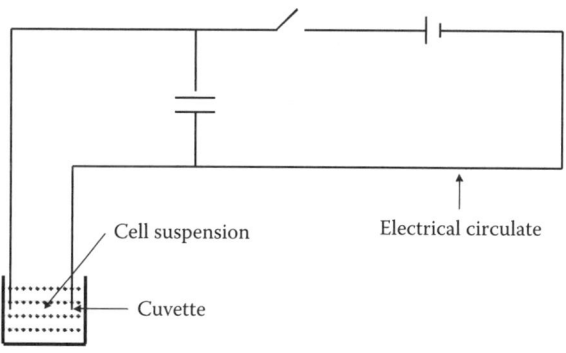

**FIGURE 13.8** An electrical circuit diagram for a simple electroporation device.

Electroporation had its inception in the 1960s with investigations into electrically induced breakdown of cell membranes (247). In 1980s electroporation was demonstrated to transfect intact cultured eukaryotic cells (248). Since then, *in vitro* transfection of animal and plant cells has become a principal application of electroporation. The success of *in vitro* transfection has led to the development of *in vivo* applications. Given the advantages of electroporation, it is a potential and promising method in the clinic setting, e.g., production of missing proteins in various deficiency syndromes, production of cytokines in the treatment of malignant cancers, or production of antigens such as vaccines to enhance an antitumoral host immune response.

Skeletal muscle is the most widely targeted tissue for *in vivo* electroporation gene delivery because it is easily accessible and able to produce secreted proteins. Muscle fibers have a long lifespan, which allows a long-term transgene expression. Thus, it is a good candidate as an endocrine tissue for expression of cytokines, growth factors or coagulation factors. An example is transfecting the tibial cranial muscle in mice with the gene encoding secreted alkaline phosphatase (SEAP). Electroporation boosted expression levels from 10- to 100-fold higher than naked DNA injection and produced long-term expression (249).

In vivo electroporation in tumor tissues is an exciting area of cancer gene therapy. Therapeutic genes including suicide genes, immune system stimulating cytokines, antiangiogenesis, and tumor suppressors can be used. For example, suicide gene therapy using HSVtk/ganciclovir technology suppressed the growth and metastasis of subcutaneously grafted mammary tumors in mice (250). Electroporation of DNA encoding cytokines such as IFN-, IL-12, and IL-18 is widely used. They have shown to reduce tumor growth and increase survival times in different tumor models.

Pretreatment of the muscle with enzymes such as hyaluronidase enhances the electroporation effect thus allowing a reduction in the field strength and consequently reducing the myofiber damage associated with electroporation (251). This raises the efficiency to the equivalent of the best that can be achieved with local delivery using the best viral vector.

Skin electroporation for increasing transdermal drug delivery have been explored. Drug delivery across skin offers advantages over conventional modes of administration. It avoids gastrointestinal degradation and the hepatic first-pass effect and has potential for controlled and sustained delivery. Using electroporation, transdermal delivery might be applicable to a broad range of drugs. Flurbiprofen, a potent nonsteroid anti-inflammatory drug, has been delivered transdermally by electroporation. Compared with iontophoresis of same amount of transported charges, flurbiprofen transport across skin by electroporation was much greater and more rapid (252).

Gene knockout is a commonly used method to study molecular functions of genes. Electroporation could be a powerful alternative for gene knockout. Recently, morpholino oligo, an efficient antisense DNA to inhibit the function of a corresponding gene, has been electroporated into developing tissues and check its effects on pattern formation (253). Electroporation is also increasingly being used for RNA transfer. mRNA has already been shown to be delivered by elecroporation (254).

In general, electroporation is a very promising technique both in gene therapy and functional genomics research as a laboratory tool. The successful use of electroporation is rapidly becoming clinically feasible for a wide range of gene correction, gene therapy and DNA vaccine applications. It may also useful for functional gennomics study via direct somatic tissue transfection.

## REFERENCES

1. Russell, S.J. (1997). Science, medicine, and the future: gene therapy. *BMJ,* 315, 1289–1292.
2. Russell, S.J., Peng, K.W. (2003). Primer on medical genomics. Part X: Gene therapy. *Mayo Clinic Proc.,* 78(11), 1370–1383.
3. Kasid, A., Morecki, S., Aebersold, P., et al. (1990). Human gene transfer: characterization of human tumor-infiltrating lymphocytes as vehicles for retroviral–mediated gene transfer in man. *Proc. Natl. Acad. Sci. U.S.A.,* 87, 473–477.
4. Flotte, T.R., Afione, S.A., Solow, R., et al. (1993). Expression of the cystic fibrosis transmembrane conductance regulator from a novel adeno-associated virus promoter. *J. Biol. Chem.,* 268(5), 3781–3790.
5. Walters, R.W., Yi, S.M., Keshavjee, S., et al. (2000). Binding of adeno-associated virus type 5 to 2, 3-linked sialic acid is required for gene transfer. *Nat. Genet.,* 24(3), 257–261.
6. Corvol, P. (1995). Liddle's syndrome: heritable human hypertension caused by mutations in the Beta subunit of the epithelial sodium channel. *J. Endocrinol. Invest.,* 18(7), 592–594.
7. Sellers, K.W., Katovich, M.J., Gelband, C.H., Raizada, M.K. (2001). Gene therapy to control hypertension: current studies and future perspectives. *Am. J. Med. Sci.,* 322(1), 1–6.
8. Takada, D., Ezura, Y., Ono, S., et al. (2003). Apolipoprotein H variant modifies plasma triglyceride phenotype in familial hypercholesterolemia: a molecular study in an eight-generation hyperlipidemic family. *J. Atheroscler. Thromb.,* 10(2), 79–84.
9. Cournoyer, D., Caskey, C.T. (1993). Gene therapy of the immune system. *Annu. Rev. Immunol.,* 11, 297–329.
10. Gilboa, E. (1990). Retroviral gene transfer. Applications to human gene therapy. *Prog. Clin. Biol. Res.,* 352, 301–311.
11. Blacklow, N.R., Hoggan, M.D., Kapikian, A.Z., et al. (1968). Epidemiology of adenovirus-associated virus infection in a nursery population. *Am. J. Epidemiol.,* 88, 368–378.
12. Stewart, P.L., Burnett, R.M., Cyrlaff, M., et al. (1991). Image reconstruction reveals the complex molecular organization of adenovirus. *Cell,* 67, 145–154.
13. Yamada, M., Lewis, J.A., Grodzicker, T. (1985). Overproduction of the protein product of a nonselected foreign gene carried by an adenovirus vector. *Proc. Natl. Acad. Sci. U.S.A.,* 82, 3597–3571.
14. Glorioso, J.C., De Luca, N.A., Fink, D.J. (1995). Development and application of herpes simplex virus vectors for human gene therapy. *Annu. Rev. Microbiol.,* 49, 675–710.
15. Bennet, C.F., Chiang, M.Y., Chan, H., et al. (1992). Cationic lipids enhance cellular uptake and activity of phosphothioate antisense oligonucleotides. *Mol. Pharmacol.,* 41, 1023–1033.

16. Ropert, C., Malvy, C., Couvreur, P. (1993). Inhibition of the Friend retrovirus by antisense oligonucleotides encapsulated in liposomes: mechanism of action. *Pharm. Res.,* 10, 1427–1433.

17. Ryser, H.P., Shen, W.C. (1978). Conjugation of methotrexate to poly(L–lysine) increases drug transport and overcomes drug resistance in cultured cells. *Proc. Natl. Acad. Sci. U.S.A.,* 75, 3867–3870.

18. Bayard, B., Bisbal, C., Lebleu, B. (1986). Activation of ribonuclease L by (2–5) (A)₄–poly(L–lisine) conjugates in intact cells. *Biochemistry,* 25, 3730–3736.

19. Mocarski, E.S., Kemble, G.W., Lyle, J.M., et al. (1996). A deletion mutant in the human cytomegalovirus gene encoding $IE_{1491aa}$ is replication defective due to a failure in autoregulation. *Proc. Natl. Acad. Sci. U.S.A.,* 93, 11321–11326.

20. Robertson, E.S., Ooka, T., Kieff, E.D. (1996). Epstein-Barr virus vectors for gene delivery to B lymphocytes. *Proc. Natl. Acad. Sci. U.S.A.,* 93, 11334–11340.

21. Moss, B. (1996). Genetically engineered poxviruses for recombinant gene expression, vaccination, and safety. *Proc. Natl. Acad. Sci. U.S.A.,* 93, 11341–11348.

22. Frolov, I., Hoffman, T.A., Pragai, B.M., et al. (1996). Alphavirus-based expression vectors: strategies and applications. *Proc. Natl. Acad. Sci. U.S.A.,* 93, 11371–11377.

23. Feng, M., Jackson, W.H., Goldman, C.K., et al. (1997). Stable in vivo gene transduction via a novel adenoviral/retroviral chimeric vector. *Nat. Biotech.,* 15, 866–870.

24. Dobbelstein, M. (2003). Viruses in therapy—royal road or dead end? *Virus Res.,* 92, 219–221.

25. Boris-Lawrie, K.A., Temin, H.M. (1993). Recent advances in retrovirus vector technology. *Curr. Opin. Genet. Dev.,* 3(1), 102–109.

26. Miller, A.D. (1992). Retroviral vectors. *Curr. Top. Microbiol. Immunol.,* 158, 1–24.

27. Buchschacher, G.L., Jr. (2001). Introduction to retroviruses and retroviral vectors. *Somat. Cell. Mol. Genet.,* 26, 1–11.

28. Pear, W.S., Nolan, G.P., Scott, M.L., Baltimore, D. (1993). Production of high-titer helper-free retroviruses by transient transfection. *Proc. Natl. Acad. Sci. U.S.A.,* 90(18), 8392–8396.

29. Bowles, N.E., Eisensmith, R.C., Mohuiddin, R., Pyron, M., Woo, S.L. (1996). A simple and efficient method for the concentration and purification of recombinant retrovirus for increased hepatocyte transduction in vivo. *Hum. Gene Ther.,* 7, 1735–1742.

30. Burns, J.C. (1993). Vesicular stomatitis virus G glycoprotein pseudotyped retroviral vectors: concentration to very high titer and efficient gene transfer into mammalian and nonmammalian cells. *Proc. Natl. Acad. Sci. U.S.A.,* 90, 8033–8037.

31. Shinnick, T.M., Lerner, R.A., Sutcliffe, J.G. (1981). Nucleotide sequence of Moloney murine leukemia virus. *Nature,* 293, 543–548.

32. Blaese, R.M. (1995). T-lymphocyte–directed gene therapy for ADA-SCID: Initial trial results after 4 years. *Science,* 270, 475–480.

33. Matano, T., Odawara, T., Iwamoto, A., et al. (1995). Targeted infection of a retrovirus bearing a CD4-Env chimera into human cells expressing human immunodeficiency virus type 1. *J. Gen Virol,* 76, 3165–3169.

34. Han, X., Kasahara, N., Kan, Y.W. (1995). Ligand-directed retroviral targeting of human breast cancer cells. *Proc. Natl. Acad. Sci. U.S.A.,* 92, 9747–9751.

35. Kasahara, K., Dozy, A.M., Kan, Y.W. (1994). Tissue-specific targeting of retroviral vectors through ligand-receptor interactions. *Science,* 266, 1373–1376.

36. Valsesia-Wittman, S., Morling, F.J., Nilson, B.H.K., et al. (1996). Improvement of retroviral retargeting by using amino acid spacers between an additional binding

domain and the N terminus of Moloney murine leukemia virus SU. *J. Virol.,* 70, 2059–2064.

37. Bonini, C. (1997). HSV-TK gene transfer into donor lymphocytes for control of allogeneic graft-versus-leukemia. *Science,* 276, 1719–1724.

38. Cavazzana-Calvo, M. (2000). Gene therapy of human severe combined immune deficiency (SCID)-X1 disease. *Science,* 288, 669–672.

39. Tannock, L., Hill, R. (1998). *The Basic Science of Oncology,* 3rd ed. McGraw-Hill Health Professions Division, 420–442.

40. Svensson, V., Persson, R. (1984). Entry of adenovirus 2 into HeLa cells. *J. Virol.,* 51(3), 687–694.

41. Greber, U.F., Webster, P., Weber, J., Helenius, A. (1996). The role of the adenovirus protease on virus entry into cells. *EMBO J,* 15(8), 1766–1777.

42. Seth, P., Willingham, M.C., Pastan I. (1984). Adenovirus-dependent release of 51Cr from KB cells at an acidic pH. *J. Biochem.,* 259(23), 14350–14353.

43. Yang, Y., Nunes, F.A., Berencsi, K., Furth, E.E., Gonczol, E., Wilson, J.M. (1994). Cellular immunity to viral antigens limits E1-deleted adenoviruses for gene therapy. *Proc. Natl. Acad. Sci. U.S.A.,* 91(10), 4407–4411.

44. Smith, T.A., Mehaffey, M.G., Kayda, D.B., et al. (1993). Adenovirus mediated expression of therapeutic plasma levels of human factor IX in mice. *Nat. Genet.,* 5(4), 397–402.

45. Schiedner, G. (1998). Genomic DNA transfer with a high-capacity adenovirus vector results in improved *in vivo* gene expression and decreased toxicity. *Nat. Genet.,* 18, 180–183.

46. Kochanek, S., Schiedner, G., Volpers, C. (2001). High-capacity 'gutless' adenoviral vectors. *Curr. Opin. Mol. Ther.,* 3, 454–463.

47. Gilardi, P., Courtney, M., Pavirani, A., Perricaudet, M. (1990). Expression of human alpha 1-antitrypsin using a recombinant adenovirus vector. *FEBS Lett.,* 267(1), 60–62.

48. Rosenfeld, M.A., Siegfried, W., Yoshimura, K., et al. (1991). Adenovirus-mediated transfer of a recombinant alpha 1-antitrypsin gene to the lung epithelium in vivo. *Science,* 252(5004), 431–434.

49. Fisher, K.D. (2001). Polymer-coated adenovirus permits efficient retargeting and evades neutralising antibodies. *Gene Ther.,* 8, 341–348.

50. Croyle, M.A. (2002). PEGylation of E1–deleted Adenovirus vectors allows significant gene expression on readministration to liver. *Hum. Gene Ther.,* 13, 1887–1900.

51. Perricone, M.A., Morris, J.E., Pavelka, K., et al. (2001). Aerosol and lobar administration of a recombinant adenovirus to individuals with cystic fibrosis. II. Transfection efficiency in airway epithelium. *Hum. Gene Ther.,* 12(11), 1383–1394.

52. Wakefield, P.M., Tinsley, J.M., Wood, M.J., Gilbert, R., Karpati, G., Davies, K.E. (2000). Prevention of the dystrophic phenotype in dystrophin/utrophin-deficient muscle following adenovirus-mediated transfer of a utrophin minigene. *Gene Ther.,* 7(3), 201–204.

53. Cerrato, J.A., Yung, W.K., Liu, T.J. (2001). Introduction of mutant p53 into a wild-type p53-expressing glioma cell line confers sensitivity to Ad-p53-induced apoptosis. *Neuro-oncol.,* 3(2), 113–122.

54. Chen, S.H., Shine, H.D., Goodman, J.C., Grossman, R.G., Woo, S.L. (1994). Gene therapy for brain tumors: regression of experimental gliomas by adenovirus-mediated gene transfer in vivo. *Proc. Natl. Acad. Sci. U.S.A.,* 91(8), 3054–3057.

55. McCart, J.A., Wang, Z.H., Xu, H., et al. (2002). Development of a melanoma-specific adenovirus. *Mol. Ther.,* 6(4), 471–480.

56. Kay, M.A., Landen, C.N., Rothenberg, S.R., et al. (1994). *In vivo* hepatic gene therapy: complete albeit transient correction of factor IX deficiency in hemophilia B dogs. *Proc. Natl. Acad. Sci. U.S.A.,* 91, 2353–2357.

57. Morsy, M.A., Alford, E.L., Bett, A., Graham, F.L., Caskey, C.T. (1993). Efficient adenoviral-mediated ornithine transcarmylase expression in deficient mouse and human hepatocytes. *J. Clin. Invest.,* 92, 1580–1586.

58. Kozarsky, K.F., Jooss, K., Donahee, M., Strauss, J.F., III, Wilson, J.M. (1996). Effective treatment of familial hypercholesterolaemia in the mouse model using adenovirus-mediated transfer of the VLDL receptor gene. *Nature Genet.,* 13, 54–62.

59. Bordignon, C., Notarangelo, L.D., Nobili, N., et al. (1995). Gene therapy in peripheral blood lymphocytes and bone marrow for ADA-immunodeficient patients. *Science,* 270, 470–475.

60. Rosenfeld, M.A., Yoshimura, K., Trapnell, B.C., et al. (1992). In vivo transfer of the human cystic fibrosis transmembrane conductance regulator gene to the airway epithelium. *Cell,* 68(1), 143–155.

61. Arcasoy, S.M., Latoche, J.D., Gondor, M., Pitt, B.R., Pilewski, J.M. (1997). Polycations increase the efficiency of adenovirus-mediated gene transfer to epithelial and endothelial cells in vitro. *Gene Ther.,* 4, 32–38.

62. Rabinowtz, J.E., Samulski, J. (1998). Adeno-associated virus expression systems for gene transfer. *Curr. Opin. Biotechnol.,* 9, 470–475.

63. Tamayose, K., Hirai, Y., Shimada, T. (1996). A new strategy for large-scale preparation of high-titer recombinant adeno-associated virus vectors by using packaging cell lines and sulfoNat.ed cellulose column chromatography. *Hum. Gene Ther.,* 7, 507–513.

64. Mamounas, M., Leavitt, M., Yu, M., Wong-Staal, F. (1995). Increased titer of recombinant AAV vectors by gene transfer with adenovirus coupled to DNA-polylysine complexes. *Gene Ther.,* 2, 429–432.

65. Malik, P., McQuiston, S.A., Yu, X.J., et al. (1997). Recombinant adeno-associated virus Med.iates a high level of gene transfer but less efficient integration in the K562 human hematopoietic cell line. *J. Virol.,* 71, 1776–1783.

66. Davidson, B.L. (2000). Recombinant adeno–associated virus type 2, 4, and 5 vectors: transduction of variant cell types and regions in the mammalian central nervous system. *Proc. Natl. Acad. Sci. U.S.A.,* 97, 3428–3432.

67. Gao, G.P. (2002). Novel adeno-associated viruses from rhesus monkeys as vectors for human gene therapy. *Proc. Natl. Acad. Sci. U.S.A.,* 99, 11854–11859.

68. Auricchio, A., Behling, K.C., Maguire, A.M., et al. (2002). Pharmacological regulation of protein expression from adeno-associated viral vectors in the eye. *Mol. Ther.,* 6, 238–242.

69. Gately, S., Twardowski, P., Stack, M.S., et al. (1997). The mechanism of cancer-mediated conversion of plasminogen to the angiogenesis inhibitor angiostatin. *Proc. Natl. Acad. Sci. U.S.A.,* 94, 10868–10872.

70. Russell, S.J. (1996). Peptide-displaying phages for targeted gene delivery. *Nature Med.,* 2, 276–277.

71. Koeberl, D.D., Alexander, I.E., Halbert, C.L., Russell, D.W., Miller, A.D. (1997). Persistent expression of human clotting factor IX from mouse liver after intravenous injection of adeno-associated virus vectors. *Proc. Natl. Acad. Sci. U.S.A.,* 94, 1426–1431.

72. Okada, H., Miyamura, K., Itoh, T., et al. (1996). Gene therapy against an experimental glioma using adeno-associated virus vectors. *Gene Ther.,* 3, 957–964.

73. Vigna, E., Naldini, L. (2000). Lentiviral vectors: excellent tools for experimental gene transfer and promising candidates for gene therapy. *J. Gene Med.,* 2, 308–316.

74. Kafri, T., Blomer, U., Peterson, D.A., Gage, F.H., Verma, I.M. (1997). Sustained expression of genes delivered directly into liver and muscle by lentiviral vectors. *Nat. Genet.,* 17, 314–317.

75. Caputo, A., Rossi, C., Bozzini, R., et al. (1997). Studies on the effect of the combined expression of anti-tat and anti-rev genes on HIV-1 replication. *Gene Ther.,* 4, 288–295.

76. Miyoshi, H., Takahashi, M., Gage, F.H., Verma, I.M. (1997). Stable and efficient gene transfer into the retina using an HIV-based lentiviral vector. *Proc. Natl. Acad. Sci. U.S.A.,* 94, 10319–10323.

77. Kay, M.A., (2001). Viral vectors for gene therapy: The art of turning infectious agents into vehicles for therapeutics. *Nat. Med.,* 7, 33–40.

78. Follenzi, A. (2002). Efficient gene delivery and targeted expression to hepatocytes *in vivo* by improved lentiviral vectors. *Hum. Gene Ther.,* 13, 243–260.

79. Kafri, T. (2001). Lentivirus vectors: difficulties and hopes before clinical trials. *Curr. Opin. Mol. Ther.,* 3, 316–326.

80. Logan, A.C., Lutzko, C., Kohn, D.B. (2002). Advances in lentiviral vector design for gene-modification of hematopoietic stem cells. *Curr. Opin. Biotechnol.,* 13(5), 429–436.

81. Miyatake, S. (1997). Herpes simplex virus as a vector for gene therapy. *Uirusu,* 47(2), 239–246.

82. Tonini, T., Claudio, P.P., Giordano, A., Romano, G. (2004). Retroviral and lentiviral vector titration by the analysis of the activity of viral reverse transcriptase. *Methods Mol. Biol.,* 285, 155–158.

83. Wolff, J.A., Malone, R.W., Williams, P. (1990). Direct gene transfer into mouse muscle in vivo. *Science,* 247, 1465–1468.

84. Schmidt-Wolf, G.D., Schmidt-Wolf, I.G. (2003). Non-viral and hybrid vectors in human gene therapy: an update. *Trends Mol. Med.,* 9(2), 67–72.

85. Davis, M.E. (2002). Non-viral gene delivery systems. *Curr. Opin. Biotechnol.,* 13(2), 128–131.

86. Desnick, R.J., Schuchman, E.H. (1998). Gene therapy for genetic diseases. *Acta Paediatr. Jpn.,* 40(3), 191–203.

87. Kumar, V.V., Singh, R.S., Chaudhuri, A. (2003). Cationic transfection lipids in gene therapy: successes, set-backs, challenges and promises. *Curr. Med. Chem.,* 10(14), 1297–1306.

88. Nishikawa, M., Hashida, M. (2002). Nonviral approaches satisfying various requirements for effective in vivo gene therapy. *Biol. Pharm. Bull.,* 25(3), 275–283.

89. Allison, S.D., Molina, M.C., Anchordoquy, T.J, (2000). Stabilization of lipid/DNA complexes during the freezing step of the lyophilization process: the particle isolation hypothesis. *Biochim. Biophys. Acta,* 1468, 127–138.

90. Kwok, K.Y., Adami, R.C., Hester, K.C., Park, Y., Thomas, S., Rice, K.G. (2000). Strategies for maintaining the particle size of peptide DNA condensates following freeze-drying. *Int. J. Pharm.,* 203, 81–88.

91. Anchordoquy, T.J., Allison, S.D., Molina, M.C., Girouard, L.G., Carson, T.K. (2001). Physical stabilization of DNA-based therapeutics. *Drug Discov. Today,* 6, 463–470.

92. Richardson, S.C.W., Kolbe, H.V.J., Duncan, R. (1999). Potential of low molecular mass chitosan as a DNA delivery system: biocompatibility, body distribution and ability to complex and protect DNA. *Int J. Pharm.,* 178, 231–243.

93. Fischer, D., Bieber, T., Li, Y., Elsasser, H–P, Kissel, T. (1999). A novel non-viral vector for DNA delivery based on low molecular weight, branched polyethylenimine: effect of molecular weight on transfection efficiency and cytotoxicity. *Pharm. Res.,* 16, 1273–12791.

94. Hwang, S., Bellocq, N., Davis, M.E. (2001). Effects of structure of -cyclodextrin-containing polymers on gene delivery. *Bioconjugate Chem.,* 12, 280–290.

95. Collard, W.T., Yang, Y., Kwok, K.Y., Park, Y., Rice, K.G. (2000). Biodistribution, metabolism and *in vivo* gene expression of low molecular weight glycopeptide poly-ethylene glycol peptide DNA co-condensates. *J. Pharm. Sci.,* 89, 499–512.

96. Kircheis, R., Blessing, T., Brunner, S., Wightman, L., Wagner, E. (2001). Tumor targeting with surface-shielded ligand–polycation DNA complexes. *J. Control Release,* 72, 165–170.

97. Oupicky, D., Howard, K.A., Konak, C., Dash, P.R., Ulbrich, K., Seymour, L.W. (2000). Steric stabilization of poly-l-lysine/DNA complexes by the covalent attachment of semitelechelic poly [*N*-(2-hydroxypropyl)methacrylamide]. *Bioconjugate Chem.,* 11, 492–501.

98. Schaffer, D.V., Fidelman, N.A., Dan, N., Lauffenburger, D.A. (2000). Vector unpaking as a potential barrier for receptor-mediated polyplex gene delivery. *Biotechnol. Bioeng.,* 67, 598–606.

99. Fisher, K.D., Stallwood, Y., Green, N.K., Ulbrich, K., Mautner, V., Seymour, L.W. (2001). Polymer-coated adenovirus permits efficient retargeting and evades neutralising antibodies. *Gene Ther.,* 8(5), 341–348.

100. Hagstrom, J.E. (2003). Plasmid-based gene delivery to target tissues in vivo: the intravascular approach. *Curr. Opin. Mol. Ther.,* 5(4), 338–344.

101. Herweijer, H., Wolff, J.A. (2003). Progress and prospects: naked DNA gene transfer and therapy. *Gene Ther.,* 10(6), 453–458

102. Lu, Q.L., Bou-Gharios, G., Partridge, T.A. (2003). Non-viral gene delivery in skeletal muscle: a protein factory. *Gene Ther.,* 10(2), 131–142.

103. Hickman, M.A., Malone, R.W., Lehmann-Bruinsma, K., et al. (1994). Gene expression following direct injection of DNA into liver. *Hum. Gene Ther.,* 5, 1477–1483.

104. Zhang, G., Vargo, D., Budker, V., Armstrong, N., Knechtle, S., Wolff, J.A. (1997). Expression of naked plasmid DNA injected into the afferent and efferent vessels of rodent and dog livers. *Hum. Gene Ther.,* 8, 1763–1772.

105. Yang, J.P., Huang, L. (1996). Direct gene transfer to mouse melanoma by intratumor injection of free DNA. *Gene Ther.,* 3, 542–548.

106. Yu, W.H., Kashani-Sabet, M., Liggitt, D., Moore, D., Heath, T.D., Debs, R.J. (1999). Topical gene delivery to murine skin. *J. Invest. Dermatol.,* 112, 370–375.

107. Kawase, A., Nomura, T., Yasuda, K., Kobayashi, N., Hashida, M., Takakura, Y. (2003). Disposition and gene expression characteristics in solid tumors and skeletal muscle after direct injection of naked plasmid DNA in mice. *J. Pharm. Sci.,* 92(6), 1295–1304.

108. Nomura, T., Yasuda, K., Yamada, T., et al. (1999). Gene expression and antitumor effects following direct interferon (IFN)-gamma gene transfer with naked plasmid DNA and DC-chol liposome complexes in mice. *Gene Ther.,* 6(1), 121–129.

109. Aihara, H., Miyazaki, J. (1998). Gene transfer into muscle by electroporation in vivo. *Nat. Biotechnol.,* 16, 867–870.

110. Fritz, J.D., Herweijer, H., Zhang, G., Wolff, J.A. (1996). Gene transfer into mammalian cells using histone-condensed plasmid DNA. *Hum. Gene Ther.,* 7, 1395–1404.

111. Zaitsev, S.V., Haberland, A., Otto, A., Vorob'ev, V.I., Haller, H., Bottger, M. (1997). H1 and HMG17 extracted from calf thymus nuclei are efficient DNA carriers in gene transfer. *Gene Ther.,* 4, 586–592.

112. Yoshida, J., Mizuno, M. (2003). Clinical gene therapy for brain tumors. Liposomal delivery of anticancer molecule to glioma. *J. Neurooncol.,* 65(3), 261–267.

113. Maurer, N., Fenske, D.B., Cullis, P.R. (2001). Developments in liposomal drug delivery systems. *Expert Opin. Biol. Ther.,* 1(6), 923–947.

114. Chonn, A., Cullis, P.R. (1995). Recent advances in liposomal drug-delivery systems. *Curr. Opin. Biotechnol.,* 6(6), 698–708.

115. Cullis, P.R., Chonn, A. (1998). Recent advances in liposome technologies and their applications for systemic gene delivery. *Adv. Drug Deliv. Rev.,* 30(1–3), 73–83.

116. Audouy, S.A., de LeiJ, L.F., Hoekstra, D., Molema, G. (2002). In vivo characteristics of cationic liposomes as delivery vectors for gene therapy. *Pharm. Res.,* 19(11), 1599–1605.

117. Liu, F., Huang, L. (2002). Development of non-viral vectors for systemic gene delivery. *Control Release,* 78(1–3), 259–266.

118. Hirko, A., Tang, F., Hughes, J.A. (2003). Cationic lipid vectors for plasmid DNA delivery. *Curr Med. Chem.,* 10(14), 1185–1193.

119. Pedroso de Lima, M.C., Neves, S., Filipe, A., Duzgunes, N., Simoes, S. (2003). Cationic liposomes for gene delivery: from biophysics to biological applications. *Curr. Med. Chem.,* 10(14), 1221–1231.

120. Miller, A.D. (2003). The problem with cationic liposome/micelle-based non-viral vector systems for gene therapy. *Curr. Med. Chem.,* 10(14), 1195–1211.

121. Nicolau, C., Papahadjopulos, D. (1998). Gene therapy: liposomes and gene delivery— a perspective. In: Lasic, D. and Papahadjopoulos, D., eds., *Medical Applications of Liposomes.* Elsevier, Amsterdam, 347–352.

122. Zhdanov, R.I., Podobed, O.V., Vlassov, V.V. (2002). Cationic lipid-DNA complexes-lipoplexes-for gene transfer and therapy. *Bioelectrochemistry,* 58(1), 53–64.

123. Almofti, M.R., Harashima, H., Shinohara, Y., Almofti, A., Baba, Y., Kiwada, H. (2003). Cationic liposome-mediated gene delivery: Biophysical study and mechanism of internalization. *Arch. Biochem. Biophys.,* 410(2), 246–253.

124. Clark, P.R., Hersh, E.M. (1999). Cationic lipid-mediated gene transfer: current concepts. *Curr. Opin. Mol. Ther.,* 1(2), 158–176.

125. El-Aneed, A. (2004). An overview of current delivery systems in cancer gene therapy. *J. Control Release,* 94(1), 1–14.

126. Dass, C.R., Burton, M.A. (1999). Lipoplexes and tumours. A review. *J. Pharm. Pharmacol.,* 51(7), 755–770.

127. Berh, J.P., Demeneix, B., Loeffler, J.P., Mutul, J.P. (1989). Efficient gene transfer into mammalian primary endocrine cells with lipopolyamine coated DNA. *Proc. Natl. Acad. Sci. U.S.A.,* 86, 6982–6986.

128. Felgner, P.L. (1995). The evolving role of liposomes in gene delivery. *J. Liposome Res.,* 5, 725–734,

129. Felgner, P.L., Gadek, T.R., Holm, M., et al. (1987). Lipofection: a highly efficient, lipid mediated DNA—transfection procedure. *Proc. Natl. Acad. Sci. U.S.A.,* 84, 7413–7416.

130. Felgner, J.H., Kumar, R., Sridhar, C.N., et al. (1994). Enhanced gene delivery and mechanism studies with a novel series of cationic lipid formulations. *J. Biol. Chem.,* 269, 2550–2561.

131. Basic, D. (1997). *Liposomes in Gene Delivery.* CRC Press, Boca Raton, FL.

132. Pector, V., Backmann, J., Maes, D., Vandenbranden, M., Ruysschaert, J.M. (2000). Biophysical and structural properties of DNAdiC(14)-amidine complexes. *J. Biol. Chem.,* 275, 29533–29538.

133. Hong, K., Zheng, W., Baker, A., Papahadjopoulos, D. (1997). Stabilisation of cationic liposome/DNA complexes by polyamines and polyethylenglycol-phospholipid conjugates for efficient in vivo gene delivery. *FEBS Lett.,* 414, 187–192.

134. Hui, S.W., Langner, M., Yzhao, Y.L., Ross, P., Hurley, E., Chan, K. (1996). The role of helper lipids in cationic liposome-mediated gene transfer. *Biophys. J.,* 71, 590–599.

135. Zabner, J., Fasbender, A.J., Moninger, T., Poellinger, K.A., Wellsh, M.J. (1995). Cellular and molecular barriers to gene transfer by a cationic lipid. *J. Biol. Chem.*, 270, 18997–19007.

136. Leventis, R., Silvius, J. (1990). Interactions mammalian cells with lipid dispersions containing novel metabolizable cationic amphiphiles. *Biochim. Biophys. Acta,* 1023, 124–132.

137. Zhou, X., Huang, L. (1994). DNA transfection mediated by cationic liposomes containing lipopolysine: characterization and mechanism of action. *Biochim. Biophys. Acta,* 1189, 195–203.

138. Zhu, N., Liggit, D., Lin, Y., Debs R. (1993) Systemic gene expression after intravenous DNA delivery in adult mice. *Science,* 261, 209–211.

139. Hong, K., Zheng, W., Baker, A., Papahadjopoulos, D. (1997). Stabilisation of cationic liposome/DNA complexes by polyamines and polyethylenglycol–phospholipid conjugates for efficient in vivo gene delivery. *FEBS Lett.,* 414, 187–192.

140. Ota, T., Maeda, M., Tatsuka, M. (2002). Cationic liposomes with plasmid DNA influence cancer metastatic capability. *Anticancer Res.,* 22(6C), 4049–4052.

141. Ren, T., Song, Y.K., Zhang, G., Liu, D. (2000). Structural basis of DOTMA for its high intravenous transfection activity in mouse. *Gene Ther.,* 7(9), 764–768.

142. Ciani, L., Ristori, S., Salvati, A., Calamai, L., Martini, G. (2004) DOTAP/DOPE and DC-Chol/DOPE lipoplexes for gene delivery: zeta potential measurements and electron spin resonance spectra. *Biochim. Biophys. Acta.,* 1664(1), 70–79.

143. Zhdanov, R.I., Kutsenko, N.G., Podobed, O.V., et al. (1998). Cholesteroyl polyethylen/propylen/imines as mediators of transfection of eukaryotic cells for gene therapy. *Dokl. Akad. Nauk.,* 361, 695–699.

144. Farhood, H., Serbina, N., Huang, L. (1995). The role of dioleoyl phosphatidylethanolamine in cationic liposome mediated gene transfer. *Biochim. Biophys. Acta.,* 1235(2), 289–295.

145. Zhdanov, R.I., Sviridov, V., Podobed, O.V., et al. (1998). In vitro and in vivo functional gene transfer with cholesteroyl polyethylen/propylen/imines. In: *6th Symposium on Gene Therapy Towards to Gene Therapeutics,* MDC, Berlin-Buch, May 4–6, 101, Abstract.

146. Kikuchi, A., Aoki, Y., Sugaya, S., et al. (1999). Development of novel cationic liposomes for efficient gene transfer into peritoneal disseminated tumor. *Hum. Gene Ther.,* 106, 947–955.

147. Plank, C., Mechtler, K., Szoka, F.C., Jr, Wagner, E. (1996). Activation of the complement system by synthetic DNA complexes: a potential barrier for intravenous gene delivery. *Hum. Gene Ther.,* 7, 1437–1446.

148. Yang, J.P., Huang, L. (1997). Overcoming the inhibitory effect of serum on lipofection by increasing the charge ratio of cationic liposome to DNA. *Gene Ther.,* 4, 950–960.

149. Li, S., Huang, L. (1997). In vivo gene transfer via intravenous administration of cationic lipid-protamine-DNA (LPD) complexes. *Gene Ther.,* 4, 891–900.

150. Templeton, N.S., Lasic, D.D., Frederik, P.M., Strey, H.H., Roberts, D.D., Pavlakis, G.N. (1997). Improved DNA: Liposome complexes for increased systemic delivery and gene expression. *Nat. Biotechnol.,* 15, 647–652.

151. Wattiaux, R., Jadot, M., Warnier-Pirotte, M.T., Wattiaux-De Coninck, S. (1997). Cationic lipids destabilize lysosomal membrane in vitro. *FEBS Lett.,* 417, 199–202.

152. Zuidam, N.J., Barenholz, Y. (1997). Electrostatic parameters of cationic liposomes commonly used for gene delivery as determined by 4-heptadecyl-7-hydroxycoumarin. *Biochim. Biophys. Acta,* 1329, 211–222.

153. Thierry, A.R., Dritschilo, A. (1992). Intracellular availability of unmodified, phosphorothioated and liposomally encapsulated oligodeoxynucleotides for antisense activity. *Nucleic Acids Res.,* 20, 5691–5698.
154. Capaccioli, S., Di Pasquale, G., Mini, E., Mazzei, T., Quattrone, A. (1993). Cationic lipids improve antisense oligonucleotide uptake and prevent degradation in cultured cells and in human serum. *Biochem Biophys. Res. Commun.,* 197, 818–825.
155. Morishita, R., Gibbons, G.H., Ellison, K.E., et al. (1993). Single intraluminal delivery of antisense cdc2 kinase and proliferating-cell nuclear antigen oligonucleotides results in chronic inhibition of neointimal hyperplasia. *Proc. Natl. Acad. Sci. U.S.A.,* 90, 8474–8478.
156. Bongartz, J.P., Aubertin, A.M., Milhaud, P.G., Lebleu, B. (1994). Improved biological activity of antisense oligonucleotides conjugated to a fusogenic peptide. *Nucleic Acids Res.,* 22, 4681–4688.
157. Jaaskelainen, I., Monkkonen, J., Urtti, A. (1994). Oligonucleotide-cationic liposome interactions. A physicochemical study. *Biochim. Biophys. Acta.,* 1195, 115–123.
158. Venanzi, M., Zhdanov, R.I., Petrelli, C., Moretti, P., Amici, A., Petrelli, F. (1993). Entrapment of supercoiled DNA into performed amphiphilic lipid vesicles. In: Gregoriadis, G. and Florence, A., eds., *Int. Conference Liposomes, Nineties and Beyond,* London, 122–123, Abstract.
159. Tarahovsky, R., Khusainova, A., Gorelov, K., Dawson, A.K., Ivanitsky, G. (1996). DNA initiates polymorphic structural transition in lectin. *FEBS Lett.,* 390, 133–137.
160. Harvie, P., Wong, F.M., Bally, M.B. (1998). Characterization of lipid DNA interaction: I. Destabilization of bound lipids and DNA dissociation. *Biophys. J.,* 75, 1040–1051.
161. Godbey, W.T., Wu, K.K., Mikos, A.G. (1999). Poly(ethylenimine) and its role in gene delivery. *J. Control Release,* 60(2–3), 149–160.
162. Lemkine, G.F., Demeneix, B.A. (2001). Polyethylenimines for in vivo gene delivery. *Curr. Opin. Mol. Ther.,* 3(2), 178–182.
163. Trubetskoy, V.S., Wong, S.C., Subbotin, V., et al. (2003). Recharging cationic DNA complexes with highly charged polyanions for in vitro and in vivo gene delivery. *Gene Ther.* 10(3), 261–271.
164. Kircheis, R., Wightman, L., Wagner, E. (2001). Design and gene delivery activity of modified polyethylenimines. *Adv. Drug Deliv. Rev.,* 53(3), 341–358.
165. Coll, J.L., Chollet, P., Brambilla, E., Desplanques, D., Behr, J.P., Favrot, M. (1999). In vivo delivery to tumors of DNA complexed with linear polyethylenimine. *Hum. Gene Ther.,* 10(10), 1659–1666.
166. Aral, C., Akbuga, J. (2003). Preparation and in vitro transfection efficiency of chitosan microspheres. containing plasmid DNA: poly (L-lysine) complexes. *J. Pharm. Pharm. Sci.,* 6(3), 321–326.
167. Read, M.L., Etrych, T., Ulbrich, K., Seymour, L.W. (1999). Characterisation of the binding interaction between poly(L-lysine) and DNA using the fluorescamine assay in the preparation of non-viral gene delivery vectors. *FEBS Lett.,* 461(1–2), 96–100.
168. Xiang, J.J., Tang, J.Q., Zhu, S.G., et al. (2003). IONP-PLL: a novel non-viral vector for efficient gene delivery. *J. Gene Med.,* 5(9), 803–817.
169. Jeon, E., Kim, H.D., Kim, J.S. (2003). Pluronic-grafted poly-(L)–lysine as a new synthetic gene carrier. *J. BioMed. Mater. Res.,* 66A(4), 854–859.
170 Luo, D., Woodrow-Mumford, K., Belcheva, N., Saltzman, W.M. (1999). Controlled DNA delivery systems. *Pharm. Res.,* 16(8), 1300–1308.
171. Brownlie, A., Uchegbu, I.F., Schatzlein, A.G. (2004). PEI-based vesicle-polymer hybrid gene delivery system with improved biocompatibility. *Int. J. Pharm.,* 274 (1–2), 41–52.

172. Ogris, M., Wagner, E. (2002). Tumor-targeted gene transfer with DNA polyplexes. *Somat. Cell. Mol. Genet.,* 27(1–6), 85–95.

173. Tseng, W.C., Jong, C.M. (2003). Improved stability of polycationic vector by dextran-grafted branched polyethylenimine. *Biomacromolecules,* 4(5), 1277–1284.

174. Boletta, A., Benigni, A., Lutz, J., Remuzzi, G., Soria, M.R., Monaco, L. (1997). Nonviral gene delivery to the rat kidney with polyethylenimine. *Hum. Gene Ther.,* 8, 1243–1251.

175. Boussif, A., Lezoualch, F., Zanta, M.A., et al. (1995). A versatile vector for gene and oligonucleotide transfer into cells in culture and in-vivo-polyethylenimine. *Proc. Natl. Acad. Sci. U.S.A.,* 92, 7297–7301.

176. Lemkine, F., Goula, O., Becker, N., Paleari, N., Levi, G., Demeneix, B.A. (1999). Optimisation of polyethylenimine-based gene delivery to mouse brain. *J. Drug Target,* 7, 305.

177. Coll, L., Chollet, P., Brambilla, E., Desplanques, D., Behr, J.P., Favrot, M. (1999). In vivo delivery to tumors of DNA complexed with linear polyethylenimine. *Hum. Gene Ther.,* 10, 1659–1666.

178. Aoki, K., Furuhata, S., Hatanaka, K., et al. (2001). Polyethylenimine-mediated gene transfer into pancreatic tumor dissemination in the murine peritoneal cavity. *Gene Ther.,* 8, 508–514.

179. Ferrari, S., Moro, E., Pettenazzo, A., et al. (1994). ExGen 500 is an efficient vector for gene delivery to lung epithelial cells in vitro and in vivo. *Gene Ther.,* 4, 1100–1106.

180. Ferrari, S., Pettenazzo, A., Garbati, N., Zacchello, F., Behr, J.P., Scarpa, M. (1999). Polyethylenimine shows properties of interest for cystic fibrosis gene therapy. *Biochim. Biophys. Acta,* 1447, 219–225.

181. Bragonzi, B., Boletta, A., Biffi, A., et al. (1999). Comparison between cationic polymers and lipids in mediating systemic gene delivery to the lungs. *Gene Ther.,* 6, 1995–2004.

182. Shi, L., Tang, G.P., Gao, S.J., et al. (2003). Repeated intrathecal administration of plasmid DNA complexed with polyethylene glycol-grafted polyethylenimine led to prolonged transgene expression in the spinal cord. *Gene Ther.,* 10(14), 1179–1188.

183. Wiseman, J.W., Goddard, C.A., McLelland, D., Colledge, W.H. (2003). A comparison of linear and branched polyethylenimine (PEI) with DCChol/DOPE liposomes for gene delivery to epithelial cells in vitro and in vivo. *Gene Ther.,* 10(19), 1654–1662.

184. Ozbas-Turan, S., Aral, C., Kabasakal, L., Keyer-Uysal, M., Akbuga, J. (2003). Co-encapsulation of two plasmids in chitosan microspheres as a non-viral gene delivery vehicle. *J. Pharm. Pharm. Sci.,* 6(1), 27–32.

185. Chiou, H.C., Tangco, M.V., Levine, S.M., et al. (1994). Enhanced resistance to nuclease degradation of nucleic acids complexed to asialoglycoprotein-polylysine carriers. *Nucl. Acids Res.,* 22, 5349–5446.

186. Houk, B.E., Hochhaus, G., Hughes, J.A. (1999). Kinetic modeling of plasmid DNA degradation in rat plasma. *AAPS PharmSci.,* 1, E9.

187. Monck, M.A., Mori, A., Lee, D., et al. (2000). Stabilized plasmid-lipid particles: Pharmacokinetics and plasmid delivery to distal tumors following intravenous injection. *J. Drug Target,* 7, 439–452.

188. Dash, P.R., Read, M.L., Barrett, L.B., Wolfert, M.A., Seymour, L.W. (1999). Factors affecting blood clearance and in vivo distribution of polyelectrolyte complexes for gene delivery. *Gene Ther.,* 6, 643–650.

189. Ishiwata, H., Suzuki, N., Ando, S., Kikuchi, H., Kitagawa, T. (2000). Characteristics and biodistribution of cationic liposomes and their DNA complexes. *J. Control Release,* 69, 139–148.

190. Dash, P.R., Read, M.L., Fisher, K.D., et al. (2000). Decreased binding to proteins and cells of polymeric gene delivery vectors surface modified with a multivalent hydrophilic polymer and retargeting through attachment of transferrin. *J. Biol. Chem.*, 275, 3793–3802.

191. Fisher, K.D., Ulbrich, K., Subr, V., et al. (2000). A versatile system for receptor-mediated gene delivery permits increased entry of DNA into target cells, enhanced delivery to the nucleus and elevated rates of transgene expression. *Gene Ther.*, 7, 1337–1343.

192. Carlisle, R.C., Bettinger, T., Ogris, M., Hale, S., Mautner, V., Seymour, L.W. (2001). Adenovirus hexon protein enhances nuclear delivery and increases transgene expression of polyethylenimine/plasmid DNA vectors. *Mol. Ther.*, 4(5), 473–483.

193. Tagawa, T., Manvell, M., Brown, N., et al. (2002). Characterisation of LMD virus-like nanoparticles self-assembled from cationic liposomes, adenovirus core peptide mu and plasmid DNA. *Gene Ther.*, 9(9), 564–576.

194. Kaneda, Y., Yamamoto, S., Hiraoka K. (2003). The hemagglutinating virus of Japan—liposome method for gene delivery. *Methods Enzymol.*, 373, 482–493.

195. Hopkins, N. (1993). High titers of retrovirus (vesicular stomatitis virus) pseudotypes, at last. *Proc Natl. Acad Sci. U.S.A.*, 90 (19), 8759–8760.

196. Cherng, J.Y., Van de Wetering, P., Talsma, H., Crommelin, D.J.A., Hennink, W.E. (1996). Effect of size and serum proteins on transfection efficiency of poly ((2-dimethylamino)ethyl methacrylate)-plasmid nanoparticles. *Pharm. Res.*, 13, 1038–1042.

197. Cherng, J.Y., Talsma, H., Crommelin, D.J.A., Hennink, W.E., (1999). Long term stability of poly((2-dimethylAm.ino)ethyl methacrylate)-based gene delivery systems. *Pharm. Res.*, 16, 1417–1423.

198. Cherng, J.Y., Schuurmans-Nieuwenbroek, N.M., Jiskoot, W., Talsma, H., et al. (1999). Effect of DNA topology on the transfection efficiency of poly((2-dimethylamino)ethyl methacrylate)-plasmid complexes. *J. Control Release,* 60, 343–353.

199. Erbacher, P., Bettinger, T., Belguise-Valladier, P., Zou, S., Coll, J.L., Behr, J.P., Remy, J.S. (1999). Transfection and physical properties of various saccharide, poly(ethylene glycol), and antibody-derivatized polyethylenimines (PEI). *J. Gene Med.,* 1 (3), 210–222.

200. Anderson, A.J., Dawes, E.A. (1990). Occurrence, metabolism, metabolic role, and industrial uses of bacterial polyhydroxyalkanoates, *Microbiol Rev.,* 54(4), 450–472.

201. Piddubnyak, V., Kurcok, P., Matuszowicz, A., et al. (2004). Oligo-3-hydroxybutyrates as potential carriers for drug delivery. *Biomaterials,* 25(22), 5271–5279.

202. Koh, J.J., Ko, K.S., Lee, M., Han, S., Park, J.S., Kim, S.W. (2000). Degradable polymeric carrier for the delivery of IL-10 plasmid DNA to prevent autoimmune insulitis of NOD mice. *Gene Ther.,* 7(24), 2099–2104.

203. Vandervoort, J., Ludwig, A. (2002). Biocompatible stabilizers in the preparation of PLGA nanoparticles: a factorial design study. *Int. J. Pharm.,* 238(1–2), 77–92.

204. Mao, H.Q., Roy, K., Troung-Le, V.L., et al. (2001). Chitosan-DNA nanoparticles as gene carriers: synthesis, characterization and transfection efficiency. *J. Control Release,* 70(3), 399–421.

205. MacLaughlin, F.C., Mumper, R.J., Wang, J., et al. (1998). Chitosan and depolymerized chitosan oligomers as condensing carriers for in vivo plasmid delivery. *J. Control Release,* 56, (1–3), 259–272.

206. Panyam, J., Zhou, W.Z., Prabha, S., Sahoo, S.K., Labhasetwar, V. (2002). Rapid endo-lysosomal escape of poly(DL-lactide-co-glycolide) nanoparticles: implications for drug and gene delivery. *FASEB J,* 16(10), 1217–1226.

207. Davda, J., Labhasetwar, V. (2002). Characterization of nanoparticle uptake by endothelial cells. *Int J. Pharm.,* 233(1–2), 51–59.

208. Zimmer, A. (1999). Antisense oligonucleotide delivery with polyhexylcyanoacrylate nanoparticles as carriers. *Methods,* 18(3), 286–295, 322.
209. Zobel, H.P., Kreuter, J., Werner, D., Noe, C.R., Kumel, G., Zimmer, A. (1997). Cationic polyhexylcyanoacrylate nanoparticles as carriers for antisense oligonucleotides. *Antisense Nucleic Acid Drug Dev.,* 7(5), 483–493.
210. Reddy, J.A., Abburi, C., Hofland, H., et al. (2002). Folate-targeted, cationic liposome-mediated gene transfer into disseminated peritoneal tumors. *Gene Ther.,* 9(22), 1542–1550.
211. Manunta, M., Tan, P.H., Sagoo, P., Kashefi, K., George, A.J. (2004). Gene delivery by dendrimers operates via a cholesterol dependent pathway. *Nucleic Acids Res.,* 32(9), 2730–2739.
212. Schwartz, A.L. (1989). The hepatic asialoglycoprotein receptor. *CRC Crit. Rev. Biochem.,* 19, 207–233.
213. Spiess, M. (1990). The asialoglycoprotein receptor: model for endocytic transport receptors. *Biochemistry,* 29, 1009–1018.
214. Wu, G.Y., Wu, C.H. (1988). Receptor-mediated gene delivery and expression in vivo. *J. Biol. Chem.,* 263, 14621–14624.
215. Wu, C.H., Wilson, J.M., Wu, G.Y. (1989). Targeting genes: Delivery and persistent expression of a foreign gene driven by mammalian regulatory elements in vivo. *J. Biol. Chem.,* 264, 16985–16987.
216. Chen, J., Gamou, S., Takayanagi, A., Shimizu, N. (1994). A novel gene delivery system using EGF receptor-mediated endocytosis. *FEBS Lett.,* 338, 167–169.
217. Ferkol, T., Perales, J.C., Eckman, E., Kaetzel, C.S., Hanson, R.W., Davis, P.B. (1995). Gene transfer into the airway epithelium of animals by targeting the polymeric immunoglobulin receptor. *J. Clin. Invest.,* 95, 493–502.
218. Buschle, M., Cotton, M., Kirlappos, H., et al. (1995) Receptor-mediated gene transfer into human T lymphocytes via binding of DNA/CD3 antibody particles to the CD3 T cell receptor complex. *Hum. Gene Ther.,* 6, 753–761.
219. Batra, R.K., Wang-Johanning, F., Wagner, E., Garver, R.I., Jr., Curiel, D.T. (1994). Receptor-mediated gene delivery employing lectin-binding specificity. *Gene Ther.,* 1, 255–260.
220. Yin, W., Cheng, P.W. (1994). Lectin conjugate-directed gene transfer to airway epithelial cells. *Biochem. Biophys. Res. Commun.,* 205, 826–833.
221. Gottschalk, S., Cristiano, R.J., Smith, L.C., Woo, S.L. (1994). Folate receptor mediated DNA delivery into tumor cells: potosomal disruption results in enhanced gene expression. *Gene Ther.,* 1, 185–191.
222. Ding, Z.M., Cristiano, R.J., Roth, J.A., Takacs, B., Kuo, M.T. (1995). Malarial circumsporozoite protein is a novel gene delivery vehicle to primary hepatocyte cultures and cultured cells. *J. Biol. Chem.,* 270, 3667–3676.
223. Hart, S.L., Harbottle, R.P., Cooper, R., Miller, A., Williamson, R., Coutelle, C. (1995). Gene delivery and expression mediated by an integrin-binding peptide. *Gene Ther.,* 2, 552–554.
224. Huckett, B., Ariatti, M., Hawtrey, A.O. (1990). Evidence for targeted gene transfer by receptor-mediated endocytosis. Stable expression following insulin-directed entry of NEO into HepG2 cells. *Biochem. Pharm.,* 40, 253–263.
225. Ross, G.F., Morris, R.E., Ciraolo, G., et al. (1995). Surfactant protein A-polylysine conjugates for delivery of DNA to airway cells in culture. *Hum. Gene Ther.,* 6, 31–40.
226. Curiel, D.T. (1994). High-efficiency gene transfer employing adenovirus-polylysine-DNA complexes. *Nat. Immun.,* 13, 141–164.

227. Thurnher, M., Wagner, E., Clausen, H., et al. (1994). Carbohydrate receptor-mediated gene transfer to human T leukaemic cells. *Glycobiology,* 4, 429–435.

228. Shibata, A., Yamamoto, M., Yamashita, T., Chiou, J.S., Kamaya, H., Ueda, I. (1992). Biphasic effects of alcohols on the phase transition of poly(α-lysine) between α-helix and β-sheet conformations. *Biochemistry,* 31, 5728–5733.

229. Carlisle, R.C. (2001). Adenovirus hexon protein enhances nuclear delivery and increases transgene expression of polyethylenimine/plasmid DNA vectors. *Mol. Ther.* 4, 473–483.

230. Iovine, M.K., Watkins, J.L., Wente, S.R. (1995). The GLFG repetitive region of the nucleoporin Nup116p interacts with Kap95p, an essential yeast nuclear import factor. *J. Cell Biol.,* 131, 1699–1713.

231. Rexach, M., Blobel, G. (1995). Protein import into nuclei: association and disso-ciation reactions involving transport substrate, transport factors, and nucleoporins. *Cell,* 83, 683–692.

232. Ludtke, J.J., Zhang, G., Sebestyen, M.G., Wolff, J.A. (1999). A nuclear localization signal can enhance both the nuclear transport and expression of 1 kb DNA. *J. Cell Sci.,* 112, 2033–2041.

233. Schirmbeck, R., Konig-Merediz, S.A., Riedl, P., et al. (2001). Priming of immune responses to hepatitis B surface antigen with minimal DNA expression constructs modified with a nuclear localization signal peptide. *J. Mol. Med.,* 79, 343–350.

234. Zauner, W., Blaas, D., Küchler, E., Wagner, E. (1995). Rhinovirus Med.iated endo-somal release of transfection complexes. *J. Virol.,* 69, 1085–1092.

235. Wu, G.Y., Zhan, P., Sze, L.L., Rosenberg, A.R., Wu, C.H. (1994). Incorporation of adenovirus into a ligand-based DNA carrier system results in retention of original receptor specificity and enhances targeted gene expression. *J. Biol. Chem.,* 269, 11542–11546.

236. Curiel, D.T., Wagner, E., Cotton, M., et al. (1992). High-efficiency gene transfer by adenovirus coupled to DNA-polylysine complexes. *Hum. Gene Ther.,* 3, 147–154.

237. Plank, C., Oberhauser, B., Mechtler, K., Koch, C., Wagner, E. (1994). The influence of endosome-disruptive peptides on gene transfer using synthetic virus-like gene transfer systems. *J. Biol. Chem.,* 269, 12918–12924.

238. Gel, J. (2003). Electroporation: theory and methods, perspectives for drug delivery, gene therapy and research. *Acta Physiol. Scand.,* 177, 437–447.

239. Chang, D.C., Chassy, B.M., Saunders, J.A., Sowers, A.E. (1992). *Guide to Electropo-ration and Electrofusion.* Academic Press, San Diego, CA.

240. Keating, A., Toneguzzo, F. (1988). Gene transfer by electroporation. *Exp. Hematol. Today,* 71–74.

241. Tsong, T.Y. (1991). Electroporation of cell membrane. *Biophys. J.,* 60, 297–306.

242. Weaver, J.C. (1993). Electroporation: a general phenomenon for manipulating cells and tissue. *J. Cell. Biochem.,* 51, 426–435.

243. Nickoloff, J.A. (Ed.) (1995). *Methods in Molecular Biology, Vol. 48: Animal Cell Electroporation and Electrofusion Protocols.* Human Press, pp. 3–38.

244. Garbriel, B., Teissie, J. (1997). Direct observation in the millisecond time range of fluorescent molecule asymmetrical interaction with the electropermeabilized cell membrane. *Biophys. J.,* 73, 2630–2637.

245. Golzio, M., Teissie, J., Rols, M.P. (2002). Direct visualization at the single cell level of electrically mediated gene delivery. *Proc. Natl. Acad. Sci. U.S.A.,* 99, 1292–1297.

246. Rols, M.P., Teissie, J. (1990). Electropermeabilization of mammalian cells. Quin-titative analysis of the phenomenon. *Biophys. J.,* 58, 1089–1098.

247. Sale, A.J., Hamilton, W.A. (1967). Effects of high electric fields on micro-organisms. 1. Killing of bacteria and yeasts. *Biochim. Biophys. Acta*, 163, 37–43.

248. Neumann, E., Schaefer-Ridder, M., Wang, Y., Hofshneider, P.H. (1982). Gene transfer into mouse lyoma cells by electroporation in high electric fields. *EMBO J.*, 1, 841–845.

249. Bettan, M., Emmanuel, F., Darteil, R., et al. (2000). High-level protein secretion into blood circulation after electric pulse-mediated gene transfer into skeletal muscle. *Mol. Ther.*, 2, 204–210.

250. Shibata, M.A., Morimoto, J., Otsuki, Y. (2002). Suppression of murine mammary carcinoma growth and metastasis by HSVtk/GCV gene therapy using in vivo electroporation. *Cancer Gene Ther.*, 9, 16–27.

251. McMahon, J.M., Signori, E., Wells, K.E., Fazio, V.M., Wells, D.J. (2001). Optimisation of electrotransfer of plasmid into skeletal muscle by pretreatment with hyaluronidase-increased expression with reduced muscle damage. *Gene Ther.*, 8, 1264–1270.

252. Perez de la Cruz, M., Eeckhoudt, S., Verbeeck, R., Preat, V. (1997). Transdermal delivery of flurbiprofen in the rat by iontophoresis and electroporation. *Pharm. Res.*, 11, 309.

253. Tucker, R.P. (2001). Abnormal neural crest cell migration after the in vivo knockdown of tenascin-C expression with morpholono antisense oligonucleotides. *Dev. Dyn.*, 222, 115–119.

254. Van Tendeloo, V.F., Ponsaerts, P., Lardon, F., et al. (2001). Highly efficient gene delivery by mRNA electroporation in human hematopoietic cells: superiority to lipofection and passive pulsing of mRNA and to electroporation of plasmid cDNA for tumor antigen loading of dendritic cells. *Blood*, 98, 49–56.

# 14 Regulatory and Compendial Issues

*Michael J. Groves, Ph.D., D.Sc.*

## CONTENTS

## INTRODUCTION

A newcomer to the biotechnology industry, probably immersed in a highly specialized aspect of the scientific subject with scant knowledge of the outside world, may be surprised to find that their work is subject to the scrutiny of outsiders, often strangers without the specialized knowledge that our young scientist is bringing to the field. This may cause confusion and, perhaps, even resentment. However, with maturity and growth, it will become evident that the external review process is not dissimilar to the review process for a scientific publication in a refereed journal. There is an advantage, however, in that the object of this review is to place a product in the marketplace and thereby derive income for the company, to everybody's benefit.

The main external review body in the biopharmaceutical and pharmaceutical industries is the Food and Drug Administration (FDA), but at some later stage it may be necessary to prepare a monograph for the quality control of a new product and this would be achieved by interacting with officials from the United States Pharmacopeia (USP) in order to have a section published in the pharmacopoeia. The purpose of this chapter, therefore, is to introduce both of these organizations and provide an understanding of their basic functions.

# THE FDA

## ORIGINS AND ORGANIZATION

Founded by Congress at the beginning of the last century in response to a perceived need to protect the public from unacceptable manufacturing practices encountered, at that time, mainly in the food industry, one finds that even today members of the organization take their function of protecting the public extremely seriously. Fortunately today the often disgusting practices found in the food industry have been corrected, although current concerns about BSE (bovine spongiform encephalopathy) passing from the cattle population to humans has shown a continuation in some very questionable areas.

Currently based in Rockville, MD, the FDA is organized under a commissioner who reports to the Department of Health and Human Resources. There is a central headquarters in Rockville with various offices around the country. The headquarters is organized into numerous functionary groups. From the pharmaceutical perspective the two most important are the Center for Drug Evaluation and Research (CDER) and the Center for Biologics and Research (CBER). Unfortunately, although these acronyms are very similar, broken down, the functions of these two groups are self-explanatory. Not equal in size, it is remarkable that the FDA has so quickly recognized the relative importance of biologically derived pharmaceutical products by creating CBER and it is to be hoped that this represents a pattern for the future.

The FDA has turned its attention to the regulatory control of the pharmaceutical industry. Following the ethylene glycol incident in the late 1930s in which this toxic solvent was used in a liquid sulfanilamide formulation, the agency was charged with ensuring that drugs and other components of pharmaceutical products were safe to use by the general public. This was achieved by only allowing a new product to go on the market following review and scrutiny of documents submitted by the manufacturer.

This process worked well, despite frustration expressed by the companies at the expense of preparing the paper work requested by the reviewers and the inevitable delays in getting a new drug product to market. However, the remit of the agency was vastly expanded when Congress demanded that the efficacy of a drug substance should also be determined. This has proven to be a major obstacle to getting a drug onto the marketplace, and this delay needs to be taken into account when a company is planning its marketing strategy.

Since the FDA does not have laboratories of its own that are capable of checking the submitted information, the data is closely reviewed by agency scientists who do not necessarily have relevant and related expertise but who are quite capable of exercising scientific judgment. In the final stages external review committees, consisting mainly of medically qualified experts in the area of interest, will review the data again and recommend if the drug should be allowed to enter the marketplace. The agency does not have to accept the recommendation although it often does.

The FDA also has another function in that, like the interaction it has with the food industry, pharmaceutical manufacturing plants are subject to inspection. The stated purpose of the inspections is to ensure that the product is safe for consumption by the general public and not contaminated at any stage of the manufacturing process. In actual fact, there are rarely instances where some dangerous contaminant can get into the product and all the inspector is doing is following the process to ensure that the literal letter of the law is being adhered to, with an appropriate bureaucratic paper trail. The main exception here is microbiological contamination which represents a major threat in the food industry but, again, is less common in the pharmaceutical arena.

It might be appropriate here to echo the mantra used by all involved in the manufacturing process, the subsequent quality control processes, and the FDA inspector:

*"If it wasn't written down it never happened!"*

Of course, this mantra is not confined to the pharmaceutical industry but now forms the underlying basis for quality control of many industrial products.

## GOOD MANUFACTURING PRACTICES AND THE INSPECTION PROCESS

In the early 1950s it seemed that the pharmaceutical manufacturing industry was composed of a number of eccentric individuals who had considerable expertise in, for example, color coating tablets but who were unwilling to share their technology. Every company had its own experts and they were jealously guarded and protected. Unfortunately difficulties arose when these individuals retired or were, rarely, promoted since there were only inadequate records or, in some cases none whatsoever. There was also no lateral movement of the technical knowledge, with inevitable problems.

FDA inspectors rapidly realized these issues were sometimes serious enough as to affect the quality of the products and started to put together a codified document in order to control manufacturing processes across the industry. Strongly resisted by industry initially, it took a while for the managers to realize that there were some advantages to be gained in sharing information. Strongly influenced by the earlier British Orange Guide, these FDA documents eventually became known as the current Good Manufacturing Practices (cGMPs). Originally written in the early 1970s, these regulations are still undergoing development in details, and are known worldwide. Many smaller countries have either adopted them wholesale or have adapted them. It can be safely said that in many ways the FDA made a major contribution to the science and technology of pharmaceutical manufacturing by insisting on this codification and this has proved to be beneficial to everyone concerned.

The emphasis is, as noted in the saying above, on the written word. Every process is written down in detail, describing what is to be produced and how. Each stage is described and signed off by the operators involved and their responsible managers before the product is taken to the next stage. Admittedly a bureaucratic procedure, it does enable external inspectors to check the whole manufacturing process from beginning to end in order to determine whether or not the product is suited for its stated purpose. Of course, the process works to the manufacturer's advantage since they are no longer dependent on individual operators; there is improved safety for the patient and fewer manufacturing errors are made which means fewer expensive product recalls.

That pharmaceutical manufacturers are now totally committed to cGMPs as a fact of life, reluctantly admitted by the newer and inexperienced biopharmaceutical technology-based companies. Indeed, there are examples of young companies going out of business because they felt that this was an unnecessary bureaucratic process that would hinder the application of their wonderful new product. This attitude was perilous in the extreme because, ultimately, the FDA has to approve the drug and its process before it can be marketed.

It is interesting to note that since the introduction of cGMPs, other codified procedures have sprung up such as Good Clinical Practices, Good Clinical Trial Practices, Good Laboratory Practices, and the like. Failure to follow these often mandated regulations can be dire, especially if something goes wrong because they were not followed.

As noted, the FDA has statutory powers to inspect and evaluate a company to determine if there are any potential threats to the health and safety of patients taking the product. This can take some interesting forms such as the use of federal marshalls to break down the front door of a plant locked to keep inspectors out. This is, perhaps, an extreme example although it appears that every inspector could tell of similar experiences gained during their professional career. However, inspection of facilities is one of the main and original functions of the FDA. The inspectors usually work in teams and operate out of offices situated near concentrations of the industry, such as San Francisco, Chicago, and New York. However, they also inspect plants overseas, such as Ireland, the UK, China, and India. The inspectors have the right to be admitted to a plant without prior notice and, as noted, even have the right to break down doors if necessary to access records and equipment. Naturally, in some quarters there has been resistance to this concept and inspectors have been threatened to the point where some are entitled to carry firearms. Whether or not these statutory powers have ever been used against members of the so-called "Big Pharma" is a moot point, especially when the companies know there is so much to be lost by opposing the inspection process. It should also be noted that inspectors have the powers to close a plant down immediately if there is a perception that it is being operated with a total disregard for protocols and written procedures or a particular area is found to be unclean or unsafe.

Although the inspectors usually turn up without advance notice, most plants are now prepared for this eventuality and internal procedures have been set up to deal with this type of contingency. A senior management personnel such as the Head of Quality Assurance is usually assigned to the task of meeting the inspectors. As soon as the inspectors identify themselves they are conducted to a meeting room and they will then discuss with their guide what they wish to see during this visit. The proceedings should be courteous and efficient but, as is the way in the real world, sometimes there are disagreements and these will be discussed in the meeting room after the inspection has been made. The senior inspector and his team will then prepare a series of observations and present these to the company management on the dreaded Form 483. If a really serious issue is raised it is usually necessary for a company chief operating officer to be present so that he can hear the issues for himself. Either way, the form has to be presented to senior management and a written response is usually required within a prescribed time frame. Some have deemed it necessary to ignore their response or delay in providing answers, but, as experience has shown, there are penalties invoked when the FDA perceive that this is happening.

Although these Form 483 observations are supposed to be confidential, only between the regulatory agency and the company, some ingenious lawyers have found that it is possible to access them through the Freedom of Information Act, and often this information is used by competitors or stock market analysts to determine just how well the company is run.

It should be noted that there are usually indicators of the seriousness of a complaint. For example, only one inspector turning up who is gone within the hour can be contrasted with a large team of as many as 20 inspectors who might take a month to investigate particular issues—clearly an indication that something is perceived to be wrong.

After the departure of the inspectors the company has a limited time to respond to the concerns expressed on Form 483 and, if the agency is not satisfied, a letter will be sent to the company warning them of the issues. Recently this letter has been followed by fines on the company, in some cases of penalties in excess of $500 million.

In fairness it has to be admitted that visits from inspectors are often highly professional. However, there have been suspicions that inspectors from certain local offices have slightly different agenda that are usually unfavorable to the industry. It is not a pleasant experience to hear an inspector openly taking credit for effectively putting a company out of business; this should not be the purpose of the inspection process.

## THE REVIEW PROCESS

When a company feels that it is ready to market, say, a new chemical entity or drug, appropriately tested and evaluated, it applies to the FDA for an approval letter. Before this can be issued several critical stages need to be successfully passed. For example, the agency needs to review and approve all of the documents relating to the initial research, testing, and manufacture of the compound before it can be tested in humans. Human testing is divided into four main stages, the first being a small trial in a limited number of patients or volunteers to make certain the drug is absorbed and is safe. The second and third stages are carried out with increasing numbers of patients to determine if the drug is efficacious against the targeted disease in a clinical setting. This is likely to result in a limited number of clinical trials to determine any effects on the general population.

Unfortunately, this is often when issues arise. To illustrate this point, let us say that 1 in 1000 patients show an allergic response to the new drug. This response is unlikely to become evident in the small numbers of patients involved in the initial clinical testing and such reactions only become increasingly likely as the number of patients exposed to the drug increases. This is why it has become important to have a fourth, postmarketing surveillance, stage. In this stage there is continual surveillance in order to monitor the way in which the drug is administered and to determine if there are any unexpected side effects, which otherwise would not have been predicted from results on the hundred or so patients involved in the earlier stages of testing.

With so much bureaucracy to support it should not be surprising to realize that the FDA is an expensive unit of the federal government paid for by the tax payer. With this in mind Congress has recently allowed fees to be imposed, allowing the companies who will benefit to provide some support.

Along the same lines, it has to be recognized that this regulatory process significantly increases the cost to the company, especially when it is required to demonstrate that the drug is essentially free of toxicity. A recent survey from Tufts University (2004) has suggested that the cost to a company of developing a new chemical entity from discovery to the market is now approaching $900,000,000. Since in many cases the company only has 5 remaining years of marketing exclusivity to recover this not insignificant amount of investment, it is hardly surprising that the marketing costs for a new drug often appear to be high. This issue will be discussed in more detail in the final chapter.

The FDA has proved to be responsive to criticism, probably more so than many other federal agencies. To give one example, in 1989 it took roughly 33 months for a market approval letter to be issued from the date of request; 10 years later this dropped to 13 months. Issues still remain, however. No new drug excipients had been approved in over 20 years because no manufacturer was willing to spend the money on the extensive toxicity testing required without commercial exclusivity or protection.

A recently appointed commissioner has noted that, in fact, the number of applications for approval of drug *and biologicals* (my emphasis) has decreased, which should be of no surprise to any student of economics. The commissioner also noted that the genomically based drugs will become even more expensive to approve and this factor alone will begin to have a significant impact on the overall cost of healthcare. However, in contrast, it should also be pointed out that the new drugs should be more effective, being targeted and precise in their application. This debate is sure to continue.

## THE UNITED STATES PHARMACOPEIA

At the present time the United States Pharmacopeia (USP) is the premier book of pharmaceutical standards throughout the world. It is enforced as the book of drug standards and is used by the FDA to define quality in the United States during inspections and reviews.

### ORIGINS AND ORGANIZATION

Historically, books of drug standards began to appear in Europe during the 14th century, especially in the small states that comprised what is now Germany and Italy. In the United Kingdom there was a London Pharmacopoeia dating from 1618, and similar books were published in Edinburgh in 1699 and Dublin in 1807. Eventually these were merged together into the British Pharmacopoeia in 1864 and this is still being published although it is now subservient to the European Pharmacopoeia.

In pre-Revolutionary America most drugs were imported from London and the standards of the London Pharmacopoeia were used. After the war nationalist feelings became dominant and there was a suspicion that imported drugs were of poor quality, especially after a sea voyage lasting up to 3 months. In addition, it was realized that many drugs being imported were actually of native American origin, and it made little sense to send materials for processing when it could be done more cheaply and more satisfactorily at home.

A group of doctors under the leadership of Dr. Lyman Spalding met in 1807 to determine if it was feasible to organize an American Pharmacopoeia. A convention was organized based on the division of the now United States into four geographical groups and the first United States Pharmacopoeia was published in Boston in 1820. A second edition was published out of New York in 1830 (twice, because of errors the first time around), and eventually it was published from Philadelphia as a national pharmacopoeia published in five yearly revisions. Now in its 28th edition (2005), the pharmacopoeia has lost its diphthong and is now published annually, with two annual supplements to ensure that the compendium is current and up to date. The National Formulary (NF) was added to the pharmacopoeia in 1975 and the book contains standards for drugs and excipients as well as simple formulae for vehicles and the like.

It should be noted that the function and format has changed over the years. For example, the 1820 edition only contained formulae for 317 drugs and preparations. Starting in 1880 the text started to contain rudimentary drug standards and a section still headed Tests and Assays.

Like many organizations today, the USP has a mission statement and this was stated in the 27th edition: the objective of a pharmacopoeia is to provide and disseminate authoritative standards for medicines, other healthcare technologies and related practices to maintain and improve public health, and to provide information for practitioners and patients. In addition, the compendial activities of the USP are also designed to support the availability of safe and good quality medicines for consumers everywhere. In parenthesis, the term "quality" is not defined here because it is intuitively understood by everyone. One working definition holds that a quality item is one suitable for a stated purpose, but this is somewhat unsatisfactory and a better definition is required.

Although the organization of the USP is obviously very complex and, by definition, expensive, it must be emphasized here that the organization is a private enterprise and is not government funded. The United States Pharmacopeia is still published by a convention which meets every 5 years and is comprised of medically and pharmaceutically qualified individuals who come as representatives or delegates from medical and pharmacy colleges across the country. Other professional organizations are represented such as veterinarians and nurses and, increasingly, representatives from the biochemical and biotechnological areas. One point often not always appreciated is the fact that foreign experts are becoming increasingly involved, especially in the area discussed under Harmonization. In many ways the USP is, in a number of senses, an international enterprise in its own right. At the time of writing it was announced that a Spanish language edition is to be published.

The convention is the engine that drives the organization, containing a nucleus of permanent staff under the direction of a chief executive and a board of trustees, rather like an industrial company but designed to be nonprofit. Income is principally derived from the sales of the publications and reference samples required for many chemical and biological assays. Expenditures are maintenance and upkeep of the buildings, salaries of the secretariat and expenses involved in bringing in members of the committees of experts as well as the organization of various conferences designed to discuss current technical issues. Since the expertise required to prepare and publish monographs on the many different types of drug in the compendium is obviously very diverse, the main part of the pharmacopeia is put together by committees of experts with different responsibilities. It is evident that the entire process of putting a pharmacopeia together is heavily dependent on interested volunteers donating their time and energy; without their contributions the whole process would become too unwieldy to function effectively. Typically a monograph for a new drug would be put together by a single member of a committee, then the committee as a whole would review the draft. The draft would then be edited by the permanent staff and published in the Pharmacopeial Forum (PF).

The PF is a bimonthly publication in which changes or additions to the pharmacopeia are published for public review and comment. This is usually the first sight of a proposal and the PF is turning out to be a very important document for those who are involved in quality control. Proposed changes are described in a page-by-page collation and, usually, there are discussion articles so that the scientific background is available to allow appreciation of the significance of present or future technical proposals.

Once a monograph has been published in PF a period of time is allowed for discussion or submissions, usually by pharmaceutical companies who might be affected. The draft monograph might then be rewritten and again published in PF to ensure that there is comfort with the proposal. After approval by the board of trustees the monograph then goes forward for incorporation into the next edition of the PF. However, even after publication the monograph is not written in stone and can be modified in detail if necessary, changes going through the same process.

## FORMAT

The format of the modern pharmacopeia remains roughly similar to those early medieval compendia that effectively listed drugs or the crude galenicals made from them in individuals monographs. Of course, these early documents were written in Latin but nowadays about the only remaining traces of Latin are in the titles of occasional botanical descriptions, which have recently made a return to official pharmacy under the rubric of Nutritional Supplements.

A modern monograph is devoted to a drug substance and starts with a concise description of the product. This is followed by instructions for packaging and storage, relevant USP reference standards, identification, and procedures for the assay of the drug or the product components. There is sufficient information for those "skilled in the arts" to be able to analyze the drug and its components, making the compendium a valuable source of information in its own right. There are also limits and simple tests for contaminants or impurities.

The addition of the National Formulary section to the pharmacopeia, followed more recently by the section of Nutritional Supplements, has resulted in a considerable increase in the numbers of substances described by an official monograph. These are mainly excipients and diluents for the NF and standards for the popular Nutritional Supplements (including my personal favorite, Chocolate).

There are two sections in the modern compendium that would not be familiar to the early pioneers, the Notices and the General Chapters. The first, described as General Notices in the text lays down the guidelines for subsequent interpretation and is required reading for any individual who picks up the pharmacopeia, providing, as it does, definitions of parameters such as temperature and methods of calculating percentages.

The most valuable information in the whole compendium, perhaps, is to be found in the various General Chapters. There are currently 1251 chapters, indicated by brackets, < >, although the observant reader will notice that the editors have held some of the assigned numbers back for later use.

These chapters contain specific information about the principles of assays defined in the text and contain what may well be the most comprehensive guide to the science of measurement in the scientific literature. Both the NF and the Nutritional Supplements have their own general chapters. Guidelines to Good Manufacturing practices are supplied, with guidance to small-scale pharmaceutical operations such as those encountered in a typical dispensing pharmacy.

It has to be said, there are a remarkable mix of topics in this section, ranging through a complete discussion of biotechnology-derived articles (<1045> and <1047>), cell and gene therapy products (<1046>), and validation of compendial (test) methods (<1225>). Specific methods for detection and measurement of suspended particulate matter are discussed in <788>, and there are detailed instructions on how to measure weights, <41>, volume <31>, or temperature <21>. Every student and, indeed, their supervisors need to be aware of what is in this valuable section; there is so much valuable information that the compilers cannot be praised enough.

It might be noted that these topics are not just confined to products on the market and in the pharmacopoeia. For example, there are only a few products currently available derived from biotechnology or from gene engineering on the market place or being the subject of a compendial monograph and yet the future needs are anticipated. This is an impressive achievement, especially when one considers that each chapter has the potential to be updated every year, thereby avoiding obsolescent information.

## HARMONIZATION

In compendial-speak harmonization is a new word rapidly becoming significant in today's world and may suggest that, in the future, there may only need to be one pharmacopoeia for international purposes. This may or may not come about but the intention is certainly there.

This whole subject came about during informal discussions during the late 1980s when it was realized that there were only three significant pharmacopoeias on the world scene: the United States, Japan, and Europe. However, looking at the three it was evident that there were discrepancies between them for substances described in compendial monographs, which were probably otherwise identical. One example was lactose, a common excipient with three different specification for the same substance in commerce, often with only minor or trivial differences. This was held to be significant in terms of world trade since a manufacturer had to label the material differently according to where it was being marketed, even if the material in the container was identical in composition.

Gradually these discussions have broadened in scope and an official body was formed under the name of The Pharmacopoeal Discussion Group, consisting of representatives from each of the three pharmacopoeias. They are currently coordinating changes in monographs for excipients and in some cases analytical methods covered in the General Chapters. The progress over the past decade has been impressive and the work is proceeding. It is possible or even likely that the influence of local practices influencing each individual pharmacopoeia will hold sway for at least a while so the need for regional compendia will continue. Who knows what will happen in the future?

It might be of interest to observe here that this logical need to harmonize on an international scale is not exactly new. In the United States there was a move in the 1860s to harmonize or join the United States Pharmacopoeia (as it was spelt in those days) with the British Pharmacopoeia, which had just been published for the first time. However, nothing came of this move for various reasons and we have had to wait for another hundred years before harmonization has come about with some unanticipated partnerships.

There can be little doubt about the international nature of business these days and anything that removes potential trade barriers is usually welcomed. This is happening and now it does not matter from where we purchase our lactose.

# 15 Some Challeges Relating to the Future of Biopharmaceutical Technology

*Michael J. Groves, Ph.D., D.Sc.*

## CONTENTS

## INTRODUCTION

Although this field of technology has been expanding rapidly over the past 5 to 10 years and apparently shows a great deal of promise for the future, it seems reasonable that we should review the progress and determine if, indeed, there is a future for pharmaceutical biotechnology. After all, the enormous potential for the good of mankind and the considerable costs involved would appear to be mutually exclusive, suggesting that some serious decisions will need to be made in the near or distant future.

To provide a hypothetical example of a potential budgetary impact, let us think about the repercussions if a cure for Alzheimer's disease were to be uncovered that required biopharmaceutical processing and separation of an active ingredient. The hypothetical cost of a single patient's drug treatment alone might well exceed $30,000 per annum; multiply this number by the 6 million or so patients diagnosed with the disease and we begin to approach direct costs that could seriously rival or exceed the military budget! Decisions will have to be made in the future about where this money will come from, especially as it would be politically unacceptable to deny the unfortunate patients this treatment.

Although this is a hypothetical example, most biotechnology-derived drugs are used to treat relatively small patient populations, so the overall demand for money from the central authorities would also remain relatively small and more easily digestible. This may be the present (2005) situation but there can be little or no doubt that the demand for biotechnology-derived drugs will increase. We will find ourselves trying to balance the budgetary impact of the costs against all the benefits associated with an efficacious and safe treatment for a disease.

## EXAMPLES OF ESSENTIAL BIOPHARMACEUTICALS

### INSULIN

In the early 1970s it became evident that there was unlikely to be sufficient naturally derived insulin from slaughtered cattle and pigs available to treat the existing and expanding diabetic population by the end of the decade. A consortium of industrial scientists succeeded in genetically engineering human insulin from bacterial sources and, after 20 years of continuous development, this remains a source of human insulin guaranteed to be pure and free of microbiological contaminants.

On the other hand, bovine and porcine insulins are still produced after over 80 years of development, especially in underdeveloped countries. These materials are relatively pure but there is always a danger of contamination with prions such as bovine spongiform encephalopathy (BSE). This danger has not been demonstrated to date but it is conceivable. Nevertheless, it remains necessary to point out that the naturally derived materials are less expensive than the genetically engineered form of the drug. Health insurance companies, perhaps wisely, have never demanded generic substitution of the bioengineered form by the natural, but such are the vagaries of the healthcare system; this option remains a possibility in the future.

### GROWTH HORMONE

Human growth hormone (hGH) or somatotropin has returned to the limelight recently when genetically engineered material became more readily available and more applications for the drug opened up.

The protein was originally derived from human cadaver pituitary glands and was used initially to treat dwarfism. However, it became evident that the material was contaminated with the prions that caused Creutzfeldt-Jakob syndrome and for many years this line of research was not followed up. Recently the protein has been produced by genetic engineering in significantly larger quantities and is now being explored for use in other applications (e.g., cosmetics).

In the United States alone there are approximately 50,000 adults with growth hormone deficiency, resulting in dwarfism and there may be as many as 6,000 newly diagnosed cases per annum. The cost of treating a child with growth deficiency is currently estimated at around $11,000–$18,000 per annum. An alternative way of looking at this issue is to note that it costs between $35,000 and $40,000 to increase the height of an affected child by one inch.

The legitimate medical demand for this drug is therefore at about $1 billion per year and this money presumably must be found somewhere in the system. A recent suggestion has been that the drug should be used to treat adults or children who are otherwise healthy but show signs of being a little shorter than average; this seems a questionable medical and ethical requirement for the drug (DeMonaco 2003).

Another aspect of this drug that is certainly highly questionable is its use for cosmetic and athletic purposes, but this issue is going to have to be resolved by the regulatory authorities.

## INTERFERON β 1a

Das (2003) made an interesting point recently when he suggested that there were currently 371 new products currently on trial against diseases such as Alzheimer's, AIDS, arthritis, cancer, heart disease, and multiple sclerosis (MS). For MS, Serono (Geneva, Switzerland) announced that their new drug Rebif®, a form of interferon 1a, would cost a patient $17,000 a year. Das therefore suggested that supply of many new protein therapeutics for a year was quite likely to exceed $10,000. In spite of this expense, the patient may have no alternatives and this would remain the situation where the disease had inadequate therapy or no treatment at all.

## OTHER EXPENSIVE DRUG THERAPIES

Looking at the numbers of bioengineered products coming on the market and the diseases that they claim to treat, it seems that the list of diseases is getting longer and perhaps more obscure. Nevertheless, there are some diseases such as rheumatoid arthritis excruciatingly painful and have little in the way of effective treatments. The use of antitumor necrosis factors (TNF), principally as a humanized monoclonal antibody, as anti-inflammatory agents has met with some success and various forms are now entering clinical practice. However, even as this is being written (2004), reports are appearing on the side effects of this therapy. In addition, one has to wonder about the long-term effects of a therapy designed to suppress the natural antitumor substance used in nature to destroy the cancers that normally appear in the body spontaneously.

## PHARMACOECONOMICS AND VALUE FOR MONEY

With concerns growing about the rising cost of drug therapies it is not surprising that a new academic discipline has arisen—pharmacoecononics. Not entirely new, the study of therapeutic outcomes from an academic perspective has remained somewhat intuitive up to now but obviously is of considerable interest to developing companies in the pharmaceutical biotechnological area interested in selling an expensive commodity to a limited number of purchasers.

The producer must demonstrate objectively that a new and expensive therapy represents good value for money, especially if the drug costs a great deal more than existing treatments. Breakthrough products that provide a treatment or cure where none has previously existed must also demonstrate value for money when it comes to reimbursement by the government or insurance agency. Beyond this point, the consumer must be convinced that the new drug is worthwhile, both in terms of alleviation of the condition and likely prolongation of life. Quality-of-life issues are also important and form part of the final evaluation equation.

Unfortunately, these issues are difficult or challenging to measure directly and the current situation involves a number of empirical or semiempirical numbers that seem to obscure the fact that there is no significant underlying science. This is controversial in the sense that some claim it is a practical or applied science. (However, Lord Rayleigh, the great Victorian scientist, claimed if something could not be measured it was not science.) Looking at patient outcomes following specific treatment some effects can be measured (time to death, for example), and there is usually a causal relationship between drug use and improvement in condition or cure of a disease. How exact numbers can be put to these effects has become controversial and there is a large literature on this subject.

To provide but one example of the many papers in the recent literature, Neumann et al. (2000) asked if pharmaceuticals were cost-effective and provided examples of current thinking in this field. These considerations will undoubtedly have an influence on the way in which Medicare and health insurance organizations calculate compensation.

The empiricism in this area is paramount but can hardly be avoided because there are no strong measurements for factors such as therapeutic outcome. One factor that is easily measured is the direct cost of the drug or treatment in terms of the cost of a daily dose. Strangely, this simple measurement does not always appear to be of concern, although there are situations where the cost has an outcome on the chosen therapy. Neumann et al. (2000) calculated a "quality-adjusted life year" (QALY) to demonstrate the benefits of, for example, treating a patient with a herpes zoster infection with acyclovir as opposed to providing no treatment at all in order to save money. On the other hand, they calculated that treating a 50-year-old Caucasian woman, otherwise healthy, with estrogens would cost $12,000 against no treatment. However, it is not entirely clear from this publication how this estimate was obtained although it obviously includes the direct cost of the drug. It might be worth commenting that the direct cost of a small molecular weight drug (<500 Da) does not usually cost more than about $6,000 per annum directly to the patient and most insurance companies and Medicare seem content with this, irrespective of other factors artificially built in to the estimated price.

The point of this discussion is to note that it does not appear to be feasible or, indeed, desirable to base a comparison of the cost of drug therapies as a whole on the simple cost of the drug per dose or per annum alone. A biotechnologically derived drug may cost a great deal of money per unit dose, but the actual cost/benefit ratio may be such that society can absorb the cost. This topic does not appear to have been played out at this point in time and it is certain there will be developments in the future.

Neumann et al. (2000) quoted an internal document dating from the Clinton administration that stated that, for every $1 spent on drug therapy, there was a saving of $3.50 in hospital spending. This number has been widely quoted but, as always, there has been little or no attempt to confirm or validate this estimate. It may simply represent the optimism of administrators dealing with a huge drug bill, aided, perhaps, by the pharmaceutical industry. Indeed, Neumann et al. draw attention to the skepticism associated with drug industry–supported studies of the economic issues of drug treatment. Most reputable medical journals have now instituted strong controls designed to ensure that the reported data is objective, thereby avoiding some of the suspicions of earlier abuse of the system.

It could be argued that pharmacoeconomics is a serious attempt to determine if a particular drug, no matter its origin, offers value for money. This issue will continue to be of interest to everyone involved in treatment of disease by biotechnologically derived drugs.

## GENERIC BIOTECHNOLOGICAL DRUG ISSUES

The respected scientific journalist W. Wayt Gibbs, writing in *Scientific American* (2003), has suggested that many of the issues associated with concerns about the high costs of individual bioengineered protein drugs will go away when the patents lapse and generic drug manufacturers become involved.

Gibbs correctly identified some of the issues when generic "copies" of established drugs became available to the public. However, he may be in error in suggesting that the cost of "biotech" or "biologics" would decrease very much for the simple reason that these drugs are not prepared in the same way as small molecular weight drugs. The new generation of drugs now available are typically large molecular weight proteins constructed within biological systems such as yeast or *E. coli* cells and the proteins need to be folded in certain critical ways to be biologically active. In addition, their properties are affected by the additional attachments of sugars and other moieties that control their solubility, stability, and availability. These syntheses are quite different from the relatively simple small chemical entities and will certainly be more difficult to replicate. There is also the problem of available facilities for making these drugs, such as the fermentation plants required to grow the biological cells. There is currently a shortage worldwide of fermentation capacity and this will also inhibit generic competition.

There are other issues associated with the generic industry that require resolution. The pharmaceutical industry at present consists of two main components: originators and generic copyists. Originators discover a drug entity, patent it, develop methods of analysis and delivery, and carry out clinical and toxicological tests before developing a marketplace. On the other hand, once the patents have expired competitors are able to do an end-run around much of this testing and research, enabling them to put the drug on the market more cheaply. This is in principle; in fact it is noticeable that many of the generic prices are only slightly less than those of the originator.

It is the patent system which encourages this competition but there have been instances where the technology has been pirated before the patent(s) expire. Since originators may have to spend up to a billion dollars before a new chemical entity enters the marketplace, it is not surprising that the entry price is high since, commercially, this investment has to be recovered. Depending on the length of the remaining patent life this investment has to be recovered before competition appears on the market place, after which, in principle, market forces take over to drive the price down.

This whole issue has become a political discussion point, and in countries like Canada there is legislation to ensure that the final price is more acceptable to the consumer. This may happen in the United States but at present there is no clear answer to the problem, especially if insurance or government agencies will not carry the cost.

Returning to biologics, it seems that the technology is so specialized that any intervention by generic drug manufacturers when the patent(s) expire is unlikely to significantly reduce the cost to the consumer, and it may well be much less than that anticipated by some journalists.

## ALTERNATIVES TO PROTEIN BIOLOGICS

It should be noted that there may be alternative drugs to those biological materials produced by gene manipulation and expression. This will be especially relevant if the high cost of bioengineered compounds becomes unacceptable to society as a whole. The point is that proteins may have unique biological activities that can be mimicked in some cases by small molecular weight compounds, in other words, conventional drugs. At present there are few examples of this type of activity but research has already been initiated—and, who knows, this may have a successful outcome.

## GENE ENGINEERING

The recently discredited "biological dogma" held that specific proteins were synthesized by specific genes which involved unique DNA and RNA sequences. It is now evident that this is by no means a dogma although there is a broad basis in fact in many situations. The popular press keeps telling us that the genes responsible, ultimately, for such and such a disease have been isolated, implying that a cure will be available within the immediate future. Unfortunately, unless the protein responsible can be isolated, identified, and produced in sufficient quantity and quality to be clinically tested, this is simply not going to happen.

An alternative to the use of purified protein would be to insert the appropriate genes into the human cells, either generally or in the specific target cells, and allow the cells to function naturally. This approach has been tried clinically in a number of diseases but not with any conspicuous success. Part of the problem here is how to insert the gene into the appropriate cell, as discussed earlier. Some viruses capable of entering cellular structures and empty viruses, containing the gene, have been tested *in vitro* and *in vivo* as well as clinically. Although the possible advantages of this approach to the treatment of some diseases have been evident for over 30 years, making the approach practical and realistic has proved somewhat elusive.

As an example, at least one of the trials was criticized after a volunteer died following poor clinical practices. In another trial three patients were successfully treated for an x-linked severe combined immunodeficiency disease but 2 years after the treatment two of the patients developed T-cell acute lymphoblast leukemia. The FDA promptly placed a ban on any further similar clinical studies.

Part of this difficulty may have been caused by insertional mutagenesis at or near the cancer-promoting LMO-2 gene, resulting in an aberrant production of the LMO-2 protein. At the time the research was initiated it was considered that the chance of a viral vector inserting itself into a gene was very small but, with hindsight, it seems that the chances are actually quite high. Much of the human DNA has no known function and for a long time this so-called junk DNA was not considered to be important other than as packing material between active centers. This viewpoint may be incorrect. Recently a retrovirus has been found to hit a gene 34% of the time, a much higher rate than would be anticipated if the process were entirely random. Since this virus had a striking preference for the initial sequences of a gene associated with turning genes on or off, the risk associated with the use of retroviruses as gene delivery systems is obviously much higher than originally thought.

Nonviral delivery systems such as positively charged liposomes are being explored but the use of viral systems requires careful consideration. There may be other viruses that could be used as safer drug delivery systems. Adenoviruses might be among this group but, in any case, they do not allow the corrective gene to be permanently incorporated into the cell and the treatment would have to be repeated at intervals.

Gene therapy using viral carriers would appear to be at a crossroad at the present time, although other options for gene delivery to cells require exploration and evaluation. The optimism of a generation ago is being replaced by cautious exploration.

## CONCLUSIONS

Although the overall theme of this chapter may seem to be pessimistic, one would not wish to close on a negative note. On the contrary, the whole of the scientific endeavor identified as pharmaceutical biotechnology must be regarded with considerable optimism and pride, both for past achievements and for future prospects. The intention, however, is to serve as a warning that we cannot proceed too quickly and caution must always be in mind. Serious and fatal errors have been made that could affect the future of the industry as a whole. Realistically it is unlikely that failure and disappointment will be avoided in the future but these can minimized if everyone involved has a clear vision of the advantages for the future of mankind. There are so many prospects for the successful treatment of disease that we cannot give up at this stage, no matter what the issues are today.

## ACKNOWLEDGMENTS

It is with considerable pleasure that I acknowledge a number of discussions I have had with Dr. Simon Pickard, Assistant Professor in the Center for Pharmaconomic Research, College of Pharmacy, University of Illinois at Chicago. Dr Pickard provided me with invaluable advice and some of his ideas have been incorporated into this present text. However, all and any errors are mine.

For clarification and more information, please contact Dr. Pickard at *Pickard1@uic.edu.*

## REFERENCES

Editorial. (2003). Juengst, E.T. What next for human gene therapy? *Brit. Med. J.,* 326, 1410–1411.

Das, R.C. (2003). Progress and prospects of protein therapeutics. *Am. Biotechnol. Lab*., Oct, 8–12.

DeMonaco, H.J. (2003). Debating the use of growth hormone in healthy but short children. Available at www.Intelihealth.com (accessed November 2003).

Gibbs, W.W. (2003). Can cells be generic?. *Sci. Amer.,* 287(11), 48.

Neumann, P.J., Sandberg, E.A., Bell, C.M., Stone, P.W., and Chapman, R.H. (2000). Are pharmaceuticals cost-effective? A review of the evidence. *Health Affairs,* 19(2), 92–109.

# Index

## A

**397**